ALGONQUI
The History of the McRae Lumber Company

Donald L. Lloyd
Foreword by Roy MacGregor

Robert D. McRae

Whitney, Ontario

Copyright © 2006 by Donald L. Lloyd

Published by Robert D. McRae, Whitney, Ontario

Also by the author:

Land Use of the Welland Canal Area. M.A. Thesis, University of Western Ontario, 1961.

The Patterns and Problems of Agriculture within Kent and Northumberland Counties, New Brunswick, Canada, 1961-1965. Ph.D. Dissertation, University of Maryland, 1971.

Algonquin Voyageurs. Board Game. 1972. © Lloyd Travel Games

Algonquin Provincial Park Fun to Learn Book. Don Lloyd, Pat Tozer, Ron Tozer, and Ann Rocchi. Illustrations by Don Lloyd. The Friends of Algonquin Park, 1989. ISBN 0-921709-46-3

Dynamic Canada: The Environment and Economy. Margaret Fagan and Donald L. Lloyd. McGraw-Hill Ryerson, 1991. ISBN 0-07-551060-X

Canoeing Algonquin Park. Donald L. Lloyd. Illustrations by D. Lloyd. Donald L. Lloyd, 2000. ISBN 0-9686556-0-2

The ownership of copyrighted material used in the text has been carefully traced and permission has been granted for its use, for which the author and publisher are grateful. The author and publisher welcome any information enabling them to rectify any error in reference or credit in subsequent editions.

CANADIAN CATALOGUING IN PUBLICATION DATA

Lloyd, Donald L., 1932–

ALGONQUIN HARVEST: The History of the McRae Lumber Company

ISBN 0-9686556-1-0

1. History of McRae Lumber Company 2. Changes in Timbering and Mill Methods
3. Algonquin Provincial Park 4. History of Park Protectionist Movement
5. Provincial Policies and Forest Industry

Layout and design by *Fortunato Design Inc.*, Toronto, Ontario

First Printing: August 2006

Cover image: Hay Lake Mill: McRae Lumber Company

Printed in Canada

*This book is dedicated to the memory of
John Stanley Lothian McRae,
founder of McRae Lumber Company;
his wife, Janet; Marjorie McRae McGregor;
and to our father and mother,
Donald M. and Helen McRae.*

JANET, BOB, JOHN AND CATHY.

TABLE OF CONTENTS

Foreword	xvii
Preface and Acknowledgements	xix
Introduction	xxii

Section I: SETTING THE SCENE			1
Chapter	1	Peopling the Ottawa, Bonnechere and Madawaska River Valleys	3
Chapter	2	Early Mills of the Whitney Area	11
		The St. Anthony Lumber Company: 1895–1909	11
		The Munn Lumber Company: 1909–1910	15
		The Dennis Canadian Lumber Company: 1912–1922	18
		The Mickle, Dyment Lumber Company: 1903–1922	21
Chapter	3	The McRae Family Background, the Early Years of John Stanley Lothian McRae, the Purchase of the Airy Mill, and Observations of Early Whitney	
		The Campbells	24
		The McRaes	25
		The Keenans	26
		The McGregors	27
		J.S.L. McRae's Early Years in Logging	27
Section II: THE OPERATIONS OF THE MCRAE LUMBER COMPANY: 1922–1980			33
Chapter	4	The Airy Mill Operations: 1922–1933	36
		Operations in the Bush	38
		Good and Plentiful Food	41
		The Nature of the Forest	46
		Jobs at the Camp	46
		Scaling, or Measuring the Logs	53
		The Haul Road	54
		The Drive by Water	63
		The Canisbay Licence and William Finlayson, Minister of the Department of Lands and Forests	71
		Accidents	73

		The Airy Mill: 1922–1932	74
		Government Regulations of the Day	80
		The End of the Airy Mill	84
Chapter	5	**The Lake of Two Rivers Operations: 1933–1942**	85
		Operations in the Woods	85
		The Saw Filer and the Blacksmith	89
		The Beginnings of Mechanization	92
		The Mill	94
		Donald McRae as Clerk	97
		J.S.L. McRae and Business Opportunities	99
		Accidents Remained a Part of Work in the Bush	99
		Lighter Times	100
Chapter	6	**The Hay Lake Mill Operations: 1942–1952**	103
		Problems in Starting Up	103
		The Product and Markets	105
		Government Regulations and Forest Regeneration	106
		Bush Camps and Woods Operations	107
		Mechanization: Tractors and Trucks	109
		The Haul	110
		Log Dumps	113
Chapter	7	**The Mink Lake Operations: 1952–1957**	118
		The Mill	118
		Sawing	119
		Government and Forestry	120
		Gravel Roads and Bush Camps	121
		Mechanization	121
		The McRae Lumber Company and the Union	127
		Cooks and Cooking	128
Chapter	8	**The Rock Lake Mill and Operations: 1957–1979**	133
		Sawing	137

A Change in the Wood Coming from the Forest and 139
  Consequent Changes in Mill Operations
The Influence of Felix Tomaszewski 140
Chips: A New Product—and Also a Gamble 141
The Selling of the English Lands 142
Bush Operations: A Time of Great Change 143
The Search for a Machine to Replace the Horse 143
Further Mechanization in the Woods: The Replacement of 148
  Gin-Poles and Jammers by Loaders
The Cutting System of the 1960s: Hot Skidding 149
Sandy McGregor: Bush Superintendent 150
The Last of the Horses 151
Bush Camps and Food 151
Donald McRae as Manager of the McRae Lumber Company 152
Donald McRae's Response to the Master Plan of 1974 and the 154
  Algonquin Forestry Authority of 1975
The Apprenticeships of Bob and John McRae 154
  Bob McRae 155
  John McRae 155
A Smooth Change in Management 157

Section III: INTERACTION WITH THE PROVINCE OF ONTARIO 159
Chapter 9 Reconciling Timber Harvesting with Recreational Use in 161
  Algonquin Provincial Park
The Frank MacDougall Years as Deputy Minister of the 161
  Department of Lands and Forests: 1941–1966
The Pembroke Administrative District 163
The Whitney Unit of the Pembroke Administrative District and 166
  the Work of Don George
Research in Algonquin Provincial Park 168
The Swan Lake Forest Research Reserve 169
  Yellow Birch Research 169
  Sugar Maple Research 171
  The Spreading of Ideas 171

Continuing Research at Swan Lake 173
The Rise of Conflicting Use Problems in Parks 173
The Roots of Protectionism for Parks in the 1950s and 1960s 174
Omand's Attempt at Planning 175
Logging in Algonquin and Quetico, and Other Issues in the 1960s 176
The Preservationist Movement and Douglas Pimlott 176
The Algonquin Wildlands League 177
Hueston's Provisional Master Plan: Algonquin Provincial Park, 1968 . . . 179
The Algonquin Park Master Plan, 1974 184
    Description of Zones in Algonquin Provincial Park, 1974 185
Changes to the Land Base of the Timber Cutting Area 188

Chapter 10 The Algonquin Forestry Authority: 189
Making a New Management System Work
Joe Bird and the Early Years of the Algonquin Forestry Authority: 191
    1975–1985
AFA Relations with the Ministry of Natural Resources 196
New Ways of Doing Business: The Interim Forest Management Undertaking 196
    Interim Forest Management, 1985 196
    Environmental Assessment Act, 1988 197
    The Crown Forest Sustainability Act, 1994 197
    The Timber Management Planning Process 198
    The Algonquin Park Independent Forest Audit: 1997–2002 199
Twenty-Five Years of Change in Harvesting the Algonquin Forest 200
The Impact of the AFA on the McRae Lumber Company 203
The Impact of Harvesting Changes at the Mills 203
The Forester and Silviculture 204
Tree Marking 205
Silvicultural Management Systems 206
The Local Citizens Committee 209
The AFA and Relations with the Public 212
The Quieting of Turbulent Times 214

Chapter 11 Whitney, Its People and Algonquin Park During and After World War II 216
Tragedy in Whitney Woods: The Billings/Stringer Deaths 216

The Post Family and the Red Cross Outpost Hospital: 1928–1978 217
"The Automobile Road" from Whitney to Lake Opeongo 219
Whitney in the 1930s 221
Government Work in the Depression 228
Frank MacDougall: Superintendent, Algonquin Provincial Park, ... 229
   1931–1941
John Stanley Lothian McRae 232
Whitney and World War II 236
Marjorie (McRae) McGregor 236
Donald McRae 238
The Great Escape 248
Helen Grace (McRorie) McRae 249
Algonquin Park: The War Years and Post–World War II 250
George Phillips: Algonquin Park Superintendent, 1944–1958 251
The Passing of John Stanley Lothian McRae 255
Whitney Today 258

Section IV: THE RECENT YEARS OF THE MCRAE LUMBER COMPANY ... 263
Chapter 12 The Modern Mills of the McRae Lumber Company 265
The Low-Grade Scragg Mill, 1973 265
The High-Grade Band Mill, 1979 266
The Weigh Scale and Scaling 268
Duncan Fisher MacGregor 270
The Fire at the Low-Grade Mill 273
Bush Operations 276
Safety 277
The Second Slasher Line 277
Gary Cannon 277
Products 280
Grading Hardwood 281
Residues 282
The Modern Sawyers 282
Prospects from the Bush 282
Bush Operations in 2005 283
The Bancroft/Minden Forest: Another Source of McRae Wood 284

Changing Markets and Perceptions	....	285
On Financing: Prudence	....	288
Relations with the Labour Force and Community:	....	288
Reward Good Work with Steady Employment		
Discharge of Community Responsibilities	....	289
Adapting to Changing Times	....	290
McRae Maxims for Continuing Success	....	291

End Notes 294

Appendices

I	Taped interviews	307
	McRae Collection of Tapes	307
	Lloyd Collection of Tapes	307
	Algonquin Park Visitor Centre Collection of Tapes	307
II	Mills of the McRae Lumber Company	308
III	Order-in-Council: Reduction in Stumpage Fees	309
IV	McLaughlin Township Timber Licence, 1937	312
V	McRae Lumber Company Employment Report, 1943	318
VI	Inspector's Report: Provincial Department of Health, August 1945	320
VII	Pembroke Administrative District: Objectives and Staffing, 1967	321
VIII	Audit Report: KBM Forestry Consultants, Thunder Bay, 1997–2002 ..	323
IX	Map of Townships In and Adjacent to Algonquin Provincial Park	327
X	Map of Boundary Changes to Algonquin Provincial Park: 1893–1993 ...	328

Glossary of Terms 329
Selected Bibliography 338
Index 340

List of Photographs

1	Log Loaders at Sproule Bay, Lake Opeongo, 1910	16
2	Montreal Motor Works Engine # 45928 at St. Anthony Lumber Company Mill, 1909	19
3	Barnhart Loader, Dennis Canadian Lumber Company, Amable Creek ...	20
4	Dennis Canadian Lumber Yard, Whitney, 1922	21
5	Airy Mill, 1917	23

6	"The Office," Basin Depot, Bonnechere River	25
7	Bush Camp at the Head of Galeairy Lake	38
8	Combined Cook and Bunkhouse, Louisa Creek, 1926	40
9	Saw Filer's Shack, Cranberry Lake, Canisbay Township, 1930–1931	40
10	Laundry Facilities, Cranberry Lake, Canisbay Township, 1930–1931	40
11	Latrine, Louisa Creek, 1926	41
12	Cookery, Cranberry Lake, Canisbay Township, 1930–1931	41
13	Bush Lunch	43
14	Bucking	48
15	Skidding a Log	50
16	The Sender or Roller Using the Cross-Haul Method for Building a Skidway	50
17	Cross-Haul Method of Building a Skidway Using a Decking Line	51
18	A Skidway	53
19	Ford Tordson Tractor Pulling a Log Sleigh Train	55
20	Loading a Tanker with A Barrel of Water	60
21	Boomed Softwood Logs, Whitefish Lake, 1930	61
22	Using Gin-Poles at Hardwood Dump, late 1930s	61
23	Falls and Log Chute Between Pen Lake and Rock Lake	67
24	Narrow-Gauge Cart Railway Between Rock and Galeairy Lakes	68
25	Alligator, Pointer and Hardwood Log Boom	68
26	Remains of McRae Alligator	69
27	The McRae Steam Tug	69
28	Boomed Hardwood Logs, Man with Pike Pole and Pointer in Background	70
29	Rigid-Boom Steam Hoist: Algonquin Park Station, Cache Lake	72
30	Train Hauling Logs to Airy from Algonquin Park Station, Cache Lake	73
31	The Airy Mill Showing the Jack Ladder Passing Under the Railway	74
32	Interior of Airy Mill Showing Conveyors	79
33	Disposal of Waste by Burning, Airy Mill, 1903	79
34	Interior of Bunkhouse, Canisbay Township, 1930–1931	81
35	Interior of Washhouse. Canisbay Township, 1930–1931	83
36	Barrienger Brake	85
37	Abandoned McRae Bush Camp on Little Madawaska River	88
38	Early Truck and Sleigh	93
39	Lake of Two Rivers Mill and Emergency Airfield, 1938	93
40	Marjorie McRae at Lake of Two Rivers	94

41	Lake of Two Rivers Mill Showing Jack Ladder	95
42	Donald McRae at the Lake of Two Rivers Lumber Yard	98
43	Dr. Post Coming up the Jack Ladder at the Lake of Two Rivers Mill	100
44	Jim and Henry Taylor	102
45	Hay Lake Mill	103
46	Gin-Pole Loading Veneer Logs at Whitney Station, 1949	105
47	Tanking Sleigh and Team of Horses	108
48	Truck Pulling Sleigh of Logs, 1946	109
49	Empty Sleighs Showing Long Horse Tongue at Pick-up Point	110
50	Loaded Sleighs Showing Binding Chains	111
51	Loading Sleighs with Gin-Pole	111
52	Loading Veneer Logs on Truck with Jammer	112
53	Gin-Pole Piling Logs at Dump	113
54	Skyline Power Jammer	116
55	Skyline Power Jammer—Lifting Mechanism	117
56	The Mink Lake Mill	118
57	GM Tandem Truck	124
58	Two-Man Power Saw	125
59	Veneer Birch Log	126
60	Gary Cannon	126
61	Tom Cannon	130
62	The Rock Lake Mill on Whitefish Lake	133
63	Alex Cenzura	135
64	Buddy Boldt at the Band Saw Carriage, Rock Lake Mill	136
65	Lorne Boldt Filing a Resaw Blade	137
66	Robert Recoskie Edging, Rock Lake Mill	138
67	Gary Cannon, Duncan MacGregor and J.S.L. McRae at Rock Lake	139
68	Donald McRae and Felix Tomaszewski at Rock Lake	140
69	Bark Conveyor, Bin and Hot Pond at Rock Lake Mill	142
70	Caterpillar Bulldozer Breaking Out a Road	143
71	International Crawler-Tractor	146
72	Timberjack Skidder	147
73	Prentice Loader	149
74	First Slasher: Tower Hill near Madawaska Lake	149
75	Bob and John McRae	154
76	Don George and Dave Harper, Louisa Flats	168

77 Yellow Birch at Swan Lake Naturally Regenerated in 1961 169
   Following Prescribed Fire and Group Selection Harvesting (ca. 1992)
78 A Discussion of Site Classification Methods at Swan Lake 170
79 First Board of Directors: Algonquin Forestry Authority 190
80 Four Key Algonquin Forestry Men 192
81 Tagged White Pine Plantation, Hogan Lake Road, June 2005 211
82 The Opening of the New Logging Museum, Algonquin Provincial Park, 213
   August 15, 1992
83 Ribbon-Cutting Ceremony at the Official Opening of the Algonquin .. 214
   Provincial Park Visitor Centre, August 1993
84 Red Cross Outpost Hospital 219
85 Aerial Photo of Whitney and Airy, 1930 222
86 Whitney Store 224
87 Railway Trestle with Airy in Background 226
88 Two Young Girls on Post Street 227
89 Henry Fuller Walking to His Blacksmith Shop 227
90 The Stirling Bank of Canada and St. Anthony's Horse Stables 227
91 McRae Alligator, 1922 233
92 J.S.L. McRae 233
93 Felix Shalla and Donald McRae at Airy, 1941 236
94 Marjorie McRae Overseas with Her Ambulance 238
95 Canadian Hockey Team at Luft Stalag III 243
96 POWs from Ontario at Stalag III 243
97 East Camp, Stalag III, Spring 1943 244
98 Flight Lieutenant Ferguson and Donald McRae 245
99 Joe Lavally, Wally Floody and Unknown American 249
100 The Catch 249
101 Ex-POWs' Hay Lake Fishing Party, 1945 249
102 Helen McRae, Donald McRae, Marjorie McRae and Jean Bradley at ... 250
    Whitefish Cottage
103 Janet MacGregor McRae 257
104 McRae Family, 60th Wedding Anniversary of Helen and 257
    Donald McRae, 2005
105 Whitney: Bridge, Shell Gas Station and TD Bank 259
106 Whitney: East from Bridge 260

107	Whitney: Ottawa Street	260
108	Medical Centre and Library	261
109	Aerial View of the Whitney Mills of the McRae Lumber Company	264
110	Low-Grade Mill	265
111	High-Grade Band Mill	267
112	Duncan MacGregor	270
113	Slasher Line, Deal Processor and Chipper	277
114	Gary Cannon, Edmund and Philip Kuiack	278
115	Jimmy Wojick	278
116	Bob McRae and Gary Cannon	279
117	Robert Cannon	283
118	Feller-Buncher	283

List of Maps

1	General Locations: Algonquin Park Area	xxii
2	Townships Where McRae Lumber Has Held Timber Licences	xxv
3	Locations Where the McRae Family's Ancestors Lived	4
4	Ottawa–Opeongo Road and Ottawa, Arnprior and Parry Sound Railway	6
5	Upper Madawaska Valley Locations	7
6	Timber Licences in Ontario, 1875	8
7	Timber Licences, 1875	10
8	Part of a Plan of Timber Limits on the Ottawa and Its Tributaries	14
9	South Algonquin Park Area Timber Limits, 1893	15
10	Airy Mill and Bush Camps: 1922–1933	37
11	Lake of Two Rivers Mill and Bush Camps: 1933–1942	87
12	Hay Lake Mill, Bush Camps and Log Dumps: 1942–1952	115
13	Mink Lake Mill: Bush Camps and Roads	122
14	Rock Lake Mill: Bush Camps and Roads	144
15	Management Units: Pembroke Administrative District, 1967	164
16	Recommendations for Zoning in Algonquin Park: Algonquin Wildlands League, March 1968	178
17	Algonquin Provincial Park Provisional Master Plan (William Hueston, 1968)	180
18	Timber Licences: Algonquin Provincial Park, 1969–1970	181
19	Algonquin Provincial Park Master Plan, 1974: Zoning	185

20	Land Use Zones in Algonquin Provincial Park: Management Plan, 1998	186
21	The Village of Whitney and Vicinity, Early 1930s	225
22	Whitney and Surrounding Area, 2005	258
23	Forest Management Units of Central and Eastern Ontario	285

List of Figures and Diagrams

1	St. Anthony Lumber Company and the Village of Whitney	12
2	Letterhead of the Mickle, Dyment Lumber Company, 1920	22
3	McRae Family Tree	28
4	McGregor/McRae Family Connections	29
5	Letter to the Honourable Beniah Bowman, Minister of Lands and Forests	31
6	Hall Stove	42
7	Draft-Style Horse Harness	45
8	Cutting Down a Tree: Corner Cutting to Prevent Splitting During Felling	48
9	A Method of Preventing a Log from Splitting on Uneven Ground	48
10	Skidding a Log	49
11	Building a Skidway by the Cross-Haul Method Using a Decking Line	51
12	Gin-Pole Used in Building a Skidway Using the Cross-Haul Method	52
13	A Method of Lessening Grade on a Bush Road	54
14	Road Roller	56
15	Patent or Otako Plow	57
16	Logging Sleigh	57
17	"Needle and Thread" Trip Stakes	58
18	Sprinkler or Tanker Sleigh	59
19	Jammer	60
20	A Variety of Hooks and a Crosscut Saw	62
21	McRae Hardwood Crib	64
22	Rafter Type Splash Dam	65
23	Method of Raising Stop Logs in a Dam	66
24	Airy Mill Layout	75
25	Hand Signals Used by Hardwood Sawyers	77
26	Letter from the Ontario Department of Health, 1937	82
27	Whitney Lumber Company Letterhead, 1947	114
28	Mink Lake Mill Layout	119
29	John A. Cockburn Letter to J.S.L. McRae	123
30	McRae Lumber Company Ltd. Letter to John A. Cockburn	124

31	Notification of Officers Elected to Executive Board of Local 2537, Lumber and Sawmill Workers' Union	127
32	Method of Binding Logs on Trucks	150
33	Basal Area of Sugar Maple Trees	207
34	Growth of Sugar Maple Under the Selection System	207
35	Uniform Shelterwood in White Pine	208
36	Supplies for Old-Growth White Pine Seed Collection	210
37	Missing in Action Notification	240
38	Personalkarte I: Donald McRae's Prisoner of War Identity Card	241
39	High-Grade Band Mill Layout	269
40	Low-Grade (Scragg) Mill Layout	275
41	Sawing Softwood and Hardwood Logs	280
42	Average Yield of Products from a Hardwood Log	281
43	Types of Residues Shipped By McRae Lumber	282

List of Tables

1	Bush Camp Buildings, Louisa Creek, 1926	39
2	Commissary Items for One Month, Cranberry Lake, 1930–1931	42
3	Blacksmith's Tools and Supplies	44
4	Expenses of Horses and Other Miscellaneous Items, 1926	44
5	Men Employed in the Bush: Organization and Monthly Wage Scale, 1930–1931	47
6	Equipment Used on the Haul	55
7	Bush Camp Lumbering Costs, 1926	63
8	Airy Mill Labour Force, 1926	76
9	Land Use Zones in Algonquin Park, 1998	188
10	AFA Allocations of Wood to Former Licensees, 1979	193
11	Percentage of Trees Species Harvested: 1975-1976 and 2003-2004	201
12	Pulpwood as a Percentage of Algonquin Forest Timber Cut	204
13	Chart of Hardwood Grades	281

Index	340
About the Author	348

FOREWORD

For a great many of us, reading Don Lloyd's ambitious *Algonquin Harvest* is a bit like being handed a time capsule containing your own life. I read the manuscript shouting—*"The big cookies at the cookery! Yes"*—and I read it at times weeping, remembering past places and people that have, too often, faded into the long history of Algonquin Park. What Don Lloyd has done is to bring those places back into focus and the people back to life. He has written a *living* history that is filled with character and drama and which, like the best of stories, carries well beyond the last page. While at first glance this might seem like the story of a single lumber operation, it is so much more than that: it is the story of an industry, the story of a province—and most of all—the story of all of us who have been touched, and in many ways formed, by Ontario's glorious Algonquin Park.

Reading this book also felt a little like drowning. Not because the words were impossible—quite the contrary, it is admirably written. And not because it is about Algonquin Park, where I have dumped my share of canoes. But drowning, because my life kept flashing before my eyes. Our father worked for McRae as did uncles and cousins from both sides of the family. J.S.L., the larger-than-life founder, was an uncle; his wife Janet—the kindest, sweetest woman any of us have ever known—was our father's sister. Irving McCormick, our mother's brother, worked in the office, as did Marjorie McRae, a cousin. Donald McRae, who took over from J.S.L., is a cousin. Alex McGregor was a cousin. Bob and John are cousins. And today, younger cousins are carrying on. We speak here, obviously, of a family business.

My first memories are of the Hay Lake mill in the early 1950s. I remember hearing more Polish than English being spoken around the cookhouse. I remember the old alligator moored in the boathouse; I recall the big friendly black Labrador dogs that had the run of the yard; I remember the creek where I caught my first speckled trout; I remember the dust that rose from the logging trucks that would choke a youngster walking the road back from the creek with a fishing pole and a can of worms gathered from lifting fallen boards back of the woodpiles.

Hay Lake was followed by the Mink Lake mill by Lake St. Peter. They kept workhorses in the barns and we used to go down there and tease the pigs that were kept in a pen down by the creek. The longest and sharpest memories, however, belong to the mill at Rock Lake, a big operation, where we were always made welcome and where the cookies they kept in the big tins up at the cookhouse seemed as big as car hubcaps. In small ways, my brothers and I grew up there, staying over often, fishing in the various lakes each summer and, year round, dropping in to visit (and of course, so our parents could hold their monthly sword fight over the paycheque). I remember one winter going in to a camp and having surely the best meal of my life, in the middle of the bush at 30 degrees below zero. It was at Rock Lake where we got to know the likes of Alex Cenzura and Felix Tomaszewski and, the friendliest of all, Gary Cannon. It was here I came to work one summer and left almost immediately when a job came through in Huntsville that would allow me to spend more time with the young woman, Ellen, I would eventually marry. I don't regret a moment of the summer spent with Ellen, but if it were possible to live two lives at once, I'd still have liked to have had a summer at the mill to look back on as well. Don Lloyd's wonderful book is in some ways making up for that squandered opportunity to get to know the mill and its people even better than I already did.

Those of us who grew up around Algonquin Park have always had to deal with the issue of logging and the environment. In our case, the divide was immediate, as our grandfather, Tom McCormick, was chief ranger, and our father, Duncan MacGregor, was one of the main men at the McRae mill. The two men never argued, as so often happens when two opposing forces meet in the cities. The ranger understood his job was to protect the forest, but he also knew that his neighbours and relatives in and around Whitney depended upon the forest for their livelihood. He knew that access to the great forests of the Canadian Shield had come about because of logging roads. He knew that the portages so treasured by those who come to this magnificent park exist, in large part, because of the loggers. But he knew, above all, that the forest needed protection, or else it would vanish. Long before "harvest" was a word applied to logging, he and others at the McRae operation were doing just that: harvesting. He treasured, in his own way, the environment and disdained any bad logging practices he came across. It is heartening, as a relative and as a frequent canoeist in Algonquin Park, to know that the McRaes were at the forefront of understanding the obligations of working such a valuable and irreplaceable—sacred—territory.

Our father and grandfather came to understand and respect each other; it can be no different for those who use the park in different ways today, tomorrow and—hopefully—in the centuries to come.

All of this has been captured by Don Lloyd: the history, the tensions, the resolutions, the future. He has written a book that speaks to those who work in the park as well as to those who play in the park. He has given a crash course in the history of logging and in the methods of logging, but he has done so in such a way that you do not feel you are learning. If the best that can ever be said of a book is that it is a *good* read, then this is a good read indeed.

It is because Don Lloyd has told a story here in *Algonquin Harvest*—a story of a family, yes, but a story, too, of an extended family that includes all of us, and one exceptional piece of land which we all have a stake in and a duty to protect.

Roy MacGregor
Ottawa, 2005

PREFACE AND ACKNOWLEDGEMENTS

It has been my good fortune to have canoed many of Algonquin Park's waters since 1945. In doing so, I have always been intrigued by the signs of logging I have found there. Occasionally I would run into abandoned camps, the remains of bridges that crossed streams, as well as logs, abandoned on drives, that still showed one end above water. Along portages, purple vetch, daisies and timothy betrayed the fact that lumbermen's horses had passed that way. Where falls occurred, the remains of dams and log chutes were common. I tried to envisage the work and life of the lumberman—mine was a romantic notion, for in reality the life of a lumberman was, and still can be, a tough and frequently dangerous way to make a living.

In 1983, The Friends of Algonquin Park was founded. Bob McRae, the president of the McRae Lumber Company, became a fellow director. Over the years we exchanged pleasantries before meetings and got to know each other. In 2000, I published *Canoeing Algonquin Park,* and I was looking for another project. I asked Bob about the possibility of writing a book on the McRae Lumber Company.

Hugh MacLennan wrote *Two Solitudes,* and in it described the cultural differences between the French Canadians and the English of Canada. There is assuredly another divide in Canada today—that being one between rural and urban ways of life in Canada. Although urban born, I have enjoyed canoe tripping to a considerable degree, but in delving into lumbering, I have had a great amount of learning to do just to scratch the surface.

Fortunately, I was able to glean a great deal from the field reports by students from the University of Toronto's Faculty of Forestry that dealt with the McRae Lumber Company for the years 1926, 1930–1931, 1948, 1949 and 1950. These reports are fine records of company operations for the dates written.

In 1986, Ian Radforth, author of *Bush Workers and Bosses* and now an associate professor of history at the University of Toronto, carried out a series of taped interviews with the McRae family, their friends and with employees, both current and retired. These interviews not only covered historical aspects of the company, but also revealed the personalities of the people involved. Radforth had started on a manuscript before having to abandon the project of completing a book for the McRaes. Professor Radforth released these tapes to the McRae family for use in any subsequent book—a generous gesture on his part that I am most happy to acknowledge here.

A delightful tape is one that was done by Mark Webber, husband of Bob's sister Janet, with Janet MacGregor McRae, wife of J.S.L. The archives at the Visitor Centre in Algonquin Park also hold tapes that relate to the McRae mills.

It has been my pleasure to also interview the McRae family members and former employees. I particularly wish to thank Merv Dunne of the McRae Lumber Company for taking me around town and introducing me to both family members and former employees.

Roy MacGregor wrote the Foreword to the book. I do not know him apart from brief conversations on the phone and an e-mail or two. His text arrived when I was having a bad time with the manuscript and his comments were a real boost. I had to put my head into the freezer for a while to make certain it didn't swell. Thank you, Roy. I just hope the finished product comes close to your most generous comments.

Many people have helped in putting this book together. Past and present employees of the Algonquin Forestry Authority (AFA) and the Ministry of Natural Resources (MNR) have been very generous with their time in explaining the history and role of the AFA. Bill Calvert, who wrote the Master Plan for the MNR, told of those very interesting times. Bernard Reynolds, a charter member of the Board of the AFA, described that organization's early years, along with Bill Brown, a retired general manager of the AFA. Don George, a retired management supervisor of all phases of forestry in Algonquin Park, has been very encouraging. Arthur Herridge, Frank MacDougall's right hand man at the Department of Lands and Forests in the 1960s and early 1970s, and Deputy Minister of Northern Affairs, 1979–1982, contributed insightful observations from the government perspective. Carl Corbett, the current manager of the AFA, Gord Cumming, its chief forester, and other AFA personnel have patiently answered many questions.

To all of the above, and for the contributions of friends George Welsh, Kevin Browne, Ron Tozer, Pat Tozer, Dan Strickland, Rachel McRae, Jamie McRae, Ernie Martelle, John Winters, John Simpson, Roderick MacKay and my wife, Joyce, and to the many others whom I have imposed upon in writing this book, my sincere thanks. I hope I have not tried your patience too much and that the book has turned out to be worth your efforts.

Writing the text was one thing. Crafting the draft into a book took the skills of two individuals who were a pleasure to work with. Dominic Farrell edited the draft manuscripts and carried out this task in a most professional manner. He is a master of correction. In a few cases I exercised an author's prerogative and disagreed with the official way of doing things. Fortunato Aglialoro of *Fortunato Design Inc.* did the maps, charts and diagrams and set up the book. These men turned a manuscript into a well-crafted book. I appreciate the skills of both, value their friendship and remain amazed that they still talk to me.

Algonquin Harvest is the story of a very successful business, managed to date by three family generations. In doing so, the McRaes have maintained the highest respect of their employees, the people of Whitney, the MNR, AFA personnel and their customers. The story is worthy of being told—I hope the book does justice to the subject.

As author, I have found the experience of researching and writing this book rewarding. Discovering the facts and putting them together has been compelling, but far more satisfying for me has been the pleasure, indeed

Preface and Acknowledgements

the excitement, in meeting so many wonderful people. The book is a story of people—hard-working and talented—all of whom have been delightful to get to know.

Jamie McRae researched the maps, ordered the photographs and picked up on items that needed correcting, as only one who grew up in Whitney and one who was born to sawdust could do. Thank you Jamie, you have been a pleasure to work with.

This book would not have been started or completed without the interest and cooperation of Bob McRae. Bob contributed many items, quietly made many suggestions and provided resource books from his library because he believes that the journey of the people of the McRae Lumber Company has been important historically and that this story should be told.

Thank you Bob for making this possible.

Don Lloyd
Canoe Lake, 2006.

INTRODUCTION

This book is a history of the McRae Lumber Company and tells of changes in its operations from 1911 to 2006. Since its timber largely comes from Algonquin Provincial Park, the company has had to deal with the consequences of the divergent views of how that marvellous tract of public land is used. The public and political battles over the Park have been exciting.

This book is also about people: the men who cut the trees with crosscut saws and axes, and the men who moved the logs to the mills by horse-drawn sleighs and alligator boats; it is a story about mechanization and the introduction of trucks and power saws, feller-bunchers, slasher lines, Deal processors and chippers. It is also the story of a town: life has also evolved in Whitney since its days as a company town where the people used well-water and oil lamps; where women, alone for much of the year, raised their children and tended wood-fired stoves. Today, Whitney is a modern town whose houses boast electric dishwashers and microwave ovens. Told also is the story of Donald McRae's 33 months as a POW in Stalag III, the German prisoner of war camp made famous by *The Great Escape*.

McRae Lumber operates as part of an

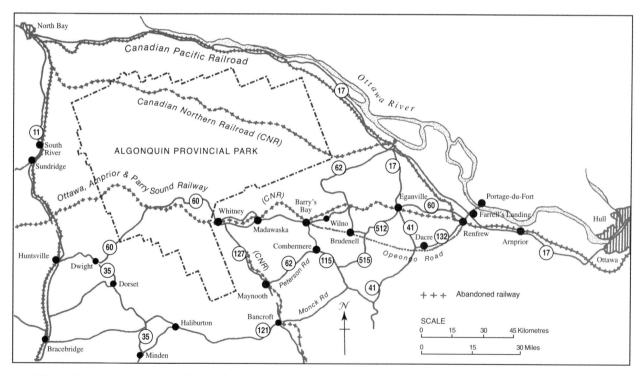

Map 1 **General Locations: Algonquin Park Area**

industry whose history in the Ottawa Valley dates back to the early 1800s, when the area was being settled. The first to arrive were the French-Canadians. They were followed by the English and Scots, who took up the better agricultural land near the Ottawa River. Numerous Irish settlers, and, later, those of German background, took up land back from the river. These settlers were followed by Poles, who settled the rocky hills to the west around Wilno, near the Opeongo Road. South of what later became Algonquin Park, English-speaking people from southern Ontario settled Bancroft, Maynooth and the neighbouring area. Men from these diverse groups still form the labour force that work in the surrounding forests.

The first McRaes arrived in what is now eastern Ontario in the 1820s, survived pioneer farming and moved into lumbering by the 1850s. Family ancestors worked for the legendary John Egan and the McLachlin Brothers, who were large-scale square-timber operators. John D. McRae owned a thriving general store and lumber mill in Eganville. The disastrous Eganville fire of 1911 nearly wiped out the livelihood of John D. McRae, but the same event launched his son John Stanley Lothian McRae (J.S.L.) into lumbering as a jobber, cutting small stands of big pine that had been ignored by the bigger companies. With fair treatment, steady employment and a paycheque that was always good, J.S.L. won the loyalty of the men who worked for him.

In 1893, Ontario passed the *Algonquin National Park Act*. This act set aside some 18 townships in the Ottawa-Huron Tract of the Nipissing District. Lumbermen were very active in the area but were permitted to stay until all the pine was removed. By 1913, legislation was enacted which permitted the cutting of all species of trees, and the name of the park was changed to Algonquin Provincial Park. The 1893 legislation, however, clearly set out conservation, recreational and scientific reasons for setting up the park.[1]

Whitney's logging history began in 1895, when the St. Anthony Lumber Company built its big mill to take advantage of the abundant sawlog-sized pine of the area, which the company was able to transport to market over J.R. Booth's newly completed Ottawa, Arnprior and Parry Sound Railway (OA&PS). Whitney was the small village built by St. Anthony to accommodate its workers. The Mickle, Dyment Company of Gravenhurst, Ontario established a smaller mill in 1903 at Airy, adjacent to Whitney.[2] It also cut mainly softwood.

By 1911, the St. Anthony Lumber Company had cut most of the pine out of its timber limits—the term used for the lands leased from the government from which it could cut timber. It sold its big mill and timber leases to the Munn Lumber Company, which was interested mainly in hardwood. Then the first conflict between recreationists and lumber interests took place—centred on Highland Inn at Cache Lake in Algonquin Park. The government of the day was pressured into buying out the Munn operation. Shortly thereafter, the Dennis Canadian Lumber Company took over the big mill in Whitney. In 1922, poor economic times led to the closing of the Dennis Canadian mill, which then burned. At the same time, when the opportunity presented itself, J.S.L. purchased the Mickle, Dyment mill, intending to cut mainly hardwood. The small village of Whitney survived as a result.

To make his purchase profitable, J.S.L. had a problem to solve: fresh-cut hardwood soon sank when placed in water, yet the only way to transport the logs from the cutting areas was by water. J.S.L. solved this problem by wiring his hardwood logs to buoyant softwood cribs and towing them with motorized boats. He was thus able to profitably harvest a large area.

McRae Lumber has always cut most of its timber in Algonquin Park: from east of Burnt Island to the west of Lake Opeongo, and south through the Park's panhandle townships of Clyde and Bruton—an area that is largely within the upper drainage area of the Madawaska River (Map 2).

The late 1960s and early 1970s brought a serious problem to the companies having timber licences in Algonquin, as preservationists/recreationists campaigned to "return Algonquin Park to nature." According to them, lumbering had no place in a provincial park. McRae Lumber, along with 11 others, had a legal licence to cut timber in Algonquin. Without a wood supply, McRae and the other companies would be out of business. A compromise was engineered by the government of the day. The Master Plan of 1974 was that compromise. The Algonquin Park Master Plan effectively separated recreationists from loggers. Zones were established where logging was prohibited, and logging was restricted in the summer months when recreationists were most numerous. The Master Plan was followed in 1975 by the creation of the Crown agency, the Algonquin Forestry Authority (AFA). The AFA was given the sole licence to cut wood in Algonquin, with the mandate to deliver wood to the previous licence holders, and it has done so.

The McRae Lumber Company prospered under J.S.L. and his son, Donald, and has continued to do so under grandsons Bob and John. In its 90-plus-year history many changes have taken place in its operations. The most dramatic change has been its evolution from a labour-intensive operation to a capital-and-machine-intensive one. Significantly, harvesting the forest is now done in a sustained-yield manner rather than by traditional high-grading.[3] Yet, as in times past, the wood is still brought from the forest by men who have to contend with winter's snow and the mosquitoes and blackflies present in the spring. Lumbering is a physically demanding and highly skilled occupation.

Today, the provincial timber industry operates under rules and regulations that are designed to produce a healthy forest of high biodiversity while also yielding quality timber. Within Algonquin Park, multiple uses continue to serve a varied constituency of patrons. Through all of these changes, the McRae Lumber Company has prospered, all the while maintaining respect in the industry, contributing to its community and being a good employer.

Talented and colourful people appear in this story: loggers and poachers, bootleggers and scientists. All have contributed to the history of the company and to Algonquin Park's continuing saga.

Introduction

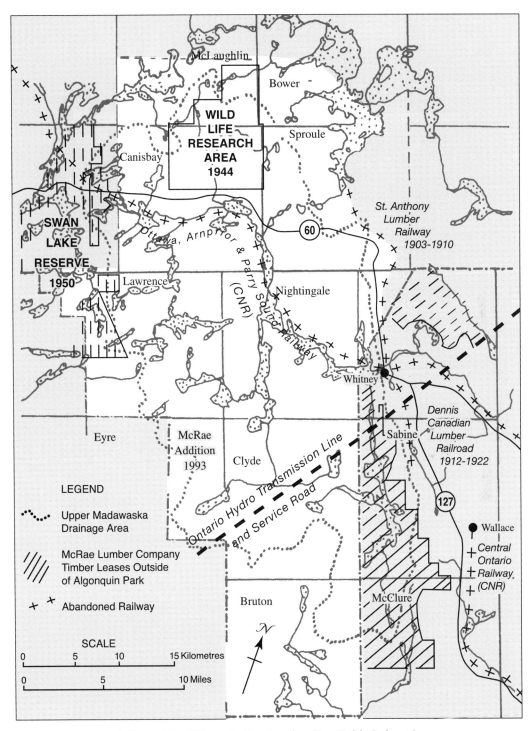

Map 2 Townships Where McRae Lumber Has Held Timber Licences

Seasoning and Preserving Timber

For the purpose of seasoning, timber should be piled under shelter, where it may be kept dry, but not exposed to a strong current of air; at the same time, there should be free circulation of air about the timber, with which view slats or blocks of wood should be placed between the pieces that lie over each other, near enough to prevent the timber from bending.

In the sheds, the pieces of timber should be piled in this way, or in square piles, and classed according to age and kind. Each pile should be distinctly marked with the number and kind of pieces, and the age, or the date of receiving them.

The piles should be taken down and made over again at intervals, varying with the length of time which the timber has been cut.

The seasoning of timber requires from two to four years, according to its size.

Musson's Improved Lumber and Log Book
New and Revised—Illustrated Edition
Based on Doyle's Rule, p. 12

Section I: SETTING THE SCENE

The settling of the Ottawa, Bonnechere and Madawaska valleys in the first half of the 1800s created a mosaic made up of French-Canadians, English, Scottish and numerous Irish. There were also distinctive German areas. Towns such as Pembroke, Arnprior and Renfrew got their start then.

Profits from the square-timber trade, begun by Philemon Wright in 1806, followed by those from the mills producing softwood lumber, stimulated commercial agriculture and the building of canals and roads. John Egan, the greatest of the square-timber barons, founded Eganville in 1853.[4] In an effort to attract more settlers, "colonization" roads were developed. The Ottawa–Opeongo Road, which opened in 1853 and ran from Farrell's Landing on the Ottawa River, was one of them. Beginning in 1859, Polish settlers took up land along it to the west, mainly near Wilno.

A population expansion also took place to the south. From English-speaking Bancroft, settlement into neighbouring areas and northward to beyond Maynooth took place. By the 1840s, sleighs carrying supplies from Maynooth used a winter road to John Egan's huge timber limit north of the Madawaska River; a branch of the road also ran westward to the Perley depot farm on Long (Galeairy) Lake. John Egan died in 1857 and several of his timber limits were sold off in an attempt to keep the leaderless business afloat. A large remnant, known as the "Egan Estate," reverted to the government and was bought by J.R. Booth in 1867. In 1893, Algonquin National Park was created out of the District of Nipissing. Booth, sensing the likelihood of increased government interference in timbering, offered to buy the entire park. Nonetheless, Booth continued to develop his holdings. To bring logs from this timber operation to market more efficiently, J.R. Booth built the Ottawa, Arnprior and Parry Sound Railway (OA&PS) in 1895. This railway spawned several other mills along its route. The St. Anthony mill and a village named Whitney were built in 1895. The Mickle, Dyment mill was built at nearby Airy in 1903.

The last significant immigration of people into the area took place in Sabine Township, south of Whitney. This influx, again of Polish people, took place after the Central Ontario Railway, which pushed northward from Maynooth to its terminus at Wallace, began operations in 1911.

The McRae family immigrated to Kenyon Township in Glengarry County in the 1820s. They survived pioneering and had enough resources to send their youngest son, John D., to McGill University. In 1872, J.D., a pre-medical student, went to Eganville and took a job as a summer school teacher, but instead of returning to school, he married the daughter of Robert Campbell, a very successful merchant of the town. Campbell took J.D. into the business and he did very well. However, the Eganville fire of 1911 destroyed the very comfortable lifestyle of J.D. McRae, as his business was largely destroyed. After the fire, three of J.D.'s sons moved to the West, but John Stanley Lothian (J.S.L.) McRae went into the timber business on his own and worked his way up the lands drained by the Madawaska River.

Starting as a small-scale timber jobber, J.S.L. gradually built up the scale of his operation as well as his good name as an employer. In 1922, both mills in Whitney were suffering from an economic downturn. The owners of the big mill, by then Dennis Canadian, closed up business; its building accidentally burned during demolition. With the pine largely cut from its limits, and not liking to cut hardwood, Mickle, Dyment sold its smaller mill at Airy to J.S.L. McRae.

Since that time, McRae Lumber has had harvesting operations and mills in the area. The company's office has always been in Airy or Whitney, and many of its employees have lived in the town and the surrounding area, thereby sustaining the community.

Chapter 1
Peopling the Ottawa, Bonnechere and Madawaska River Valleys

Philemon Wright established Wrightsville (now Hull, Quebec) in 1800 with five families and some 25 axemen. In 1806, he made his first trip down the Ottawa and St. Lawrence rivers to Quebec City with a raft of square pine and oak staves, which he hoped to sell at the timber coves of Quebec City.[5]

Up to this date, trade in timber from British North America to Britain had been largely restricted to expensive items such as masts. These could cover the high cost of shipping across the Atlantic Ocean whereas cheaper wood was usually supplied to Britain over the much shorter distance from the Baltic Sea ports. Wright was lucky, for his arrival at Quebec City coincided with the wartime blockade of the Baltic Sea by Napoleon. This action cut Britain off from its usual source of timber, which was vital for the wooden ships of its navy and merchant marine. To counter this situation, Britain instituted high import duties, termed "colonial preferences," on non-colonial timber. These duties effectively raised the value of British North American wood by making it more competitive, thus compensating for the much higher shipping costs to Britain. These duties, which remained in force for some 40 years, triggered the development of the Canadian timber industry.

By 1814, Wright and his sons, employing 200 men in the bush, were cutting, squaring and sending to Quebec as many as 20 rafts of pine a year.[6] The assault on the pine forests of Canada had begun in the Ottawa Valley as well as in the Maritimes and up the St. Lawrence into the Great Lakes.

With the end of the Napoleonic Wars in 1815, Britain's economy slumped. The following years saw immigration into the still-forested lands upstream from Montreal along the St. Lawrence by demobilized soldiers, displaced tenant farmers and others from the British Isles. At this time, the ancestors of the present-day McRaes pioneered at St. Elmo, Kenyon Township, in Glengarry County. In 1825, Colonel By arrived to supervise the digging of the Rideau Canal which opened in 1832. With the canal finished, the Irish navvy labour employed by Colonel By to dig the canal became a gang of ruffians, one that terrorized men moving timber down the Ottawa River for several decades. The "slab town" made from the discarded outer boards cut from pine logs, located at the junction of the Rideau and Ottawa Rivers, became known as Bytown (later Ottawa) (Map 3).

By 1855, when Bytown was renamed Ottawa, the area was shipping thousands of "sticks" of squared timber to Britain, and its hydraulic-powered mills were turning out

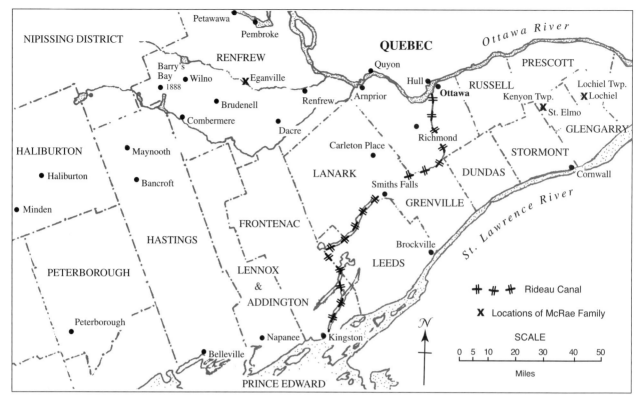

Map 3 Locations Where the McRae Family's Ancestors Lived, 1888

Farther up the Ottawa River, toward what became Renfrew and Pembroke, a mosaic made up of English, Scottish, and numerous Irish settlements was established on the better agricultural land. Intermixed were occasional French-Canadian communities. However, timbering, not agriculture, quickly came to drive the economy of the Ottawa Valley.

Algonquin Park was not established until 1893. Whitney was established in 1895.

some 250 million board feet of sawn lumber destined for the developing urban markets in the northeastern United States.[7]

In the 1850s, John Egan was the largest square-timber operator in the Canadas, having licences that covered over 2,000 square miles in Canada East (now Quebec) and Canada West (now Ontario). A timber licence gave the holder the right to cut timber on a specific piece of land, subject to certain rules, while the Crown continued to own the land. Egan's large mill was at Quyon, upstream from Ottawa in Canada East. He also had a smaller, two-saw mill at Milldam on the Little Bonnechere River in Canada West. Nearby was a large depot farm at Clancy Lake that supplied oats, potatoes and hay to the bush camps. This area was destined to become part of Algonquin Park nearly 40 years later. Egan died in 1857, and, in an effort to keep the business going some timber licences such as one on the Little Bonnechere River were sold off. Alex Barnet bought the one which included the depot farm at Clancy Lake. In 1867, after the remaining Egan licence had reverted to the Crown because of non-payment of dues, J. R. Booth purchased what had become known as the "Egan Estate" timber limit on the Madawaska river for the then huge price of $45,000 (Map 5).[8]

John Egan had bought the "Fairfield Farm" property at the fifth chute of the Bonnechere from James Wadsworth, who had purchased the property from Grégoire Bélanger, the original owner. Eganville was surveyed and given its name in 1853. The gristmill and stores located in the town and the large timber operations in the surrounding area attracted farmers. Among the townspeople, who included merchants and timber operation supervisors, were the ancestors of the current owners of the McRae Lumber Company.

From the beginning, agriculture in the Ottawa Valley, unlike that in much of the remainder of southern Ontario, was commercially oriented, producing cash crops of hay, oats, potatoes, pork and beef, as well as providing horses for the timber companies. In winter, farmers of the area and their sons often joined the lumberjacks in cutting the pine, or else they worked as teamsters with their oxen and horses.[9]

Commercial interests in the towns and in the lumber industry had a common interest in promoting better transportation. Canals were the first focus. Beginning in the 1820s, the idea of a canal joining the Ottawa River with Georgian Bay was entertained, but the idea was abandoned when the reality of the nature of the intervening country, one of rock ridges and thin soil, was revealed by surveys. Developing roads became the new strategy.

In 1853, through the office of the Ministry of Agriculture of the Canadas, a program of colonization roads was launched. This effort was designed to open up farmland to the sons of southern Ontario farmers and, so, keep them in the colony, rather than having them emigrate to the American frontier. Roads were planned and eventually pushed into the Madawaska Highlands. The Ottawa–Opeongo Road ran westward from Farrell's Landing on the Ottawa River.[10] This road (usually referred to simply as the Opeongo Road) was supposed to reach Georgian Bay, but it terminated, instead, a short distance to the west of Barry's Bay, near what became the village of Madawaska (Maps 4 & 5).

To populate these areas, the government in 1856 began advertising free lands to settlers from Britain and continental Europe. The *Free Homestead Act* of 1868 further stimulated immigration. The Opeongo Road attracted German settlers to Sebastopol and Brudenell townships before difficult terrain and infertile lands discouraged further westward development by that group of settlers (Map 4).[11]

By the 1920s, the sons and grandsons of the pioneers of this area were working as loggers in the local bush camps and mills. In the aftermath of World War II, numerous displaced persons came to Canada. Many of these had served in the British armed forces. A large number of these were Polish people who did not want to return to Poland. Some of these men turned up in the Whitney area and worked for a time for the McRae Lumber Company. A few of them settled in Whitney.

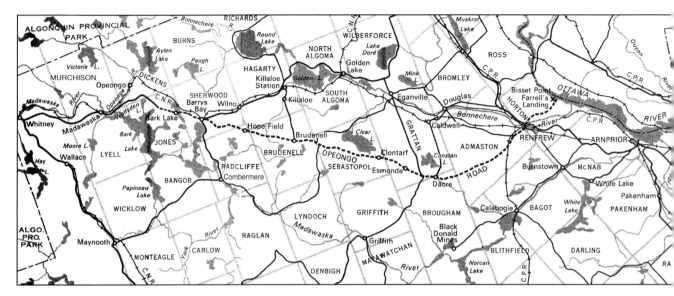

Map 4 Ottawa–Opeongo Road and Ottawa, Arnprior and Parry Sound Railway

Beginning in 1859, several thousand Polish families settled along the Opeongo Road around Wilno in Hagarty Township. In the summer, they struggled with their subsistence farms on the hilly, stony land; in winter, they worked in the timber camps of the area. Their settlements spread westward to Barry's Bay. The names of the villages and roads in the area, such as Paplinski, Luckasavitch, Coulas and others, preserve their origins.

In the 1840s, a winter road from Maynooth to the north was used to supply John Egan's huge timber limit on the Madawaska. Thomas Coghlan had a boarding house and stables for 40 horses just south of Mackenzie Lake to serve this road. If supplying the Perley farm on Long (Galeairy) Lake, the sleigh train stopped overnight at Dunn's in Murchison Township before turning west in the morning for the depot at Farm Bay (Map 5).[12] Coghlan's prosperous business ended when J.R. Booth, who purchased the Egan Estate timber limits in 1867, constructed the Ottawa, Arnprior and Parry Sound Railway. The OA&PS Railway reached Whitney in 1895.[13]

Source: Used by permission of *Canadian Geographic Journal*.

Ottawa–Opeongo Road
Ottawa, Arnprior and Parry Sound Railway

facing page:
Map 5 Upper Madawaska Valley Locations

By 1875, the area that was to become Algonquin Provincial Park was covered with timber licences. In the upper Madawaska Valley, A.C. Kelly, a speculator, held a huge area that included lands from around Source Lake to Long (Galeairy) Lake. Alexander Fraser held the southeast side of Lake Opeongo. The "Egan Estate" was one of the licences owned by J.R. Booth. Note that W. M. Mackey had succeeded Alex Barnet on the Bonnechere River by this time.

Chapter 1: Peopling the Ottawa, Bonnechere and Madawaska River Valleys

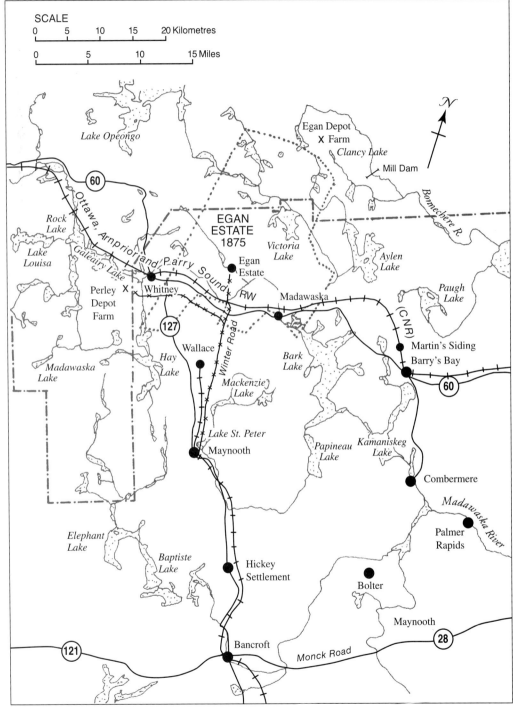

Map 5 **Upper Madawaska Valley Locations**

ALGONQUIN HARVEST

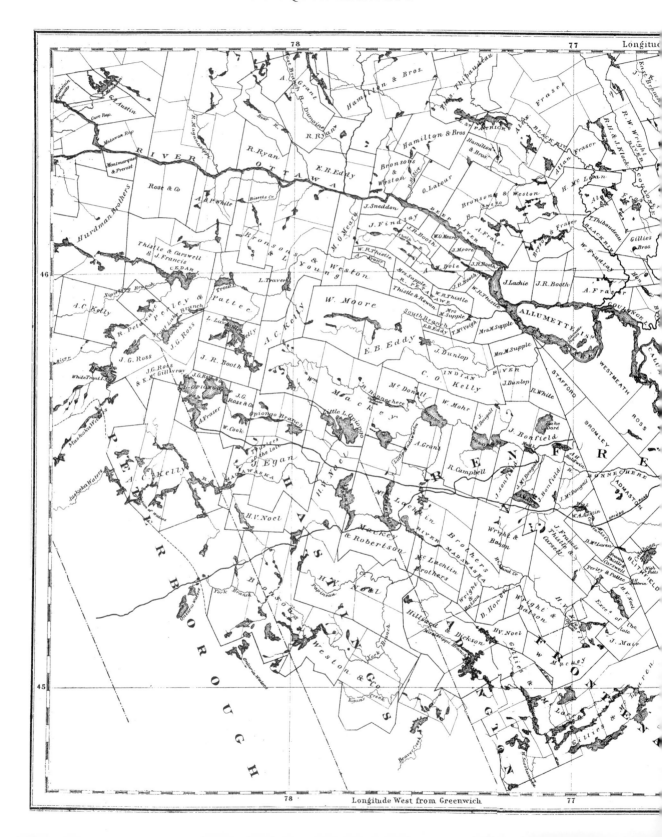

Chapter 1: Peopling the Ottawa, Bonnechere and Madawaska River Valleys

Map 6 **Timber Licences in Ontario, 1875**
Source: A.J. Russell, *The New Standard Atlas of the Dominion of Canada*, Montreal and Toronto: Walker and Miles, 1875, 157.

ALGONQUIN HARVEST

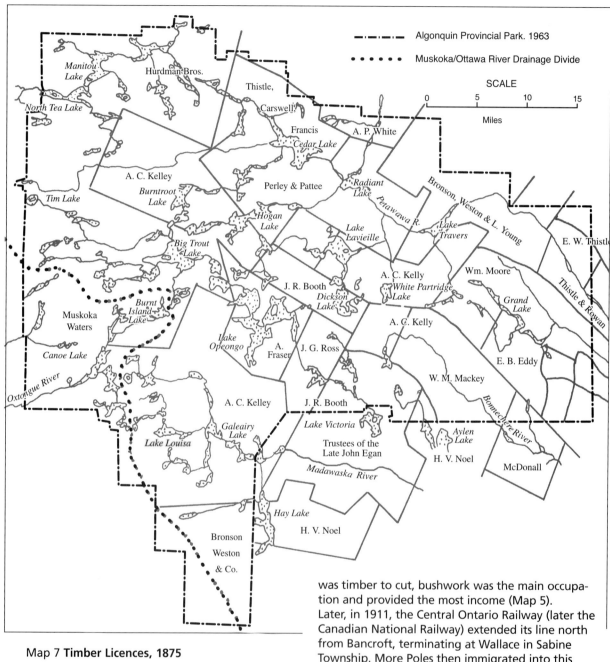

Map 7 Timber Licences, 1875

To the south and east of Maynooth, through Boulter, McArthur Mills and Hickey Settlement, scattered English-speaking settlements developed. However, farming on the rocky hills of the Canadian Shield was never really successful. As long as there was timber to cut, bushwork was the main occupation and provided the most income (Map 5).

Later, in 1911, the Central Ontario Railway (later the Canadian National Railway) extended its line north from Bancroft, terminating at Wallace in Sabine Township. More Poles then immigrated into this area. No towns were established in Sabine; only a community centre and a cemetery remain today to mark the settlement. Luckasavitch Lake is the only mapped reminder of the Polish heritage of this area.
Source: Redrawn from Russell, 1875.

Chapter 2
Early Mills of the Whitney Area

The St. Anthony Lumber Company: 1895–1909

Square-timber operators such as Perley and Pattee had already cut the easily accessible stands of large pine from the area in the 50 years or so before Booth's Ottawa, Arnprior and Parry Sound Railway reached the east end of Long (Galeairy) Lake in 1895. The St. Anthony Lumber Company of Minneapolis, Minnesota built its mill at this site in 1895 and established the village of Whitney in the same year. The site was an ideal location for a mill, for the headwaters of the Madawaska River still held immense numbers of sawlog-quality pine, and the railroad, the Ottawa, Arnprior and Parry Sound Railway, was available to carry the finished lumber to markets.

In the spring of 1895, tracklayers of the OA&PS Railway reached Long Lake to find a small dam and the nearly completed mill building of the St. Anthony Lumber Company.[14] In the previous fall, some 100 men and horses had proceeded from Barry's Bay up the railway right-of-way to start building the mill.[15] It took 14 years for St. Anthony to cut the pine from its limits, to put the logs up one of three jack ladders and into the saws of the big mill.

St. Anthony operated a large mill with up-to-date technology. Sawing began on July 25, 1895. *Canada Lumberman* reported in September 1895 that:

the mill building consists of the main building, 88 x 208 feet; its floor is without posts, the roof being supported by a truss; the shingle and lath mill, 48′ x 52′ and the sorting shed, 32′ x 276′. The lumber mill contains 3 band saws and a gang saw, with 2 six- and 1 four-saw edgers and 2 eleven-saw trimmers. The engine and boiler house, 81 x 82 feet, is built of brick, stone and iron. A 900-horsepower Allis Corliss engine and a battery of eight boilers, 60 inch[es] by 20 feet produce power for the mill.

The mill is outfitted with all the latest and best labour-saving machinery such as steam feed, steam flippers and kickers required to handle logs and lumber. The logs are not touched by hand from the time they enter, 'til they are sawed. Sawdust is used for fuel and the furnaces are fed automatically. The surplus and mill refuse is consumed in a burner 30 feet in diameter and 90 feet high.

The lumber is sorted automatically and is taken from the mill on small cars, the facilities being such that two single horses can haul it away. In the lumber yard, there are ten miles of small railway as well as five miles of standard gauge track to accommodate cars, to load lumber for shipment over the Ottawa, Arnprior & Parry Sound Railway.[16]

The mill was said to have a capacity of some 200,000 board feet per ten-hour day and

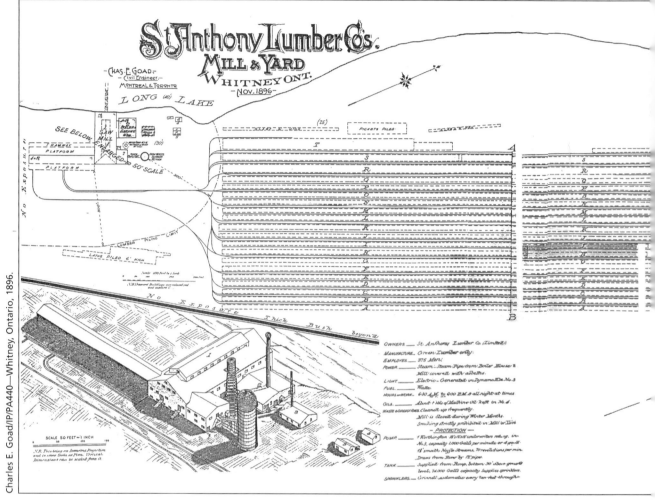

Figure 1 **St. Anthony Lumber Company and the Village of Whitney**
Source: By permission of the Library and Archives Canada.

thereby to rank with the largest mills on the continent, but actual production was less. Production did reach 50 million board feet in 1901 and 1902, but fell to less than 25 million board feet in 1909. An output of 110,000 board feet per day was closer to average. Products of the mill were pine lumber, lath and shingles.[17]

The St. Anthony Lumber Company had its beginnings in Minneapolis, Minnesota in 1886. With dwindling timber supplies in northern Minnesota, the company moved across the border into Ontario to continue its business. St. Anthony brought with it experience in the use of logging railways.[18]

The building of the mill saw the beginning of Whitney as a company town. Edwin Canfield Whitney was the general manager of

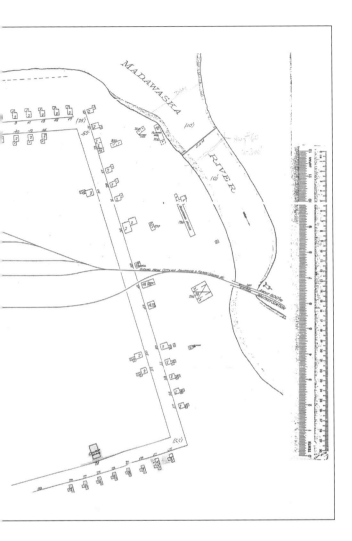

the mill, but the village was named after his brother, James P. Whitney, a Liberal politician who became premier of Ontario in 1915. The St. Anthony Lumber Company intended to build a model community. Thus no outsiders were allowed on company property, "the sale of whiskey was strictly prohibited, and drunkenness was punished with instant dismissal."[19]

The men were initially accommodated in the company's boarding houses, but the company promised that

cottages will be provided for the married men, to each a quarter of an acre of ground will be attached, which the men will be obliged to cultivate. Some fifty cottages will be erected. It is the intention to gather together an industrious and thrifty community, which will be characterized by sobriety and industry. There are 340 men employed at the mill and about 500 in the woods during the winter.[20]

In 1893, timber limits for the mill were held in the name of Arthur R. Hill of Saginaw, Michigan, a major backer of the St. Anthony mill. The limits stretched westward from Long (Galeairy) Lake through Whitefish, Rock, Lake of Two Rivers and Cache to Source Lake. The areas around Pen and Mackenzie lakes to the south, and Lake Louisa to the west, were also part of the licence (Maps 8 & 9). Logs were brought to the mill by this water system. But St. Anthony had a big problem. The pine was largely scattered throughout this upland area and there were few large stands of good quality trees. In consequence, the pine was difficult to cut economically and bush camps had to be relocated frequently. After ten cutting seasons they had to find a new source of trees.

In 1903, a 15-mile rail line from Whitney to Sproule Bay on Lake Opeongo was constructed by St. Anthony Lumber, thereby giving access to limits around the lake, including those formerly held by Alexander Fraser (Maps 6 & 7). The railway and steam hoists, capable of loading 11 flat cars in 90 minutes, made possible the efficient movement of logs from the company limits to the mill in 24 hours. Previously, logs from Lake Opeongo had to be driven down the Opeongo River to the Madawaska and then down the Ottawa River, a trip that took one whole season of open water. Accessing the Opeongo limits gave St. Anthony six more years of milling.

In 1909, the St. Anthony Lumber Company sawed its last log at Whitney after 14 years of operation. Like other lumber companies of the time, it closed up when the economical harvesting of the pine on its limits was finished. These years also marked the end of pine as the big cut in Ontario. Thereafter, spruce became the lumber of the construction trade.

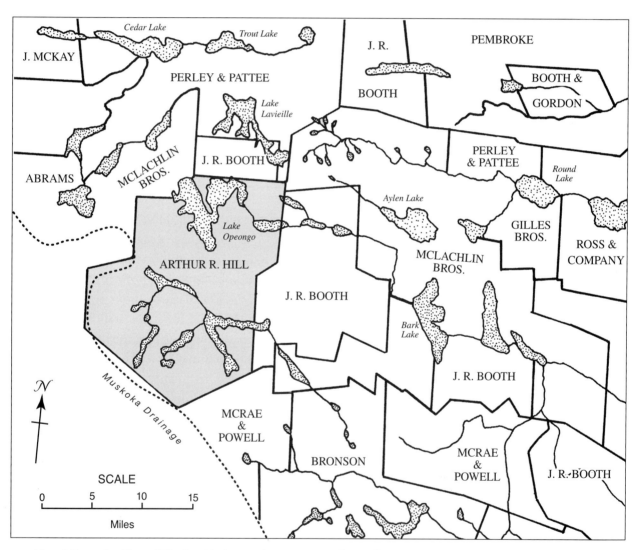

Map 8 **Part of a Plan of Timber Limits on the Ottawa and Its Tributaries**
Source: A. Charest, *Plan of Timber Limits on the Ottawa River and Tributaries, 1893.*

Chapter 2: Early Mills of the Whitney Area

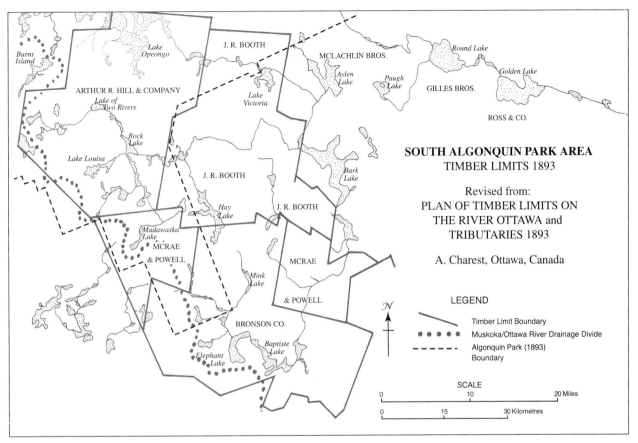

Map 9 **South Algonquin Park Area Timber Limits, 1893**
Source: Redrawn from Charest, 1893.

The collapse of the pine industry was sudden and the cut of pine has not returned to prominence, despite the fact that large areas of the province are capable of producing this valuable tree. "In 1908, the lumber industry cut about one billion board feet of red and white pine.[21] Today [2000], only 114,965,000 board feet of pine are cut in Ontario."[22]

The Munn Lumber Company: 1909–1910

The St. Anthony Lumber Company was fortunate that a buyer was interested in the hardwoods remaining on its limits. When Algonquin Park was created in 1893, timbering had been taking place in the area for over 50 years. The government had no intention of stopping it; indeed, it wanted to continue receiving fees from timbering licences and to continue employment in the industry. With

Photo 1 Log Loaders at Sproule Bay, Lake Opeongo, 1910
The date of this photo indicates that this was a Munn Company operation.
Source: R.J. Taylor. APM 3298

pine running out, changes in the *Park Act* of 1900 permitted the harvesting of "not only pine, but also spruce, hemlock, black and yellow birch, cedar, black ash and tamarack, until 1930."[23] Note that black and yellow birch trees are the same species. This act was clarified in 1913 with the result that all species were open to cutting.[24]

J.B. Tudhope, a Conservative MPP from Orillia, who needed a hardwood supply for his carriage factory, took over the Munn Lumber Company, which had operated in Haliburton. He then bought St. Anthony. However, Tudhope's Munn Company soon found itself the focus of the first confrontation between the nature-loving tourists and the logging industry in the history of Algonquin Park. Munn, unfortunately, chose to begin operations at Cache Lake, the site of both the Park headquarters and the Highland Inn, a resort hotel for wealthy tourists.

Algonquin Park headquarters was originally based on Canoe Lake, where in 1894 a substantial log building—21' x 28', with a hewn timber floor and a scoop roof—was built.

> The location seemed ideal because it was on the main waterway from the Muskoka District (Oxtongue River), through the Park to the Petawawa system; the Gilmour Lumber Company, Trenton, had started

building a tote road from Dorset to South Tea Lake, where they planned to establish their main depot; it would not be far from the proposed railway (OA&PS) and the Algonquin Park staff could get mail and supplies regularly from both of these sources.[25]

With the failure of their long log drive to their mill at Trenton on Lake Ontario, the Gilmours decided to locate a mill operation at Canoe Lake adjacent to the newly constructed OA&PS. Having park headquarters in the middle of a lumber mill was an untenable situation. A location across the lake from the mill was considered, but Cache Lake was eventually chosen.

Ottelyn Addison commented on the new headquarter's site and its construction.

> There is no record as to why Cache Lake was chosen for the permanent Park headquarters. The land might have been cleared for a depot during the building of the railroad. The inspectors and officials who came in to check construction progress were impressed by the beauty of the lake with its many wooded islands. The surrounding hills had never been burned or lumbered—a vivid contrast to Canoe Lake. It was a disadvantage to have headquarters off the main waterways but probably the scenic beauty was the deciding factor. The lumber and supplies for the new buildings arrived from the (J.R. Booth) "Egan Estate" mill in November, 1896.[26]

Despite the authority of the above quote, both Perley and St. Anthony had undoubtedly cut the Cache Lake area for pine in past years and raised its water level by a dam, but there had been enough intervening time for the lake to regain its beauty. In contrast, the scenery on Canoe Lake was dominated by a shoreline of dead trees that had been killed by high water held up by Gilmour's Tea Lake dam. In addition, considerable areas had been burned and the pine cut. Tom Thomson painted this desolate landscape in dramatic style, but others sought more pristine scenery.

In 1898, G.W. Bartlett, a former foreman of the J.R. Booth Lumber Company, was appointed park superintendent. He had a broad mandate which included tightening up the administration of the Park.

> I became Superintendent of the 3,000 square mile tract [Bartlett exaggerated here, as in his time Algonquin Park was about half that size]. The job included the position of Postmaster, Commissioner in the High Court of Justice, Police Magistrate, and Chief Coroner in the District of Nipissing. The premier of the Province told me that the Park had been a blot on the Government and asked me to make it a credit.[27]

In 1906, the Algonquin Park railway station opened at Cache Lake; this was followed by the establishment of the Highland Inn in 1908.[28] Highland Inn quickly became a very popular resort.

At the turn of the last century, timber companies were virtually a law unto themselves on their licensed land—at least many behaved that way; Munn certainly did. Slash was left in the bush, as was normal, but timber was also often cut to a small diameter, especially hemlock, whose bark was used as a source of tannin for tanning leather. According to its licence, Munn had every right to harvest timber. But it certainly chose the wrong place to start.

> By early May, 1910, the company's logging operations had commenced on the hills around Cache Lake. At first hemlock was taken. The lumbermen then stripped off its bark to be used in the tanning industry, and the remaining slash was left to spoil in the woods. On May 9th, Park Superintendent George Bartlett, concerned about the matter, wrote to Toronto to report that trees in front of his residence at park headquarters were falling fast. It would not have been so bad if only the mature timber was carefully taken out, he declared, but they were cutting hemlock down to the size of stovepipe.
>
> Over the next two weeks the situation intensified. Then, the Munn Lumber Company began cutting the birch trees at the park headquarters, and as people watched from the veranda of Grand Trunk's Highland Inn, some angry things were being said. A desperate George Bartlett again wrote to Queen's Park, urging to get the logging operations terminated. Five trees had fallen in the time it took to write the letter.[29]

In the end, the government decided to buy out the Munn Company. The letter of agreement with the government stated,

> The company agrees to accept $290,000 for the complete surrender of all its rights, pine timber and everything else, over an area of 350 square miles. It also agrees to remove its mills (at Whitney), take up the iron on its logging railway (from Whitney to Sproule Bay on Lake Opeongo), and thus the territory will come back to the Crown absolutely free from all rights of cutting, etc.[30]

This land was soon leased again, first to Dennis Canadian and later to McRae Lumber Company.

The Dennis Canadian Lumber Company: 1912–1922

The Dennis Canadian Lumber Company was organized in September 1912 and it took over the 150,000 acres of timber limits, the sawmill and other lumbering equipment of Cameron & Company of Ottawa. It was located just east of the Park in the Madawaska Valley, where, in 1912–1913, Cameron operated some eight bush camps, a railway from Aylen Station to a mill on Bark Lake, and a mill where the Canada Atlantic (formerly OA&PS) crossed the Opeongo River. The parent company, A.L. Dennis Salt & Lumber Company of Michigan, was a very successful hardwood lumbering firm. Dennis Canadian took over the former St. Anthony/Munn mill and limits on January 1, 1913, along with a 25-year lease on the hamlet of Whitney and all the buildings connected with it.[31] Dennis Canadian had to first solve the problem of how to get a large number of hardwood logs to its mill, a problem caused by the fact that these logs floated poorly.

A large portion of the timber being hardwood makes it necessary to adopt some method of lumbering other than floating the logs down the streams, as heretofore has been done. For this reason, a logging railroad is under construction from Whitney that eventually will extend about twenty-five miles southeasterly from that point. Eight miles of road, six miles of main line and two branches of about a mile each, are at present under construction and will be completed this season. The line will be extended from year to year as necessary to secure the log supply for the mill [Map 2].

The line is being constructed of standard gauge and 56-lb. steel is being used. The main line is being put in shape to make a first class roadbed, but not so much expense will be made on the branches. The first two miles of road out from Whitney are through very rough country and a good deal of rock had to be moved, but after striking the timber limits, which are reached two miles from the mill, the country is very favourable for railroad building and lumbering.

The railroad equipment has not been delivered as yet, excepting a few flat cars

Photo 2 **Montreal Motor Works Engine # 45928 at St. Anthony Lumber Company Mill, 1909**
The building in the background was a boarding house run by Sandy Haggart. Today, to the right of the engine would be the post office, while behind and to the left of the engine would be the LCBO.
Source: APM 2317

and one 50-ton Shay geared locomotive. A second locomotive, steam loaders and logging cars will be added before the beginning of log hauling in November. It is intended to keep the mill running throughout the year, cutting as largely as possible on hardwood in the winter and on softwood during the summer. The softwood from about ten thousand acres can be landed on Hay Lake during the winter and driven to the mill with very little cost for driving.[32]

Canada Lumberman noted:

This is one of the first attempts that have [*sic*] been made in Ontario to use the logging railroad for the purpose of taking out hardwood for any considerable distance, but the methods being adopted are the same that have been successfully employed in many of the States for years and some such means will have to be used to reach the bodies of hardwood at a distance from the railroads, where the pine has been floated down the streams.[33]

Bob McRae commented that "the problem with Dennis Canadian's strategy was the quality hardwoods were scattered. It was going to take too big an area to supply the mill by railroad. It would require too much railroad track."[34]

Dennis Canadian did realize this problem. They tried to enlist another wood user that was not dependent upon high-quality timber and thus have the rail lines used efficiently. *Canada Lumberman* reported on the issue:

Photo 3 Barnhart Loader, Dennis Canadian Lumber Company, Amable Creek
Evidence of the bush railway lines built by Dennis Canadian extends south from Whitney into the Hay Lake area.
Source: R. Van Meter. APM 4303

In order to utilize the enormous quantity of timber left on the lands after the removal of the sawlogs and which is fit for cordwood, an agreement has been made with the Standard Chemical Iron and Lumber Company of Toronto, whereby the Chemical company will erect within the coming year a large plant for the manufacture of charcoal, wood alcohol and by-products, using all the refuse from the sawmill for fuel and getting their cordwood from the limits on a stumpage basis. They intend to cut thirty thousand cords of wood during the coming season, which they will load on the cars of the lumber company for delivery to the factory yard. The erection of this factory with the necessary cottages for the workmen and the wood cutting operations will mean a large increase in the amount of business and labour activity at Whitney.

A shingle and tie mill is now being erected to take care of the cedar timber on the limits, and will soon be in operation. A large stock of cedar was put in with the other timber last winter and sorted into separate booms to await the erection of the shingle mill.

A general store is run in connection with the operations and does a very large business under the management of O.E. Post. The management of the business is in the hands of L.A. Van Meter, who was superintendent of the A.L. Dennis Salt & Lumber Company's operations in Michigan for eight years.[35]

The outbreak of World War I ended expansion efforts at Whitney.[36] In 1922, the big Dennis Canadian mill finally closed down and, as it was being dismantled, an accidental fire destroyed it completely. Many of the workers took the next train out of town, leaving their company houses behind.[37]

> Mr. O.E. Post had managed those houses and the mill for Dennis Canadian. In the following years the Ontario Government took over the townsite and leased it back to Mr. Post. At one point Post paid $1,100 a year to rent the town of Whitney. Those who remained experienced two decades of "hard scratching," heading into the Great Depression with no unemployment insurance.[38]

The Mickle, Dyment Lumber Company: 1903–1922

The Mickle, Dyment Lumber Company owned the smaller softwood mill at Airy, north of the Madawaska River. Two band saws, a band resaw, four shingle saws and a lath machine comprised the equipment at this mill. Some 100 men were employed to run the mill, which cut mainly pine.[39]

In comparison to the St. Anthony/Munn/Dennis Canadian operations, the Airy mill was a small one: in 1906 it sawed 6.5 million board feet, while, in 1907, *Canada Lumberman* reported production at the mill of 3.5 million board feet.[40] In contrast, the St. Anthony mill produced some 50 million board feet in 1901 and in 1902.[41]

There were three generations of Charles Mickles. Grandfather Charles pioneered in Puslinch Township on the Elora Road north of Guelph, but his main business was in timber: cutting and milling the pine of the Greenock Swamp. The product went mainly to the United States—via Dundas, Lake Ontario and

Photo 4 Dennis Canadian Lumber Yard, Whitney, 1922
When the Dennis Canadian mill burned, the lumber in the yard was sold. Later, Whitney expanded across the area.
Source: McRae Lumber Company

the Oswego Canal to New York City. When this cut was finished, the Mickles moved to Yakasippi in Bruce County and cut there. Father Charles moved to Gravenhurst, partnered for a short time with the Tates and then bought them out, in partnership with Nathaniel Dyment of Barrie. The Mickle, Dyment Company prospered, having at various times mills in Gravenhurst, Severn Bridge, Fenelon Falls, Whitney and at Thessalon on the north shore of Lake Huron, as well as working plants in Toronto and Brantford.

Son Charles S. Mickle, or "Charlie," ran the Whitney operation from 1903 until 1921. The combination of an ageing father wanting Charlie's help with the larger Gravenhurst operation and a downturn in the market in 1922, together with the fact that the softwood was running out on its Airy limits, led Mickle, Dyment to sell to J.S.L. McRae. Charlie then took over the management of the Mickle, Dyment Gravenhurst operations. Father Charles died in 1929 and the company was reorganized with Symon Dyment as milling president and Charlie as vice-president in charge of operations. A highway traffic accident claimed the life of Charlie in 1933. The company ceased large-scale operations in 1937.

Figure 2 **Letterhead of the Mickle, Dyment Lumber Company, 1920**
The title of the mill at Airy varied, as was noted previously.
Source: McRae Lumber Company

Charlie's father made interesting observations on the evolution of mill saws.

> The first mills were all the old gate type. The whole carriage moved up and down. Then came the "Muley" mills. In them it was only the saw that moved and they were thought to be a great improvement, for they gave greater speed. There was also a siding mill for cutting the cants made by the "Muley" and it was automatic both as to feeding back and forth and as to setting different thicknesses regulated by a dial. A curved bar solidly fastened to the floor and a knuckle on the carriage slid

over this bar in the forward movement of the carriage which in turn set the board. On the return of the carriage the knuckle doubled back when encountering the curved iron, there was some danger in cutting the last plank of the board running into the dogs. I was running this saw once when this occurred and it ripped the saw all to pieces and I made a dive for the sawdust hole in the floor. My lumbering days would have stopped right there if there had not been room for me to get down that hole. The pullover lever, still often used in portable mills, later overcame this.

Then came the circular saws. The circular saws gave place to the single cut band saw. Each of these was an improvement. The circular saw put the speed of production up seven or eight times, I should think, and the band saws increased it again. The first band saw cut only when the log was moving one way. Then they made them with teeth on both sides and they cut as the log moved both ways.[42]

Photo 5 Airy Mill, 1917
Note the high carrying platforms on the carts. When loaded and moved to the drying yard, the heavy boards could be off-loaded downward. This was much easier than lifting the heavy timber upward onto the drying piles.
Source: McRae Lumber Company

CHAPTER 3
The McRae Family Background, The Early Years of John Stanley Lothian McRae, The Purchase of The Airy Mill, and Observations of Early Whitney

The McRae family, owners of the McRae Lumber Company, is descendant from four immigrant families, the McRaes, the Campbells, the McGregors and the Keenans, all of whom came to Upper Canada (Ontario) shortly after 1820, following the Napoleonic Wars (1798–1815). The major motive for coming was the prospect of owning land, as opposed to staying in the British Isles and working for others. All four families, while not wealthy, brought resources enabling them to obtain land immediately. Three of the family ancestors of J.S.L. and his wife, Janet (Campbell) McRae, were Presbyterian Scots: the McRaes, the Campbells and the McGregors. The fourth, the Keenans, were Wesleyan Methodist Irish. All sides had ingrained the values of strong ties to family, clan and church; the ethic of hard work; and scrupulous fairness in dealings with employees and in business.

The Campbells

Robert Campbell was two years of age when the family arrived from Scotland. They found the land in Glengarry covered with an untouched forest made up largely of hardwood, with oak predominating. The soil was mainly based on heavy marine clay, and was very difficult to cultivate. Robert Campbell and his brother Peter left the family farm in Lochiel Township, Glengarry County, and worked for John Egan, the very successful Ottawa Valley square-timber entrepreneur. Both brothers did well. Robert quickly rose to become manager of John Egan's Bonnechere operations, and held that position between 1838 and 1857, when Egan died. Robert, in his own name, had a timber berth on the Bonnechere River at Brennans Creek near Golden Lake and "reportedly was the first man to exploit the timber of the Pine River off the Little Bonnechere River."[43]

When Richards Township was surveyed in 1861 by Robert Hamilton, there was "a house and stable on Lot 24, Concession 5, occupied by Peter Campbell, the only settler in this township at that time. Since Peter and his brother Robert had worked for John Egan, Campbell may have first occupied this site while in Egan's employ."[44]

By 1860, Robert Campbell was established in Eganville, where he owned a lumber business, Robert Campbell and Son. He then acquired a grist and lumber mill at Douglas under the name Campbell and McNab. At Eganville, he served as municipal councillor and Justice of the Peace; he later ran and was defeated as a Liberal candidate in Renfrew South. He ran again in 1882, winning that time, and was re-elected in 1887, after which he soon died. His son, Robert Campbell (1853–1934), followed his father into both business and politics, being elected as a Liberal MP in Renfrew South in the 1890s.

The McRaes

Duncan McRae married Margaret Munro and emigrated from Scotland in the 1820s. They pioneered land at St. Elmo in Lochiel Township, Glengarry County. Duncan McRae stayed on the farm in St. Elmo where, although not plentiful, there were enough resources to send John Duncan (J.D.), the youngest of his 11 children, to McGill University in Montreal, where he was enrolled in pre-medicine for two years.

J.D., during a summer break from his medical studies, came to Eganville to teach school. He gave up university and married Robert Campbell's youngest child, Isabella (Cissie), in 1872. Robert Campbell, then 59, set J.D. McRae up, under his tutelage, as a merchant in Eganville. J.D. did well and expanded his interests into the lumber business. In 1904, J.D. bought the Knight's sawmill at Eganville and property at the fourth chute of the

Photo 6 "The Office," Basin Depot, Bonnechere River
This cabin was used by J.D. McRae when timbering upriver into the Paugh Lake area with the Golden Lake Lumber Company.
Source: APM 5743

Bonnechere. Timber, cut around Paugh Lake, was driven down a tributary to the Bonnechere River and fed both mills. In 1908, J.D. used a cabin known as "The Office" at Basin Depot on the Bonnechere River as a bush office. This cabin still stands and is reputed to be the oldest structure in Algonquin Provincial Park.

A. Charest's map of timber limits in the Algonquin Park area shows two limits held by McRae and Powell. One limit was centred on Madawaska Lake and the other was south of the Conroy Marsh on the York River. Both were located in headwater areas from where it would have been difficult to drive logs to a mill. Perhaps both McRae and Powell were initially speculating on these leases, but much later J.D.'s son J.S.L. cut the Madawaska Lake area (Maps 8 & 9).

John D. McRae and Cissie had five children: Robert Campbell, George Duncan, Ethel, John Stanley Lothian (J.S.L.), and Leonard (Fig. 3). J.S.L. was four years old when his mother died in 1892. Shortly afterwards, his father married Helena Wolfe, a young music teacher. There were no children from this union. While J.S.L. and his younger brother Leonard lacked nothing in the way of material wants, the relationship between the two boys and their stepmother was distant at best.

In 1908, the Golden Lake Lumber Company was incorporated with its head office in Eganville. The charter was in the names of M.J. O'Brien and J.A. O'Brien of Renfrew and J.D. McRae and his two older sons, R.C. and George D. of Eganville. The Golden Lake mill was then enlarged.[45] J.D. was trying to establish his older sons as Robert Campbell had done for him. J.D. had done very well up to this point in time, but shortly after events turned against him.

On July 9, 1911, a fire destroyed most of Eganville. Losses in the town were assessed at some $450,000, with insurance covering only $100,000. J.D. McRae lost his lumber yard and general store, worth $12,000 together. Insurance covered only $2,000 in total. Those losses did not include the debts of customers carried on the books of the general store. Many people, wiped out by the fire, simply left town, leaving their unpaid store accounts behind. Included in the losses from its fire was a $1,500 loss for the office of the Golden Lake Company, which burned as well. The high losses in the town and the low insurance payouts were a result of the fact that Eganville was judged to have practically no fire protection. Insurance premiums were consequently very high, and people and industries could not afford adequate coverage.[46]

The Golden Lake Lumber Company continued to expand, however, and sawed over 11 million board feet in 1912. But the mill then burned. This ended J.D.'s involvement in lumbering. O'Brien survived the destruction of the Golden Lake Lumber Company because his money was apparently invested in the timber limits, which did not burn. J.D. McRae continued ownership of Egansville's electric plant, but he had no capital to aid his sons. Seeing no opportunity in Eganville, J.D.'s two older sons and the youngest, Leonard, left for lumbering in the Peace River District of Alberta. J.S.L. chose to remain in the general area, where he slowly developed his own timbering business. As the forest in the area was harvested, the general cutting movement was upstream along the watershed of the Madawaska River. In less than a decade, J.S.L. found himself in the Whitney area.

The Keenans

Both John Keenan and his wife, Margaret Hogg, were from the good farming area of Nepean, near Ottawa. Despite this, John left farming for lumbering and in time became a foreman for Robert Campbell on the Bonnechere. The 1891 census of Canada located the Keenan family on Round Lake, where they operated a stopping place with boarding facilities and stables.[47] This site was very close to the one begun by Robert Campbell. In 1897, John Keenan patented Lots 26 and 27, Concession 5, while he continued to work for other lumber companies and serve as postmaster—a post he kept until 1911.

The McGregors

John McGregor was born in Scotland and emigrated to Burnstown, McNab Township about 1853. Shortly thereafter, he married Janet Fisher, a widow with two young children. The family home was in Eganville. Seven other children followed, with Donald (Dan) being born in 1856. Dan left the farm for timbering and rose to be the "walker" for the McLachlin brothers. Walkers—also known as timber cruisers or bush superintendents—are the men who assess the type and quantity of timber on prospective or licensed timber lands for a company or the government.

While on a trip to the company licences on the upper Ottawa River in 1912, Dan McGregor suffered two strokes and died at 55 years of age. His wife, Annie, took charge of the family. For some reason, not known now, she added the "a" to MacGregor. She believed strongly in education for her children, who, as a result, all achieved success in their careers. William Keenan, the eldest son, survived World War I and became a well-known judge in the upper Ottawa Valley. Kenneth became an engineer and worked on the hydro-aluminum Kitamat/Kemano project in British Columbia. Duncan Fisher, the youngest son, and the subject of *A Life in the Bush,* a delightful family biography by his son Roy MacGregor, plays a significant part in the remainder of this story.[48] The eldest daughter, Janet Agnes, married J.S.L. McRae—her next-door neighbour in Eganville. The other two children were girls of whom little is known.

J.S.L. McRae's Early Years in Logging

J.S.L. McRae was 21 at the time of the Eganville fire. It appears that after leaving school at about 14 years of age, he joined in the family business. He would have been a participant in the bush camps of the Golden Lake Lumber Company, felling, skidding and driving logs to the mill. At the mill, he would have experienced the sawing of lumber, the grading of boards and the marketing of lumber. So to speak, he served his apprenticeship in lumbering.

In 1911, J.S.L. was cutting timber on a farm lot on Bark Lake out of Barry's Bay where Murray Brothers had a mill. He was also operating his first mill at Martin's Siding (Map 5).

Duncan MacGregor, J.S.L's brother-in-law, remembered:

> The Barry's Bay Lumber Company used to cut his logs. He got his start on an old farmer's lot on Bark Lake, by Barry's Bay. It had a great stand of timber on it … I guess you could say that Murray helped him out. Then he went to Madawaska for three or four years. From this widow-woman he bought limits and a small mill there. And then after that he bought the mill here [Whitney]. He got a big contract from the Pembroke Box Company [Map 4].[49]

Donald McRae, J.S.L.'s son said, "Dad had the mill at Martin's Siding at Madawaska. But when he first went there, Murray sawed our logs."[50]

Figure 3: McRAE FAMILY

Duncan McRae	**Robert D. Campbell**	**John McGregor**	**John Keenan**
b. 1801, Scotland	b. 1818, Scotland	b. 1813, Scotland	b. 1835? Nepean
d. 1880	d. 1877	d. 1870	d. 1901 ?
m. c1828, Glengarry	m. 1846, Glengarry	m. 1847, Bathurst Dist	married
↓	↓	↓	↓
Marg. Munro	**Elizabeth Lothian**	**Janet Fisher (widow)**	**Marg. Hogg**
b. 1811, Scotland	b. 1821, Glengarry	b. 1821, Scotland	b. 1846 ? Nepean
d. 1890, Glengarry	d. 1899	d 1888	d. 1912?
h. Kenyon, Glengarry	h. Eganville	h. Burnstown, Renfrew C.	h. Nepean
11 children,	5 children	2 + 5 children	m. 1893, 4 children,
youngest	youngest	second	eldest
↓	↓		

John Duncan ← m. 1872 → Isobella (Cissie)   Donald (Dan) Fisher ← m. 1893 → Annie
b.1849   b. 1853?   b. 1856,   b. 1870,
d. 1920   d. 1892?   d. 1912   d. 1949

h. Eganville   h. Eganville
↓

5 children
1. Robt Campbell
2. George Duncan   7 children
3. Ethel
4. John Stanley Lothian ← m. 1914? → Janet Agnes
   b.1889. d. 1969   b. 1896, d. 1981
5. Leonard

———————————————— h. Eganville → Whitney ————————————————

Donald McGregor	John (Jackie)	Marjorie Lothian
b. 1915 d. 2006	d. 2.5 years	b. 1922 d. 2004
↓		↓
m. 1945		m. 1979
h. Whitney		h. Whitney
Helen Grace McRorie		Alexander McGregor
b. 1914		b.1925 d. 2005

Janet Elizabeth	Robert Donald	John Duncan	Catherine Anne
b. 1946	b. 1947	b.1950	b. 1953
m. 1970	m. 1973	m. 1979	m. 1978
h. Toronto	h. Whitney	h. Peterborough	h. Toronto
↓	↓	↓	↓
Mark J. Webber	Rachel M. Micklethwaite	Valerie R. McLeod	Michael Freeman
No children	1. Jamie Campbell	1. Jennifer Rawes	1. David John
	b. 1978	b. 1980	b. 1984
	2. Michael David	2. Donald Cameron	2. Sarah Kathleen
	b. 1981	b. 1982	b. 1987
	3. Margaret Kaitlyn	3. Duncan McLeod	
	b.1986	b. 1984	
		4. Gillian Anne	
		b. 1988	

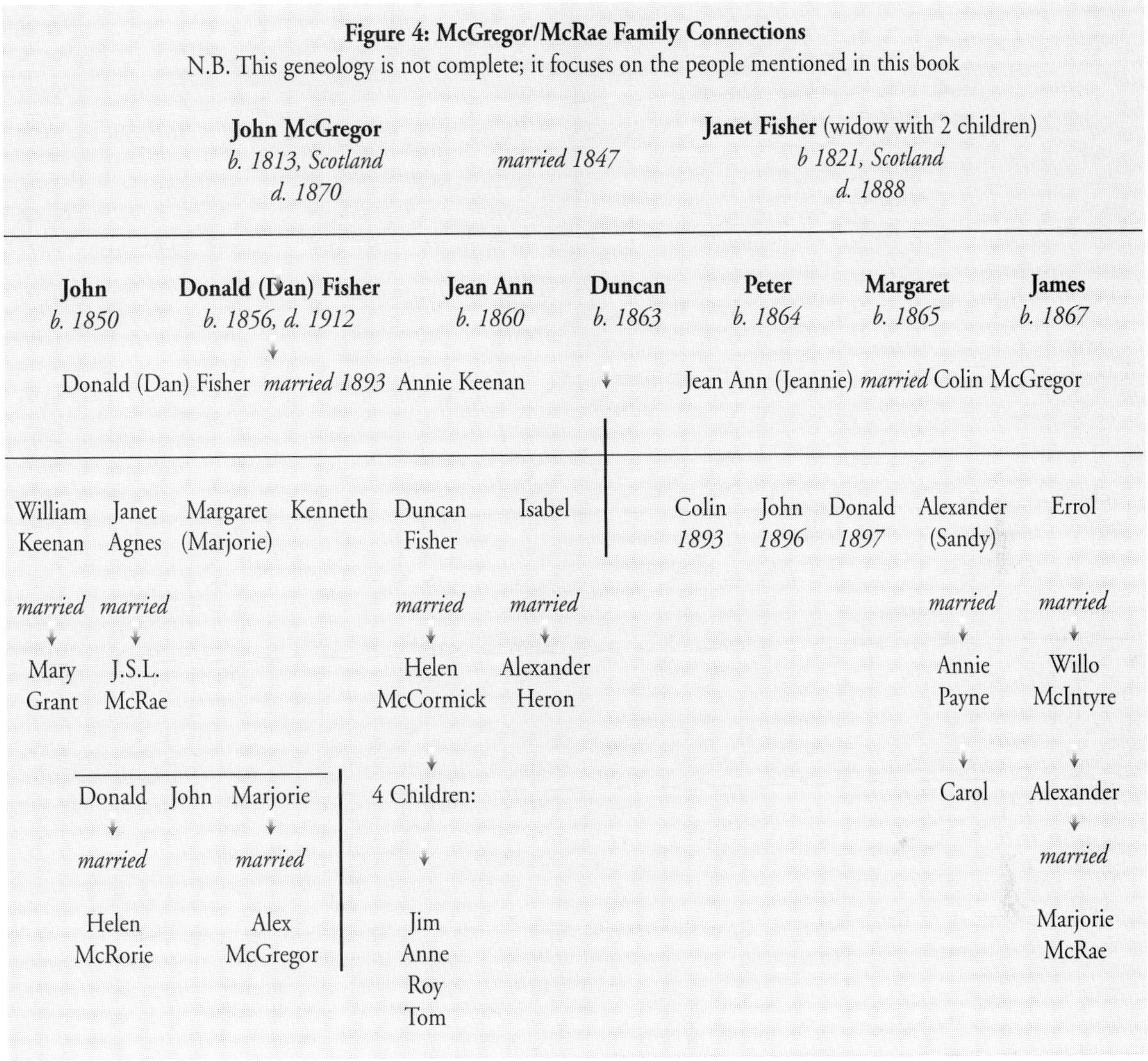

Chapter 3: The McRae Family Background

Figure 4: McGregor/McRae Family Connections
N.B. This geneology is not complete; it focuses on the people mentioned in this book

John McGregor		Janet Fisher (widow with 2 children)
b. 1813, Scotland	married 1847	b 1821, Scotland
d. 1870		d. 1888

John	Donald (Dan) Fisher	Jean Ann	Duncan	Peter	Margaret	James
b. 1850	b. 1856, d. 1912	b. 1860	b. 1863	b. 1864	b. 1865	b. 1867

Donald (Dan) Fisher *married 1893* Annie Keenan   Jean Ann (Jeannie) *married* Colin McGregor

William Keenan	Janet Agnes	Margaret (Marjorie)	Kenneth	Duncan Fisher	Isabel	Colin 1893	John 1896	Donald 1897	Alexander (Sandy)	Errol
married	*married*			*married*	*married*				*married*	*married*
Mary Grant	J.S.L. McRae			Helen McCormick	Alexander Heron				Annie Payne	Willo McIntyre
	Donald John Marjorie			4 Children:					Carol	Alexander
	married married									*married*
	Helen Alex			Jim						Marjorie
	McRorie McGregor			Anne						McRae
				Roy						
				Tom						

Opposite: Figure 3 **McRae Family Tree**
Source: Janet McRae Webber. 2005

Janet McRae, J.S.L.'s wife, remembered the early days, especially the cadging. A cadge crib is a raft of logs about 25-feet square. It was used to tow a boom of logs, with the logs cadging (or hitchhiking) a ride. To do so, a long rope tied to the cadge crib was rowed out in the direction they wanted to move the logs. An anchor was then dropped to the bottom. A winch on the cadge crib, usually turned by a horse, was used to wind up the rope, moving the cadge crib forward and towing along behind the attached boom of logs. It was a deadly dull experience. Janet experienced cadging once, and explained to her granddaughter why she never went again.

I lived at home after I was married for a while, and then we got a little house—rented it. I used to be alone at night. Take a windy night, they'd have to bring the logs down. And how they brought the logs down, they put them on a scow. And there was a two-team of horses. They had no engines. The horses were on the scow, and they had ropes, and they kept going around and around on this scow, and they kept pulling and pulling, and the rope would go up like that …Your grandfather [J.D. McRae] had logs back of Barry's Bay there; well, he gave the logs to Dad. [These logs may have come from the McRae/Powell lease on the York River—see Map 9.] They cadged the logs, with the horses on a winch, and brought the logs with a fair wind down the Kaminiskeg. No engine!

But I remember, I said this night … Oh, I didn't like staying alone because I wasn't used to it—at home with the family and everything. And Dad'd be out night after night on this old scow. I said, "I'm going to go." And he said, "Yeah, you can go with a blanket." And there was another lad with him. So I went. Well, I never went again, I'll tell you! I was sick and tired of it. You'd be there from seven o'clock in the evening, and never get back. Oh, we were all night long. I guess it was eight o'clock in the morning before we brought the logs up. The scow wouldn't be any bigger than double the size of this room if it was that big, and then there was a railing all around it. And the horses were trained to just walk around and walk around. They had a winch they'd pull. And they'd watch the logs, and they'd have to keep them straight.[51]

The beginning of the McRae Lumber Company can be established by a copy of a bill sent to G.M. Mason for 1"x 4" lumber in the amount of $241.90. It was under the letterhead John S. McRae, Lumber Manufacturer, Barry's Bay, Ontario, and dated October 22, 1914. In another document, a letter to Campbell & Johnson, Wholesale Lumber, Toronto, from 1915, J.S.L. was looking for payment, and Campbell & Johnson were delaying because the market was slow.[52]

In 1916, the Grand Trunk Railway (formerly the OA&PS) billed J.S.L. for shipping two horses from Eganville to Madawaska, and Dr. Foster, the veterinary surgeon from Renfrew, billed John S.L. McRae of Barry's Bay $214.00 for horse medicine and tonics in 1919. This large quantity of horse medicines indicates that McRae's outfit was increasing considerably in size.[53]

By 1919, J.S.L. was in the Whitney area. Here he was cutting pine, hemlock and spruce, including spruce boom logs, on a Dennis Canadian timber lease, paying only the required government stumpage. In the fall of 1919, he was setting up a "shanty operation" as evidenced by an order for "long clear" pork from the William Davies Company Limited of Toronto, which was followed by an order to Dunlop & Company Limited of Pembroke, Hardware Merchants, for cooking utensils, stove pipes and roofing. The order included the basic iron products required by a blacksmith. This purchase was broadened to other essential shanty items later in the fall.[54]

Chapter 3: The McRae Family Background

In 1920 and 1921, documentation shows that J.S.L. put considerable wood through the Mickle, Dyment Lumber Company mill at Airy. Invoices from 1921 also show evidence of cutting in a second area, where J.S.L. dealt with the Murray and Omanique Lumber Company of Barry's Bay. This account billed J.S.L. for the labour and equipment used in towing logs, sawing wood and loading freight cars. J.S.L. must have had a woods operation adjacent to Kaminiskeg Lake. In 1922, he negotiated with Rathburn Company of Deseronto for the transfer to him of timber limits on some lots in McClure and Sabine townships. These limits still remain in the hands of the McRae Company.[55]

```
The Honourable Beniah Bowman, 1922
         Minister of Lands & Forests,
           Parliament Buildings,
              TORONTO, ONTARIO.

Dear Sir:-   Re Berths Nos. 3 and 4 - Comprising Concessions
             9 to 14 inclusive, Township of McClure,
             (excepting Lots East of Railway, viz: Con-
             cession 14 - Lots 5,6,7,8,9); and Concessions
             1 to 5 inclusive, Township of Sabine, (except-
             ing the Lots east of Railway, viz: Concession 1 -
             Lots 27,28,29,30; Concession 2 - Lots 27, 28,
             29 & 30 excepting N.E.Cor., Concession 3 -
             Lots S.Pt.31), being part of Licenses Nos. 54
             and 56. Crown Timber Licenses for Season 1920-
             1921.

        The Rathbun Company, of Deseronto, has agreed to transfer
to us the timber rights in respect of the lands above noted.

        We understand that upon production of a transfer of the
timber rights in respect of the above noted lands in form satis-
factory to you, you will permit the approval of such transfer.

        We understand that we are from time to time and at all
times to be bound by Order-in-Council approved by the Lieutenant-
Governor of the Province of Ontario, March 6th, 1907, and that we
must abide by the provisions of the same so far as in your opinion
same are applicable to the above noted lands.

        We further understand that we must at all times accept
such interpretations as you may place upon such Orders-in-Council
as regards the above noted lands.
```

Figure 5 **Letter to the Honourable Beniah Bowman, Minister of Lands and Forests**
Source: McRae Lumber Company

Albert Perry, a long-time resident of Whitney and friend of J.S.L. McRae, said,

> I remember when McRae first came here [Whitney]. He didn't own the mill right away. Mickle was still operating when he came. McRae bought a little limit above Hay Lake—they called it Rainy Lake [the name was changed to Drizzle Lake in 1933]. That's where he got a start up there. He used to bring his logs down to Mickle's here, and they'd cut the lumber at so much a thousand. He did that for about two years, and then Mickle pulled out and McRae got a chance to buy the mill. I think he got a good deal on it.
>
> He didn't have much of a limit, but he started takin' out hardwood. But he had a hard time to get them from Rainy Lake. They wouldn't float. They'd lose lots, sink down. So they had to peel them. And maybe they would dry out quite a bit by the time he'd bring them down. They had to bring them down to the foot of Hay Lake, and then run them down the creek, into this lake [Long Lake; it was renamed Galeairy Lake in 1923].[56]

J.S.L. learned from this experience and developed a method of moving hardwood logs by cribbing them. He would soon apply this idea when he bought his own mill.

Midway through 1921, Canada's inflationary boom collapsed. A period of sharp deflation and deep recession began that lasted until 1924. In 1922, Dennis Canadian closed down due to depressed lumber prices and Mickle, Dyment took the opportunity to sell to J.S.L. McRae. The purchase of the mill included the limits under licence to Mickle, Dyment. As Albert Perry said, "Mickle sold out because his timber limit ran out. He only cut softwood: hemlock, spruce, balsam and pine. He wasn't a hardwood man."[57]

J.S.L. also picked up limits formerly under licence to Dennis Canadian. In 1922, J.S.L. McRae bought the mill at Airy. From that date until the present, the McRae Lumber Company has been the major industry in the community.

When I go into the woods in sharp, frosty weather I carry a few cotton rags in my pocket, and I fold one of them across the point of the wedge. With this precaution there is no danger that the wedge will fly out, at a touch, as it is likely to do without it.

Musson's Improved Lumber and Log Book
New and Revised—1905
—Illustrated Edition

Section II: The Operations of The McRae Lumber Company: 1922–1980

The language used in logging is unique to the business. It developed from the roots of the industry in the early 1800s in the Ottawa Valley. Many of the terms and much of the equipment are no longer in use in modern lumbering—they have faded away, as have the teams of horses that hauled the sleighs loaded with heavy logs. When used in this book for the first time, these terms are explained in the text. Definitions of these terms are also included in a Glossary of Terms at the end of the book.

Land in Ontario is held in two ways. Titled land is either owned or leased by individual people or businesses—they have title to a specifically described parcel of land: for example, a house lot in a town or city, or a larger area such as a farm. The remaining, untitled land in the province is owned by the Crown, and most of this area today comes under the jurisdiction of the Ministry of Natural Resources.[58] In 1999, inventoried productive forest in the province amounted to some 40.3 million hectares, of which 65 per cent, or 26.2 million hectares, was Crown forest. The Ministry of Natural Resources (MNR) issues timber leases to lumber companies from this land.[59]

Timber leases are made to forest industry companies in large part to generate income for the government; the fee charged is termed stumpage. The MNR assesses or estimates the volume of timber to be cut under licence and determines its value in a procedure known as cruising. In this procedure the land is walked over, the trees are sized and counted by species, and a value is established. The land parcel to be leased is then put up for tender. The lumber companies also "walk the bush," to assess the timber. Terms and conditions, spelled out in the lease commonly include items such as the lower diameter limits or size of trees permitted to be cut, and areas that have been withdrawn from the lease area for the purposes of watershed protection or game preservation, or because the timber has been released to other companies.[60] The agreement between the lumber companies and the government is a lease for a term of years.

In the nearly 60 years from 1922 to 1980, the bush and mill operations of the McRae Lumber Company underwent enormous changes. Both the bush and mill operations used by McRae would have been typical, however, of those used throughout the industry in eastern Canada. Where McRae used unique procedures, these exceptions have been noted.

In 1922, the Airy operations were labour intensive, both in the bush and at the mill. As McRae Lumber moved its mills to new sites, less labour and more capital were used in the business.

Highway 60 was put through Algonquin Park as a "Great Depression" make-work project between 1933 and 1935. Another such project was the clearing of an emergency airfield at Lake of Two Rivers in 1935. The highway led to an ever increasing number of tourists visiting the Park and eventually to a conflict between loggers and recreationists/protectionists. The aircraft landing field was seldom used.

While all of the above events were significant, the impact of the changes in the management and use of Algonquin Park on the operations of the McRae Lumber and the other mills having licences in the Park was most dramatic. As a result of a very public campaign by recreationist/protectionist groups, all timber licences in Algonquin Park were cancelled in 1974 with the passing by the government of the Master Plan. The following year, the Algonquin Forestry Authority, a Crown agency, was issued a single timber-cutting licence for Algonquin, along with the responsibility of supplying to the previous licence holders the wood necessary to keep their mills open.

The McRae Rock Lake mill (1957–1980) lasted some 24 years. During that time, many changes took place. Provincially mandated tree marking became the rule: trees, called stems by foresters, that were formerly ignored were marked, and had to be sold somehow. The poorer wood was cut into small pieces or chips, which were shipped out to be made into paper. Construction of gravelled all-weather bush roads now preceded cutting; and the last of the company bush camps, a set of double trailers, was closed in 1974, forcing daily commuting by the men. Mechanization of bush operations also proceeded, and eventually the Timberjack skidder replaced the horse in skidding logs out of the bush. The advent of all-weather road construction and the use of truck transportation to move logs or stems to the mill over a longer distance meant that the trees were moving to the mill rather than the reverse. Diesel electric power replaced steam in powering the mill. With the death of J.S.L. in 1969, Donald and his sister Marjorie became owners of the company with Donald managing it.

The closing of the Rock Lake mill and the move to the Whitney site was not made because McRae had run out of trees to cut. The Algonquin Provincial Park Master Plan of 1974 required that mills be removed from the Park, and there was increasing pressure to carry this out. McRae Lumber Company established two new mills, a low-grade mill in 1973 and a high-grade band mill in 1979 on a site in Whitney. Donald McRae then retired, and management of McRae Lumber Company was passed over to his sons, Bob and John.

CHAPTER 4
The Airy Mill Operations: 1922–1933

The Airy mill that J.S.L. McRae acquired in 1922 was a much larger operation than what he had run previously. While J.S.L. was very much the "hands on" owner, he needed a reliable, experienced core of men. He turned to men he knew and with whom he had previously worked to serve as key personnel.

Sandy McGregor, the walking boss, or bush superintendent, was a distant relative of J.S.L.'s wife, Janet. Albert Perry had worked for Dennis Canadian and Mickle, Dyment, as had his father. Duncan MacGregor, J.S.L.'s brother-in-law, joined McRae in 1929, and in time became supervisor of the yard crew and responsible for grading the lumber. Paul Shalla, an old friend and one of the first employees from the early Barry's Bay days, was the barn boss. Paul's son, Frank, started working for J.S.L. at the Drizzle Lake limit before the purchase of the Airy mill. To run the mill, J.S.L. kept Ben McBride, the office manager, and Jed Buckman, the millwright, both of whom had worked for Mickle, Dyment in the same capacity.

These men formed the basic core of workers and spent most of their working lives as employees of the McRae Lumber Company. In an age and in a business of intermittent employment, McRae kept these workers on full-time. They repaid J.S.L. with loyalty and the capacity to draw other good workers to the firm.

The operations of the McRae Lumber Company during the Airy mill period (1922–1933) give a good picture of bush operations and milling at the end of the pine period, even though the Airy mill was producing a much higher percentage of hardwood than it had been under Mickle, Dyment.

In 1923, during the depressed market conditions that aided him in buying the Airy mill and the Mickle, Dyment limits, J.S.L. McRae applied for an additional timber licence to cut hardwood in Nightingale and Lawrence townships on part of the old Munn limit. In 1925, the government complied,

> knowing that 400 people in the village of Whitney depended on the McRae Lumber Company for their livelihood …To ensure strict supervision over the company, Deputy Minister Cain instructed his district forester in Pembroke to reach "a thorough and distinct understanding" with McRae as to the trees to be extracted. Cain's policy for Algonquin sought to balance a utilitarian, conservationist ethic with a philosophy of economic development.[61]

Two basic operations were necessary: work in the bush, where the logs were cut and then moved to the mill; and the sawing of the logs into lumber at the mill.

Chapter 4: The Airy Mill Operations: 1922–1933

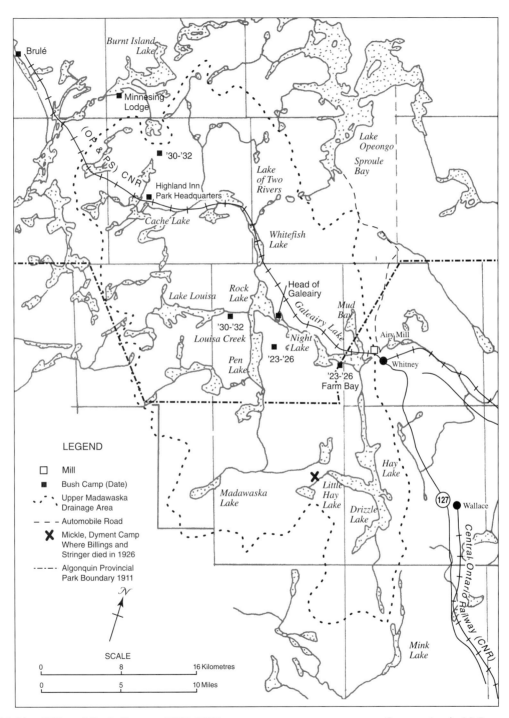

Map 10 **Airy Mill and Bush Camps: 1922–1933**   **Source:** Jamie McRae

Operations in the Bush

From 1922 to 1925, bush camps spread down Galeairy Lake and then to Lake Louisa by 1926 (Map 10). Bush camps normally operated and supplied a log dump for only two or three years. A log dump is a collection of hardwood logs brought from the cutting areas to the edge of a lake. Over the years, the company had a number of log dumps: on Galeairy Lake, at Hay Creek, Farm Bay and Forest Bay, as well as at a number of locations on the north side; on Rock Lake, at the Pen Lake portage; on Hay Lake, at Hay Creek; and on Whitefish, on the southwest side. With open water by the end of April, the dumps were taken down and the hardwood logs cribbed and hauled by tug to the mill at Airy. The softwood logs were boomed while on the ice and were towed to the mill after the hardwood logs later in the summer.

Much planning had to go into setting up the bush camps from which the men would walk to the cutting areas each day. The first step was to cruise the licence by walking through it to get the lay of the land. Decisions had to be made when establishing the cutting plan, such as which area to cut first, and what sequence to cut the ones following to give the greatest efficiency. This was followed by decisions as to where to locate the bush camps and the tote road that would be used to bring in the needed supplies for the camps. The most important decision involved where to locate the winter haul roads on which the sleigh loads of logs would be moved out of the bush on their way to the mill. The bush superintendent, Sandy McGregor, carried out the cruise, along with J.S.L. himself in the early days.

Photo 7 **Bush Camp at the Head of Galeairy Lake**
Donald McRae is the young lad in the foreground. Sandy McGregor, the bush supervisor, is behind Donald, and J.S.L McRae is to the right in suit and tie.
Source: McRae Lumber Company

Bush camps were like small villages. They required considerable planning and continuous effort to supply them. These camps were commonly built near the end of the summer season so as to be ready for the men coming to camp to cut the logs in the early fall. A typical bush camp for 100 men and 30 horses occupied a clearing of some 1.5 acres. It would be close to a good, clean water source. The stables and latrines would be located downslope from the potable water source.

The camp components listed below would be typical of Canadian bush lumber camps of the time.

The main supply route or tote road con-

Chapter 4: The Airy Mill Operations: 1922–1933

Table 1 Bush Camp Buildings, Louisa Creek, 1926

1	combined cook and bunkhouse	2 stables
1	office and van	
1	root house	1 storehouse
1	blacksmith shop	
1	meat house	1 pigpen
2	latrines	

Source: M. A. Adamson, W.E. McGraw, D. McLaren and D.M. Parker, "McRae Lumber Company, Lake Louisa, Nightingale Township," *U of T Logging Report 56*, 1926.

nected the Lake Louisa camp to the Rock Lake station on the CN Railway. At the Rock Lake train station, there was a company storehouse, supplied from stock at the Airy mill site. This storehouse held hay and other non-perishable goods. In early fall, most of the hay and provisions were brought across the lake by scow and tug to the beginning of the tote road that led to Lake Louisa. To tote means to carry supplies. A team of horses used the tote road and pulled a supply wagon or sleigh loaded with food and other supplies along it to the bush camp. In winter, the tote road ran over the ice from Rock Lake station and connected to the beginning of the land-based tote road into the camp on Lake Louisa. The tote road was fairly rough in character as little grading was done on it. Small stream crossings were bridged with corduroy, or logs, laid side by side in the stream at right angles to the road. The water simply flowed through the bridge of branches that filled up the channel of the stream, enabling the wagon to cross over. The tote road was entirely separate from the haul road, which was used to move the cut logs out of the bush. The present-day canoe portage between Rock Lake and Lake Louisa follows much of the old tote road route.

Bush camps were used for only two or three years and were then abandoned, because in that time the men had cut all the trees that could be reached within a reasonable walking distance. The camp buildings were usually built of low-value trees at hand such as balsam, poplar and spruce. Spaces between the logs were filled or chinked with moss. Low-grade, second-hand lumber was used for the gable ends and flooring. To build a camp at that time took 25 men one month and cost $3,000.

A separate cabin housed the office used by the foreman and the clerk. The records of the camp, including invoices for incoming supplies, the records of the cut and the work type of each man, were kept by the clerk under the supervision of the foreman. These men slept in this building, as did the government scalers when they were present in camp. A medium-sized Quebec heater gave warmth and a gallon iron pot was provided for warm water. The van, or camp store, was also kept in the office and was typical of those found in isolated lumber camps at this time. Articles for sale by the camp clerk included tobacco, clothes (underwear, mitts, socks and mackinaw jackets), footwear, matches, candies, drugs, stamps and stationery. The value of the stock kept on hand was about $2,200, with prices comparing favourably with those off-site. The amount purchased by each man was charged against his wages and deducted from his total earnings. Catalogues from Eaton's and Simpson's were also available.

Photo 9 **Saw Filer's Shack, Cranberry Lake, Canisbay Township, 1930–1931**

More windows were in this building, giving the saw filer better light for his work. Note the saws hanging on the outside wall.

Source: J.E. Bier and A. Crelock, "McRae Lumber Company, Cranberry Lake, Canisbay Township," *U of T Logging Report 75*, 1930–1931.

Photo 8 **Combined Cook and Bunkhouse, Louisa Creek, 1926**

The combining of cookhouse and sleeping quarters was not done after the time of the Airy mill. With wooden buildings, the risk of fire was always present and to lose both sleeping and cooking quarters in a single blaze would be a complete disaster.

Source: M.A. Adamson, W.E. McGraw, D. McLaren, D.M. Parker, "McRae Lumber Company, Lake Louisa, Nightingale Township," *U of T Logging Report 56*, 1926.

The blacksmith shop, saw filer's shack and bunkhouses were also equipped with Quebec heaters. Stables, supply shacks, pigpens and a root house were other structures at the camp.

Photo 10 **Laundry Facilities, Cranberry Lake, Canisbay Township, 1930–1931**

Any personal laundry that was done took place on Sunday. Keeping clothes clean was not an easy task at this time.

Source: J.E. Bier and A. Crelock, "McRae Lumber Company, Cranberry Lake, Canisbay Township," *U of T Logging Report 75*, 1930–1931.

Chapter 4: The Airy Mill Operations: 1922–1933

A story is told of one old teamster in another camp who didn't believe in changing his clothes.

> He figured that if he did, he would catch cold. So, as his shirt or socks became worn out and literally fell, or rotted, off him, he just bought new ones and pulled them on. "Tis said that in the spring when he finally got around to taking a bath, he had four shirt collars around his neck, and around each leg were seven red tops from his work socks!"[62]

Photo 11 **Latrine, Louisa Creek, 1926**
Two of these latrine structures would be required for a camp of 100 men. They and the stables were located downslope from the well or other potable sources of water.
Source: Source: M.A. Adamson, W.E. McGraw, D. McLaren, D.M. Parker, "McRae Lumber Company, Lake Louisa, Nightingale Township," *U of T Logging Report 56*, 1926.

Photo 12 **Cookery, Cranberry Lake, Canisbay Township, 1930–1931**
The cookery was whitewashed and kept very clean.
Source: J.E. Bier and A. Crelock, "McRae Lumber Company, Cranberry Lake, Canisbay Township," *U of T Logging Report 75*, 1930–1931.

Good and Plentiful Food

Men doing hard physical labour required large amounts of food—and it had to be good. Nothing would upset a crew more than to have a poor cook. The cook's wages in the camp, which were second only to the camp foreman, indicated his importance. McRae camps enjoyed the reputation of having the best food in the region. The head cook looked after the pastry items, including the bread. The bull cook had to be a strong man for he was the one who carried all the needed water from the well or "water hole" to the cookery, as well as the wood for the stoves. He also looked after the meat, including butchering the pigs as needed. A chore-boy washed up the utensils and dishes, kept the room swept up, maintained the wood supply for the stoves and helped the cooks.

J.S.L. watched costs, but bought the best food in good quantity. Cooks, valued by both management and the workmen, were employed for many years.

A list of supplies for 100 men for one month is given below. J.S.L. worked this out and put in the orders himself. His regular presence at the bush camps through the 1920s and into the 1930s ensured that conditions and meals were very good.

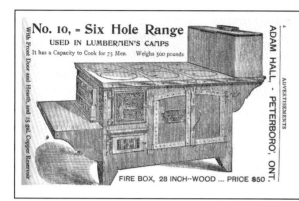

Figure 6 Hall Stove

The cookery was provided with two, six-hole Hall stoves, each having a 14-gallon water tank and 24-inch ovens. Stoves of this type, introduced into bush camps in the 1880s, greatly increased the range of food that could be prepared and thereby improved the diet of bush workers.

Source: G.D. Garland, comp. *Glimpses of Algonquin: Thirty Personal Impressions from Earliest Times to the Present.* Whitney [ON]: The Friends of Algonquin Park, 1989.

Table 2 **Commissary Items for One Month, Cranberry Lake, 1930–1931**

Meats

700 lbs.	pork	
100 lbs.	pork sausage	
8 sides	beef	
500 lbs.	lard	
400 lbs.	butter	
80 lbs.	fish	

Vegetables

40 bags	potatoes	
30 bags	cabbage, turnips and carrots	
7 bags	beets	
2 bags	peas	
86 lbs.	onions	

Canned Goods

8 dozen tins	pumpkin
9 30-lb. pails	apple filling
9 cases	rhubarb
12 30-lb. pails	jam
4 gallons	corn syrup
4 gallons	vinegar
5 1-gallon jars	pickles

Liquids

5 quarts	vanilla extract
5 quarts	lemon extract
5 gallons	molasses

Dried Goods

30 bags	flour	2 lbs.	savoury	2 lbs.	cloves
80 lbs.	tea	2 lbs.	cream of tartar	25 lbs.	soap
8 bags	sugar	5 lbs.	pastry spice	5 lbs.	mustard
12 cans	lye	7 lbs.	ginger	75 lbs.	table salt
36 boxes	Royal yeast cakes	4 lbs.	allspice	7 lbs.	pepper
7 25-lb. boxes	raisins	4 lbs.	cassia	50 lbs.	rice
4 25-lb. boxes	macaroni				
12 lbs.	prunes				

Source: J.E. Bier and A. Crelock, "McRae Lumber Company, Cranberry Lake, Canisbay Township," *U of T Logging Report 75*, 1930–1931.

Chapter 4: The Airy Mill Operations: 1922–1933

At this time, beef commonly came from local farms. Wilno supplied butchered beef but some also came from Canada Packers. Beef was taken to the camps as soon as the weather got cold. Pork was fresh, as 20 live pigs were kept at the camps. The pigs were the camp's recycling system also, as they were fattened on cookhouse leavings, then butchered as required. The pigs were sheltered in a rough lumber lean-to attached to the barn.

Potatoes, carrots, onions, turnips and parsnips were also purchased locally if possible. Canned tomatoes came from Trenton or Canada Canners, while butter came from Eganville.

The cooking staff baked 30 loaves of bread and 48 pies per day for one camp. Since flour was inexpensive, the cook was expected to supply pastries in as many forms as possible. Other common baked goods were gingersnaps, plain or raisin cookies, iced cakes, donuts and tea biscuits. Many puddings were also prepared, such as cornstarch, rice and cottage. The average cost per man for food was between 36 and 55 cents per day.

Even out in the bush, where the men were cutting, good quality meals were provided and appreciated.

Duncan MacGregor remembered:

> Gosh that was great—the hot lunches. You'd come to a certain place, just at noon hour, horses and men'd come to a little place in the bush with the logs around to sit on. And ya warmed up the beans, or sometimes beef stew or salty pork—always beans. You always had fresh bread. The cinnamon buns were loaded with lots of cinnamon![63]

Photo 13 Bush Lunch
Bernie Stubbs is on the left talking to Donald McRae. The young lad at the far right is likely Felix Shalla.
Source: McRae Lumber Company

The saw filer had his own cabin, with extra windows to provide the light that this specialized craftsman needed. A good saw filer, along with a good cook, was critical to the well-being of a camp. Cutting crews were paid on the basis of how many trees they felled, bucked and moved to a skidway. Crews had two saws. They brought the one in that they had used that day and took the one in the morning that had been filed. Extra saws were kept on hand, since falling trees frequently broke them. While the saws of the felling crews were filed daily, those of the road crews were filed every third day. Road crews worked on the haul road that was used to move logs out of the bush to dumps prior to open water in the late spring.

In 1926, eight cutting crews were used, each using a 5-foot #13 Simonds saw, and one road crew using a hardwood saw of the above type as well as a #22 Simonds saw for softwoods (Fig. 20). These saws evolved from those developed by Shurley & Dietrich of Galt, Ontario in 1874.[64] They incorporated raker teeth that removed chips and sawdust as the cut was being made. This made possible the use of crosscut saws to fell trees using a horizontal cut. Previously, the saws would bind and jam, and were useful only in bucking or cutting logs using a vertical cut that let the sawdust fall out of the cut. Bucking was the term used for sawing of logs to a desired length, usually 16 feet.

The blacksmith and handyman were housed together, thus avoiding the construction of another building. Both men were very important to the success of a lumber camp. The blacksmith shoed the horses using sharp-shod shoes for the iced haul road and dull shoes for skidding logs in the bush. Both men built the double-bunked sleighs and maintained equipment such as the patent plow and tanker sleighs (Figs. 15 & 18). They also made cant hooks, pups, bitch and pig's foot hooks (Fig. 20). Their shop contained the following tools and equipment along with some miscellaneous equipment for odd jobs and repairs.

Most of the horses were owned by McRae

Table 3 **Blacksmith's Tools and Supplies**

1 forge	chisels	1 anvil
crosscut saws	draw knives	road crawlers
2 vises	saws	wedges
1 drill	axes	cant hooks
several hammers	tongs	decking chains & pulleys
several sledges	pincers	logging chains
rasps	shoeing tongs	files
nails	road grubbers	horseshoes
wrenches	shovels	

Source: M.A. Adamson, W.E. McGraw, D. McLaren and D.M. Parker, "McRae Lumber Company, Lake Louisa, Nightingale Township," *U of T Logging Report 56*, 1926.

Table 4 **Expenses of Horses and Other Miscellaneous Equipment, 1926**

Stable Equipment:			**Additional Equipment:**	
2	blankets per team	$7.00	1	wagon
1	pail per team	.75	16	sleighs
12	lanterns	$24.00	1	water wagon
1	harness for each team	$80.00	1	jammer
	forks, shovels, brooms etc.		1	tug
2	motorboats			
25	towing cribs			

Source: M.A. Adamson, W.E. McGraw, D. McLaren and D.M. Parker, "McRae Lumber Company, Airy Mill," *U of T Logging Report 56*, 1926.

Chapter 4: The Airy Mill Operations: 1922–1933

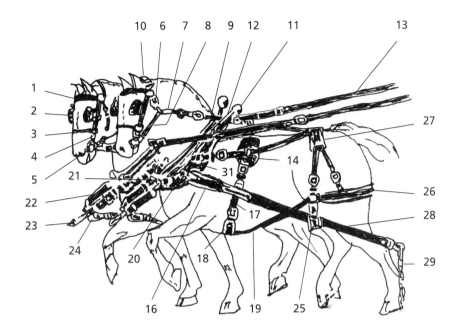

1	Brow Band and Winker	16	Flare
2	Blind	17	Billet
3	Nose Band	18	Belly Band
4	Bit Straps	19	Hold Back
5	Bit	20	No. 919 Connector
6	Gag Swivel	21	Hame Strap
7	Rein Front	22	Breast Strap
8	Throat Latch	23	No. 1616 Snap
9	Side Check	24	Martingale
10	Head Stall or Bridle Top	25	Lazy Strap or Trace Carrier
11	Hame Ball	26	Breeching Seat
12	Hame	27	Breeching Top
13	Lines or Reins	28	Trace
14	Back Bands	29	Heel Chain

Figure 7 **Draft-Style Horse Harness**
Source: Redrawn from an advertisement in *Harness, Saddlery and Baggage*, 1902.

but some were provided by local farmers along with their labour. They weighed from 3,200 to 3,400 pounds per pair or team, and cost approximately $300 per team. A working horse doing intensive labour requires 2 to 3 per cent of its body weight in food per day. Feed consisted of one bushel of oats per team per day, and 34 to 50 pounds of hay per horse per day. All of this was toted into the bush camp in 150 pound bales. Each team was also given four ounces of salt per week. The cost for hay in 1926 was between 75 cents and $1.00 per horse per day, while the average cost per team, including other feed, was $3.50.

Harnesses for each team cost $80 and were readily purchased in the 1920s and thirties. Twenty years later, they were difficult to obtain. Tractors had replaced horses on the farm, while trucks were being used in cities for deliveries.

The Nature of the Forest

The forest on the McRae timber limits in Nightingale, Lawrence and Clyde townships had been harvested twice previously. First, Perley and Pattee, with a depot farm on Farm Bay, Long (Galeairy) Lake, cut and squared the largest pine.[65] Then St. Anthony cut sawlog-sized pine. An old St. Anthony bush camp was on Lake Louisa near the beginning of Louisa Creek.[66]

The timber on the lower slopes of the area and on the more poorly drained soils had stands consisting of evergreens—black and white spruce, eastern hemlock, balsam fir and eastern cedar. The white pine was scarce because most of it had been cut. These are all evergreen trees and reproduce by seed developed in cones. The wood of these trees is light in weight and it floats well. As a group these trees are classed as softwoods.

In Algonquin, hardwoods are (and were) much more widely distributed. In general, sugar maple occurs everywhere, except immediately along lakeshores. Yellow birch is most common on the damp, lower slopes. Beech tends to be found in higher, drier areas. These trees reproduce by means of male and female flowers that develop seeds. Deciduous trees lose their leaves each fall. The wood of these trees is hard and heavy in weight because its cell structure is dense. Because of this, hardwoods float poorly. Freshly cut logs of these trees cannot be driven down rivers or moved in booms across lakes without great loss, as many logs would sink.

Most of the timber cut was sound and straight, but the hardwood was slightly overmatured, with some dead tops and heart-rot present. The average tree contained about two logs, varying in size from 10 to 18 feet, with 16 feet being the usual length.[67]

Jobs at the Camp

In 1930–31, McRae employed 12 to 14 cutting gangs of five men each, who worked from September until Christmas or shortly after. Twenty men cut the roads directly under a buckbeaver who was boss of the haul road. Each camp would have a foreman, clerk, cookery gang and a filer.

At the Airy bush camps each of the log-making crews was comprised of two log makers (sawyers), one trail cutter or swamper, one skidway man and one teamster. The teamster, the most experienced man, was in charge of the crew and he skidded the logs out of the bush to the skidway where the logs were piled. The teamster would determine the location of the skidway and haul the logs to it. Trees to be cut were selected by the head sawyer, the most experienced of the log-makers. The notching, felling and log-making were done by the two sawyers.

Axes used by the felling crews were of the common or poleaxe type. Frequently, hammer

Chapter 4: The Airy Mill Operations: 1922–1933

Table 5 Men Employed in the Bush: Organization and Monthly Wage Scale, 1930–1931

General Manager
Walking Boss/Bush Boss

Clerk		Foreman	
($70)		($115)	
Cook	Saw Filer	Road Foreman	Teamsters
($75)	($50)	($30)	($30)
Flunkeys	Blacksmith	Road Crew	Swampers
($26)	($60)	($20 to $22)	($26)
Handyman			Sawyers (log makers)
($40)			($30)
Barn Boss			Rollers
($30)			($26)

Note: Flunkey was a term usually used for a uniformed servant or waiter. There were certainly none of these in a bush camp. Here the term referred to unskilled labour that ended up with the most menial of jobs.
Source: J.E. Bier and A. Crelock, "McRae Lumber Company, Cranberry Lake, Canisbay Township," *U of T Logging Report 75*, 1930–1931.

axes, with a larger blunt head opposite the cutting edge, were used on skidways for driving the dog or pup spike on the end of the decking chain into the log. Axes weighed on average $3\frac{3}{4}$ pounds. Saws were the hardwood type: $5\frac{1}{2}$ feet long. Other tools included a measuring stick, eight feet-three inches long, tongs with a 24-inch reach for skidding logs, a cant hook, an 80-foot decking line, grab hooks and steel blocks (pulleys).

The first step in felling a tree was to make an undercut on the side where the tree was to fall, giving consideration to the lean of the tree, how to best avoid lodging (or getting caught up) in surrounding trees, and how to best access the skidding trail (Fig. 8). If possible, trees were felled uphill. When limbed and topped, a log in this position would be easier to haul away. The saw undercut was chipped out with an axe. Then, a saw cut was made from the side opposite to and slightly above the undercut until the tree was about ready to fall. The saw was then removed and the tree was wedged over. The directional felling of a tree was a learned skill and an art. It was also dangerous and accidents did occur.

After felling, the tree stems would be measured and sawn into log lengths in a procedure known as bucking. A swamper limbed the logs and prepared trails for skidding, a roller piled the logs on a skidway, and the teamster, using one team of horses, skidded the logs and decked or piled them on the skidway, by chain and gin-pole when the skidway pile of logs became too high to be piled by the roller (Photos 14–18 and Figs. 10, 11 and 12).

Logs varied in length and according to defects, taper and crook. Taper refers to the difference in thickness from the base of the log to the top. Crook is the amount of curve in

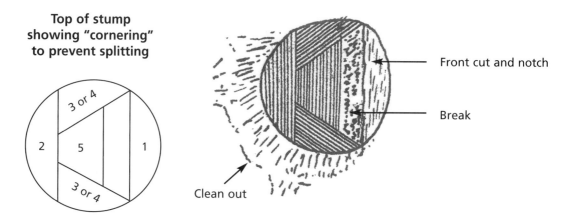

Figure 8 **Cutting Down a Tree: Corner Cutting to Prevent Splitting During Felling**
Source: R.L. Snow and C.W.R. Day, "Shier, Clintock Twp., Haliburton," *U of T Logging Report 59*, 1926.

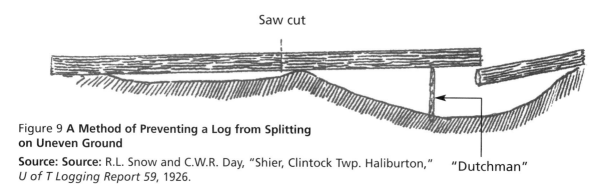

Figure 9 **A Method of Preventing a Log from Splitting on Uneven Ground**
Source: Source: R.L. Snow and C.W.R. Day, "Shier, Clintock Twp. Haliburton," *U of T Logging Report 59*, 1926.

Photo 14 **Bucking**

Two men using a crosscut saw felled the tree and cut it into 16-foot logs if possible.

Source: J.E. Bier and A. Crelock, "McRae Lumber Company, Cranberry Lake, Canisbay Township," *U of T Logging Report 75*, 1930–1931.

Chapter 4: The Airy Mill Operations: 1922–1933

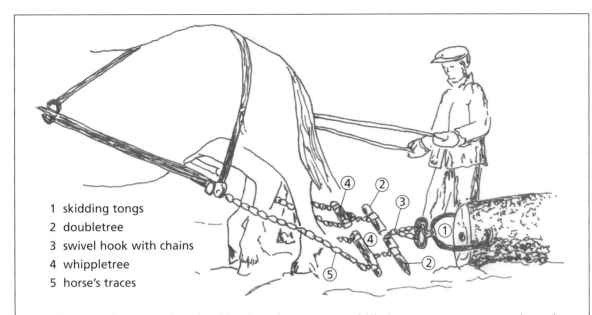

1 skidding tongs
2 doubletree
3 swivel hook with chains
4 whippletree
5 horse's traces

Skidding the logs was the job of hauling the logs from where they were felled out to the skidway. The teamster hooked onto a log with skidding tongs. He then took the chain on the doubletree and connected it to the tongs with a swivel hook. The doubletree was then connected using short chains to the two whippletrees which in turn were connected to the horses' traces that connected to the horses' collars. The teamster had to walk beside the log as it slithered through the snow. This was hazardous for he could easily have a foot or a leg jammed between the log and a stump or tree.

Figure 10 **Skidding a Log**   Source: D. Lloyd

the log. Unlike pine or other conifers, which are quite uniform, there is a considerable variety of shape in maple, birch and other hardwood trees. About 40 pieces of hardwood or 60 pieces of softwood were skidded and piled per day per crew. The skidway or landings, where piles of logs were built up along the haul roads, held up to 250 logs.

A single horse, or two together in a team, skidded one log at a time to the skidway located on the haul road. With a single horse set-up, tongs were fastened to the logs and the tongs were then chained to the whippletree. Chains coming from the horse's traces attached to the ends of the whippletree. The other ends of the traces attached to the horse collar. When the horse moved forward and pushed into the horse collar, the log moved. With a pair of horses, a doubletree was used. The ends of the doubletree chains attached to the middle of each whippletree belonging to each horse (Fig. 10 & Photo 15).

Skidding trails were, on average, 600 feet long, with a quarter of a mile being the maximum. Skidding trails, planned by the teamster and cut out by the trail makers, were plotted so as to run down slope, avoiding all possible obstructions and poor footing for the horses

Photo 15 **Skidding a Log**
The teamster had to walk beside the log as it slithered through the snow. This was hazardous for he easily could have a foot or leg jammed between the log and a stump or tree.
Source: J.E. Bier and A. Crelock, "McRae Lumber Company, Cranberry Lake, Canisbay Township," *U of T Logging Report 75*, 1930–1931.

while making it possible to get the largest number of logs from a given area.

Adjacent to the haul road were the skidways—where logs were piled at the end of skidding trails. Skidways were sloped, if possible, toward the haul road so gravity could be used to aid in decking the logs. To begin the skidway, two skid logs were placed some ten feet apart, perpendicular to the haul road, and levelled. The skid logs were notched at the end nearest the haul road. Skidways were built from the side furthest away from the haul road toward the road. The first layer of logs was rolled onto the skidway by a man called the roller using a cant hook.

Photo 16 **The Sender or Roller Using the Cross-Haul Method for Building a Skidway**

A cant hook is a tool consisting of a handle about four feet long. At one end of the handle is a "dog," or snibbie, a metal fitting with two teeth. A hook, which is free to open and close, is attached to the handle about 16 inches from the dog. The roller would place the dog on the log to be moved with the hook hanging down under and just touching the log. A lifting movement of the handle would push the hook up and into the log and further upward pressure on the handle would cause the log to roll.
Source: J.E. Bier and A. Crelock, "McRae Lumber Company, Cranberry Lake, Canisbay Township," *U of T Logging Report 75*, 1930–1931.

Chapter 4: The Airy Mill Operations: 1922–1933

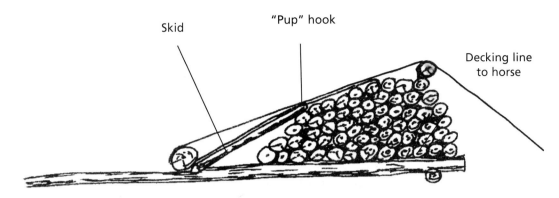

Figure 11 **Building a Skidway by the Cross-Haul Method Using a Decking Line**
Source: Redrawn from Joe Mason. *My Sixteenth Winter: Logging on the French River.* Cobalt [ON]: Highway Bookshop, 1974.

For a simple skidway, the cross-haul method was used to add subsequent layers. Two short skids, a pulley and a long decking line were used. The decking line had a sharp hook called a "pup" which was driven into a log on top of the skidway after being passed over and around a log. As the horse or team of horses pulled the other end of the decking line the log was rolled up the short skids. On passing the pup, it pulled out. A pair of men called rollers, each using a cant hook, placed the logs.

Photo 17 **Cross-Haul Method of Building a Skidway Using a Decking Line**

The skid logs are shown in this photograph. The log in the foreground is moving down the skids. The horse does not show since it is off to the right at the end of the decking line. This method of piling was both slow and dangerous, for if the spiked pup slipped, the log was free to bounce down the pile and crush any worker in the way.

Source: J.E. Bier and A. Crelock, "McRae Lumber Company, Cranberry Lake, Canisbay Township," *U of T Logging Report 75*, 1930–1931.

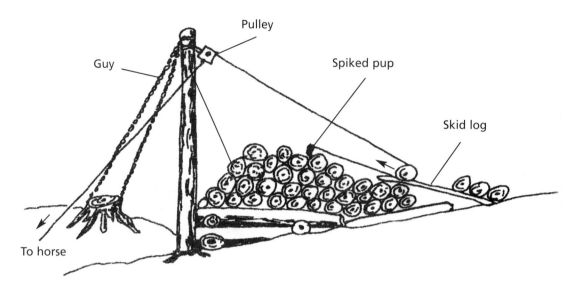

Figure 12 **Gin-Pole Used in Building a Skidway Using the Cross-Haul Method**

A long pole (gin-pole) was guyed securely in front of the skidway. A chain went through a pulley block, and over and around the log to be moved. A spiked hook, termed a pup, was driven into a log directly in front of the space where the new log was to go. The other end of the chain was hooked to the doubletree of the team of horses. The log moved up the skids as the team of horses moved forward.

Source: Redrawn from T. Wayne Crossen. *A Study of Lumbering in North Central Ontario 1885-1930 with Special Reference to the Austin-Nicholson Company.* Toronto: Ontario Minsistry of Culture and Recreation/Ministry of Natural Resources, 1976.

A variation in building a skidway involved the use of a gin-pole. A pup hook is driven into a log on the pile, with the decking line passed around the log. The decking line is then passed through a pulley that has been attached some 15 feet up a tree or pole used as a gin-pole. A gin-pole is simply a convenient vertical pole, be it a convenient tree or guyed single pole. From the pulley, a grab hook attaches the decking line to a small chain attached to the doubletree of the waiting harnessed horse. A doubletree is a pole to which leather traces from the horse collar are attached. At the centre of the pole was a ring to which a grabhook could be attached. Again, the dog is pulled out of the log when the moving log passes over it. The gin-pole method could build a higher pile of logs than the simple cross-haul method (Fig. 12).

The gin-pole method, which could be used with a single horse or team, required at least three men. Gin-poles were used to load sleighs in the bush and to pile up logs at the dumps. Skidways averaged 60 logs, but some had as few as 15 and others as many as 250 logs. The logs were moved from the bush skidways down the haul road and piled in thousands in dumps. With open water in the late spring, a gin-pole gang would take down the hardwood piles and begin the movement of the logs, usually by water, to the mill.

Chapter 4: The Airy Mill Operations: 1922–1933

Photo 18 **A Skidway**
Brought from where they were cut, these logs are waiting to be loaded on sleighs and taken down the haul road to the dump built from the logs taken from many skidways.
Source: M.A. Adamson, W.E. McGraw, D. McLaren and D.M. Parker, "McRae Lumber Company, Louisa Creek, Nightingale Township," *U of T Logging Report 57*, 1926.

Scaling, or Measuring the Logs

Scalers were Department of Lands and Forest technicians who measured the logs on the skidways. Based on the records they made, the company would pay a fee to the government called stumpage. The scalers used a scaling stick, which gave the board feet contents of each diameter class of log, a crayon and a scaling hammer. The logs on the skidways were usually 16 feet long and were piled with one end even, thus eliminating any difficulties for the scalers in determining the lengths of the logs. The scaler and his assistant stood on opposite ends of the pile of logs. Since the assistant scaler frequently could not visually identify the log to be measured, a hammer was used to hit the log being marked; the impulse sent down the log could be sensed by his partner and the log thus identified. The diameter of the log on each end was then called out. All defects on the end of the log at which the assistant scaler was located were explained to the head scaler. He then allowed for the defects from the scale to be deducted from the small end diameter of the log. In other words, he discounted the value of the log because of its defects. The assistant scaler then tallied or recorded the information on a government form. The log scaled was marked with a blue crayon. Between 700 and 800 logs were scaled each day. A record of the number of logs cut, skidded and hauled, and the numbers at each skidway was kept in a government logbook.

The scaler drew on his experience of the type and nature of all defects in order to make the proper allowance. For defects such as checks and shakes, the diameter was usually decreased, while for crooks and sweeps a deduction was made in the length of the log. Check refers to the separation of wood between spring and fall growth rings, while shake refers to a separation in growth rings in pine and hemlock.

Government scalers were employed to do the scaling in the 1920s, but the company paid them. Scalers were paid $5.50 per day and helpers, $4.50. This was a fine wage for the time. The Doyle Rule, Ontario's standard log scale, was used. The scaling cost was six cents per thousand board feet, with the money going to the government. The only check the company kept on the scaling was the clerk's log count. Each night the teamster of each cutting crew reported to the clerk the number of logs cut and skidded and the location of the skidway. At the earliest opportunity, the clerk checked the number of logs, marking them with a red crayon. If there was a discrepancy, the government scalers repeated their work. Any wide variations in the two results necessitated scaling for a third time.

When most of the cut was made up of smaller logs, the Doyle Rule underestimated the harvest to the benefit of the mill owners (i.e. the output at the mill was larger than the cut recorded in the bush). The Doyle Rule worked reasonably well when measuring big pine, but underestimated the wood in smaller logs. However, yellow birch and sugar maple have a large percentage of heartwood, which is unsuitable for manufacturing the best grade of lumber. This feature appreciably diminished the large overrun given on logs by the Doyle Rule. The use of the Doyle Rule persisted until 1946, when it was replaced by a metric system.[68]

With the logs scaled, the moving of the logs from the bush skidways to the dump was begun.

The Haul Road

The average hauling distance for a haul road was 2½ miles, with a maximum length of three miles. A day's haul consisted of 300 logs, and each team made three trips.

The ice on any lakes to be crossed had to be at least 20 inches thick to bear the weight of horses and loaded sleighs. To ensure this, the road track would be plowed clear of most of the insulating snow, which permitted the penetration of freezing temperature and the thickening of the ice. Ruts for the sleigh runners would also be watered to form ice. The ruts were made by the Otako plow, as explained later.

The hardwood was dumped on land, while the softwood, pine and spruce, was dumped on the ice. Before the ice melted, the softwood logs were surrounded by a log boom.

After Christmas, some of the men would be put on road preparation. The men employed in haul road construction and maintenance were the older men or those inexperienced in woods work. Occasionally, grading or smoothing out of the haul road's surface or grade was done. Simple "bridges" might be constructed to even out grades or if a creek had to be

Figure 13 A Method of Lessening Grade on a Bush Road
Minor variations in road gradients and small stream-crossings were commonly dealt with as shown below.
Source: R.L. Snow and C.W.R. Day, "Shier, Clintock Twp. Haliburton," *U of T Logging Report 59*, 1926.

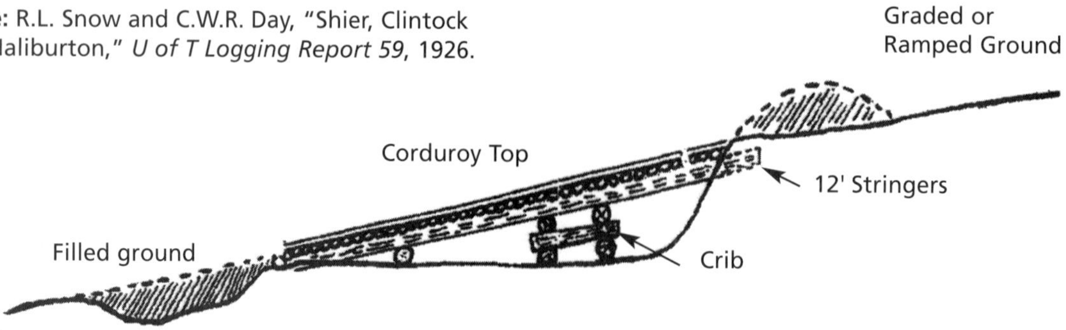

crossed. Since the road was only to be used for two or three years, rough construction, as long as it was sturdy, was quite satisfactory. The haul road was a winter route that commonly followed the water—across lakes, ponds and marshes. Snow and ice made the route passable. In spring, summer and fall, the haul road would be virtually impassable over long stretches.

Photo 19 Ford Tordson Tractor Pulling a Log Sleigh Train
In 1926, a Ford Tordson tractor was the first machine used in the bush by McRae. It was used to "break out roads." This would include removing stumps and rocks. The tractor was found to be superior to the horse for this purpose.
Source: M.A. Adamson, W.E. McGraw, D. McLaren and D.M. Parker, "McRae Lumber Company, Louisa Creek, Nightingale Township," *U of T Logging Report 57*, 1926.

Duncan MacGregor observed that, big as they were, the winter roads could hardly be found in the summer.

> You had to have 25 or 30, sometimes 50 men cuttin' the roads. You had to cut right down to the roots. It took a lot of men to make all those roads. You'd go there in the summer time and you wouldn't know there was a road there. Right down in the marshes and creeks. And they were big roads.[69]

Traces of abandoned tote and haul roads can be detected in the bush today. One indicator is the presence of clover, timothy and purple vetch—plants not native to Algonquin. The seeds of these species were scattered, undigested, in "road apples" deposited by passing teams of horses many years ago.

Table 6 Equipment Used on the Haul

16	sleds	4	hooks per team
16	teams	4	binding chains per sled
	decking lines	1	wrapper chain
	skid chains		

Source: M.A. Adamson, W.E. McGraw, D. McLaren and D.M. Parker, "McRae Lumber Company, Louisa Creek, Nightingale Township," *U of T Logging Report 57*, 1926.

Donald McRae said the haul usually began after the Christmas break and continued until the middle of March.

> Once you got four or five inches of snow, you had the team on to flatten it out so it'd make ice. Well then, after Christmas you'd go after your main road to get it in shape. You'd start, maybe January 5th through 10th, and then you'd be ready to go. The odd time you maybe got the sleigh haul started at the end of December. Most of the men would go home for Christmas, but they'd be back in three or four days. You'd have all the logs you wanted by Christmas—I mean

what you planned for. The minute the haul started you stopped all the cutting. Well, it took a while to change over, because for hauling you shod the horses sharp; for skiddin' you had dull shoes on them. At each camp there was a blacksmith to do the job.[70]

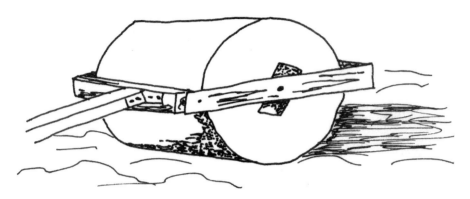

Figure 14 **Road Roller**
A crew of four was detailed for road maintenance. To avoid delays in hauling operations, plowing, rolling and sprinkling were done at night, torches being used for illumination. The compacting of snow on the road was accomplished by rollers weighing about a thousand pounds each. One team of horses was usually sufficient to pull a roller.
Source: Redrawn from Crossen, *A Study of Lumbering in North Central Ontario 1885-1930*.

Two types of plows were used to maintain the road: the V-plow and the Patent (or Otako) plow.

The V-plow was made from two heavy pine timbers, each measuring perhaps twelve feet in length, six inches in width and twelve to fifteen inches in height. These pieces were fastened together at one end by a flexible hinge to form a wide "V." About eight feet back from the front hinge were fastened two smaller brace timbers. Each brace timber was hinged to one of the two larger timbers and the two brace timbers overlapped in the centre. By adjusting the pins or steel binding holding the brace timbers together, the width of the "V" could be altered to conform to the width of the road and the amount of snow to be removed.[71]

Opposite page, top
Figure 15 **Patent or Otako Plow**
The Patent or Otako plow consisted of two solidly built sleighs connected by a heavy timber down the centre to which the flanges and wings were attached. Just behind the front bobsleigh were two wings which could be adjusted to take the crown off the road. In the front of the rear bobsleigh were two winged metal blades, which could be moved up or down, to cut grooves in the snow-packed road and clear snow outside the track. The cut grooves, which were later iced, guided the runners of the log sleighs. At the rear of the plow were two outside, rear-mounted wing-plow blades that pushed snow further off the road.[72]
Source: Redrawn from Crossen, *A Study of Lumbering in North Central Ontario 1885–1930*, 203.

Chapter 4: The Airy Mill Operations: 1922–1933

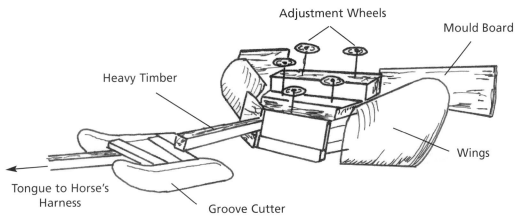

Figure 15 **Patent or Otako Plow**

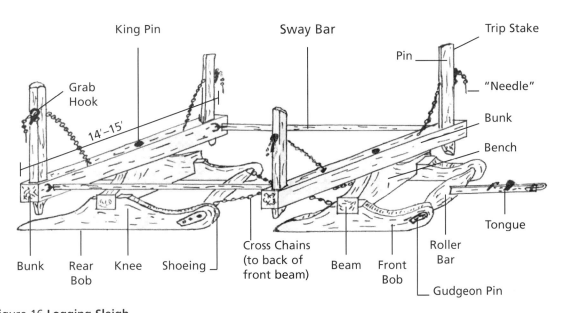

Figure 16 **Logging Sleigh**
Source: L. Cahill, D.W. MacGregor and S.W. Lukinuk, "Shier, Clinton Township, Haliburton County," *U of T Logging Report 121*, 1949.

All sleighs in use were built by the company's own blacksmiths and handymen, and were constructed especially for hauling 16-foot logs. The bobs were 6 feet wide with an 11-foot bunk. The bobs were the sleigh's runners and the bunks were where the logs rested. The rear bob was attached by cross chains from the roller bar of the rear bob to the bench of the front bob. The sleighs were equipped with a dual-purpose tongue: a short one for truck-hauling and an extension for team-hauling. The short tongue was hooked directly to the front roller bar and braced with flat irons. The extension tongue rode over the short tongue where a flat iron "U" held it on the short tongue. The end of the extension slid under a clevis pin on the short tongue where they were bolted together. This bolt was chained to the short tongue to prevent loss (Fig. 16).

On the end of the short tongue was a ring for hooking on to the hitch of the trucks. This hitch could not become unfastened unless the catch was pushed down against a spring that held it up.

The bunks were held in position by sway bars which kept the two bobs tracking. Shoeing on the sleigh's runners was half-round iron, which was the best for iced snow roads as it had less friction. However, it had a tendency to slew, especially on lake-roads.[73]

The capacity of these sleighs was on average

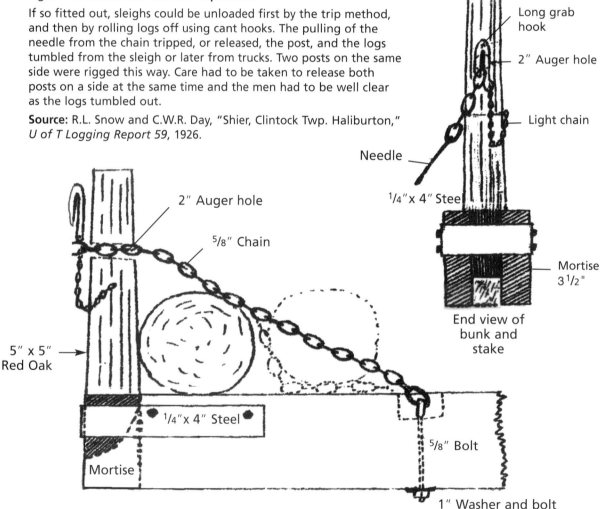

Figure 17 "Needle and Thread" Trip Stakes

If so fitted out, sleighs could be unloaded first by the trip method, and then by rolling logs off using cant hooks. The pulling of the needle from the chain tripped, or released, the post, and the logs tumbled from the sleigh or later from trucks. Two posts on the same side were rigged this way. Care had to be taken to release both posts on a side at the same time and the men had to be well clear as the logs tumbled out.

Source: R.L. Snow and C.W.R. Day, "Shier, Clintock Twp. Haliburton," *U of T Logging Report 59*, 1926.

Chapter 4: The Airy Mill Operations: 1922–1933

12 birch logs or 20 softwood logs. Each team hauled three loads per day over a distance of six miles per round trip. This sleigh shows four "needle and thread" stakes. While they speeded up the unloading of the logs, they were dangerous to use.

The haul road was iced, which made for less friction and easier pulling for the team of horses. This very cold job was usually carried out at night to ensure freezing, and since the log-sleighs were not on the road at this time.

The sprinkler tank was fitted with curved runners and shafts at either end. "This meant that the tanker men didn't have to turn their sleigh around when they ran out of water and had to go back for more. Instead they merely unhitched the horses, brought them around to the back end of the sleigh, hitched them up again and started back the way they had come."[74]

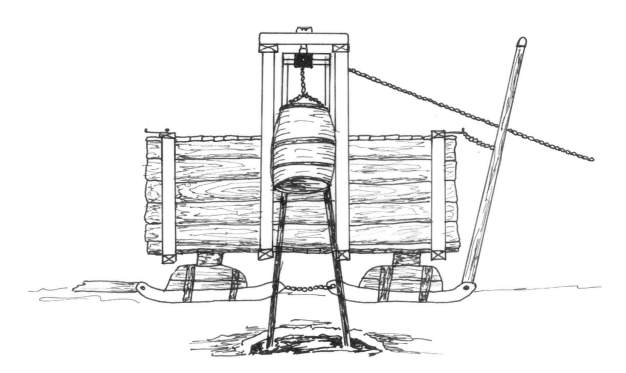

Figure 18 **Sprinkler or Tanker Sleigh**
A sprinkler consisted of a rectangular, watertight wooden tank mounted on a hauling sleigh. Tanks were filled at a lake or stream, where a ladder or skid was placed, sloping, against the side of the tank. By means of a cable attached at one end of a barrel and passing through a pulley at the top of the tank and down to a team of horses at the other side, water was drawn up and emptied into the tank. A team of horses was required to pull the heavy sprinkler. The water was guided into the sleigh tracks where it froze. The ice provided less friction and the horses could pull more logs than if the road were simply topped with packed snow.

Source: *Algonquin Park Logging Museum: Logging History in Algonquin Provincial Park.*

Photo 20 Loading a Tanker with a Barrel of Water
Source: L. Cahill, D.W. MacGregor and S.W. Lukinuk, "Shier, Clinton Township, Haliburton County," *U of T Logging Report 121*, 1949.

In order to be loaded, a sleigh was driven between the jammer and the skidway. The cable that ran through the base pulley was attached to a team of horses that pulled, lifting the log into the air and over the sleigh. Six men were needed in this operation. Two senders drove the pups into the ends of the log and guided it over the sleigh. Two tailer men used cant hooks to roll logs from the back of the skidway up to the senders, while a top loader on the load positioned the log on the sleigh. A teamster directed the horse that provided the power.

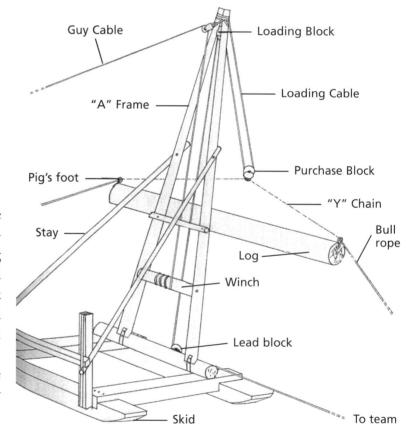

Figure 19 Jammer
Loading of logs on to the sleighs was done at the skidways by a crew of six men using a block and tackle arrangement called a jammer.
Source: Redrawn from L. Cahill, D.W. MacGregor and S.W. Lukinuk, "Shier, Clinton Township, Haliburton County," *U of T Logging Report 121*, 1949.

The side jammer was a more efficient apparatus than the gin-pole for piling logs. Two long poles, supported by braces and guyed, were topped by a block and tackle, while from a second block and tackle hung two steel cables with hooks called pups. The pups were driven into the log to be moved. A third pulley was at the base of the pole.

Chapter 4: The Airy Mill Operations: 1922–1933

Photo 21 Boomed Softwood Logs, Whitefish Lake, 1930
Part of the log boom surrounding the logs shows on the left side.
Source: McRae Lumber Company

dump. The hardwood logs were dumped on land near the shore and piled in high skidways by a jammer. A jammer had two upright poles joined at the top where there was a pulley. The jammer was free-standing. The softwoods were dumped and boomed on the ice. The winter's cut of approximately 40,000 hardwood logs and 25,000 softwood logs would be at the log dumps by spring.[76]

Softwood such as spruce, hemlock and pine could be floated to the mill. These logs were first dumped on lake ice in winter and then surrounded by large boom logs that were chained together. When open water came in the spring, powered pointer boats or "alligators" towed the logs across the lakes. The logs were then driven down a river to the next lake where the logs were again boomed and the procedure repeated until Galeairy Lake and the mill was reached.[75]

Bob McRae recalls that

> Dad [Donald McRae] tells of the McRae Lumber winter dump on Whitefish Lake, which he remembers from his boyhood. The logs were cut from one camp [probably Speckled Trout Lake] during the winter in Lawrence Township, then hauled by horse-drawn sleighs to the Whitefish

Photo 22 Using Gin-Poles at Hardwood Dump, late 1930s
At the dumps, the hardwood piles would be taken down by a gin-pole gang for cribbing. This method of piling and taking down dumps was slow and dangerous.
Source: McRae Lumber Company

Cant Hook
The cant hook has a hook and a dog on a handle. It is used to roll logs.

Pup Hook
The pup hook or dog is attached to the end of the decking chain that is used to move logs onto skidways.

Bitch hooks are used to tighten the chains over a load of logs on either a sleigh or truck.

Pig's Foot
The pig's foot hook is driven into the log to be lifted by the jammer. Rope attached to these hooks enable the bull rope handlers to guide the log to the waiting top loader.

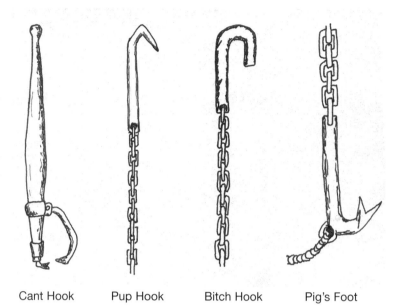

Cant Hook   Pup Hook   Bitch Hook   Pig's Foot

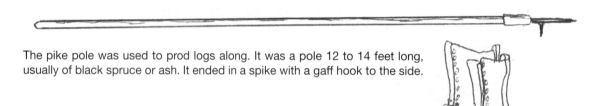

The pike pole was used to prod logs along. It was a pole 12 to 14 feet long, usually of black spruce or ash. It ended in a spike with a gaff hook to the side.

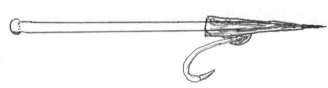

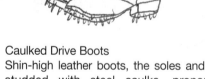

Caulked Drive Boots
Shin-high leather boots, the soles and heels studded with steel caulks, pronounced "cocks," enabled a river driver to run over floating sawn logs.

The peavey on a five-foot stock had a spike at the end. The thumb-like hook moved up or down—not sideways. The peavey could be used both to prod logs along or to lever them into the water using the side hook.

The introduction of pairs of raker teeth to the saw by Surley and Deitrich of Galt, Ontario in 1874 made possible horizontal sawing without binding.

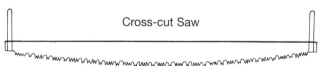

Cross-cut Saw

Figure 20 **A Variety of Hooks and a Crosscut Saw**
These hooks were made by the blacksmith and his helper.
Source: D. Lloyd

Chapter 4: The Airy Mill Operations: 1922–1933

J.S.L. kept close track of the expenses of his operation. Note that the costs listed below are given per thousand board feet. This of course was the measurement of potential boards in a standing tree, in a log and in the output of a mill. This common denominator enabled the mill manager or owner to quickly see where his expenses were relative to production, or as J.S.L. might have said where an item or labour "was costing me debt."

Table 7 **Bush Camp Lumbering Costs, 1926**

Costs per Thousand Board Feet for the Whole Operation:	
Cost of log making	$ 0.76
Cost of skidding	2.69
Cost of hauling	2.80
Cost of scaling	.06
Total	$6.51

Note: Overhead expenses are not included

Source: M.A. Adamson, W.E. McGraw, D. McLaren and D.M. Parker, "McRae Lumber Company, Louisa Creek, Nightingale Township," *U of T Logging Report 57*, 1926.

The cost of feeding the men in the bush is not included in the table given above. Since the men were paid on a piecework basis, calculating the cost of felling the trees and cutting the stems into log lengths was simple. The cost of skidding would have included the expenses of the horses, as did the costs of the haul and the piling of the hardwood logs adjacent to the lake from which they would be towed in the spring. The cost of scaling was a fee set by the government.

The Drive by Water

The success of McRae's financial gamble in purchasing the Airy mill hinged on getting hardwood logs from the dumps to the mill efficiently, without the loss of logs. The McRae limits were in the townships of Nightingale and, later, Lawrence and Clyde. These limits were far beyond horse-haul sleigh range to the Airy mill. Galeairy, Rock, Whitefish, Pen and McKenzie's (Clydegale) Lakes, along with wooden dams and chutes at McKenzie's, Pen and Galeairy lakes and on Hay Creek brought the cutting areas within range of the mill, if the logs could be floated (Map 10). The traditional way of moving logs to a mill was to drive them: logs were dumped into the rivers and floated down stream with the current. When a lake was met, there was no current to move the logs. To cross the lake, the logs were gathered together inside a large loop of connected logs called a log boom and dragged across the lake with a cadge crib and later a tug or alligator. McRae used this system to move softwood logs.

Hardwoods, particularly yellow birch and sugar maple, were McRae's main cut. Logs of these types quickly sank and were lost. Peeling the hardwood logs and letting them dry before floating and booming them was time-consuming, expensive and only marginally successful. J.S.L. solved the problem of how to move hardwood logs over water by wiring the logs to floater cribs.

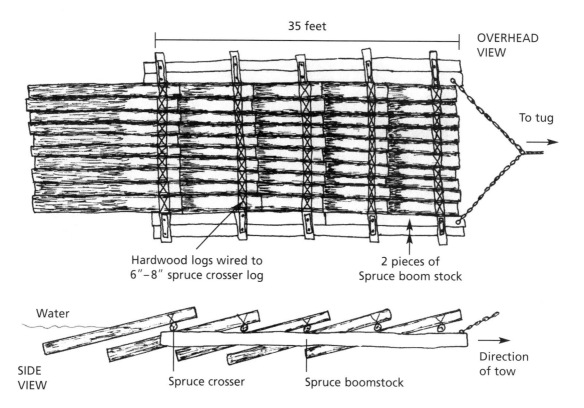

Figure 21 **McRae Hardwood Crib**
Source: D. Lloyd

McRae devised a crib of softwood which floated well, and the hardwood logs were wired to crossers of the crib. In 1926, a crew of about six men was used to build the cribs, the cost being about $20 each, of which $15 was for labour, the rest being for material.

Donald McRae explained how McRae moved hardwood logs across water.

> When the ice went out, they would begin to tow all the logs down to the mill on Galeairy. The hardwood logs would be lifted from the skidways with a jammer and skidded to the lake by horses. They were then cribbed and towed down to the mill first. The hardwood crib, 35 by 25 feet, was made of two main floaters of 35-foot spruce logs with crossers of balsam or hemlock 25 feet long. The hardwood logs were wired onto the crossers. Each log was wired onto the leading crosser. Perhaps ten logs were wired to this crosser. In this fashion, they would wire 60 to 70 logs to a crib.

> A large pointer boat with a large gas motor would tow two or three cribs together down Whitefish Lake, through the river and down Rock to the dam between Rock and Long [Galeairy] lakes. Dad thought that the gas boat would tow five cribs at once. The cribs had to be narrow enough to pass through the river, and between the piers of the railway bridge at Whitefish Lake.

Chapter 4: The Airy Mill Operations: 1922–1933

The company had a narrow-gauge rail track to portage the logs around the Rock Lake dam [Photo 24]. Each crib was towed up to the train cart, and a jammer with a gas-powered winch lifted each log from the boom onto the cart, which would hold ten logs. Then the cart was pulled along the track by a gas-powered winch. A jammer then transferred the logs to a waiting crib in Long Lake. Eight cribs twice a day would be filled with logs and towed to the mill at Airy. The pine logs would be boomed rather than cribbed, and pulled down Rock Lake, put through the sluice at the dam, log by log, and reformed into booms on Galeairy. The tugboat would go up Galeairy every day to pull the cribs or booms down to the mill.

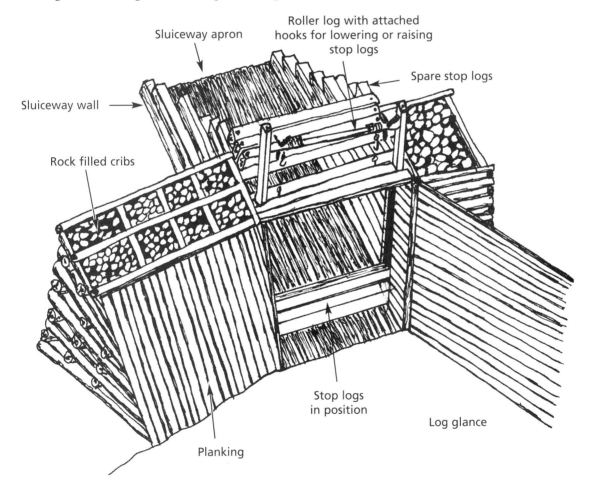

Figure 22 **Rafter Type Splash Dam**
Source: S. Brackenbury, W. Cleaverley and F.T. Collict, "Canadian International Paper Company, Manawaki, Quebec," *U of T Logging Report 123*, 1950.

One steam tug was used together with two gas launches of a two-cylinder type. The hulls of the boats were company-built and the engines cost $400 each for the launches and $1,000 for the steam tug. The gas boats were used to pick up stray logs, mainly from the softwood booms, and to pull the hardwood cribs into position for connecting, towing and positioning for the jack ladder at the mill. The cost of towing was about $18 for 600 logs.[77]

Philemon Wright, the founder of Hull, took the first raft of squared timber down the Ottawa River and St. Lawrence River to Quebec City in 1807. At that time there were no log dams on the rivers, and the rafts of timbers had to be completely taken apart, the logs run through the rapids and then reassembled.

Ruggles Wright, the son of Philemon, constructed the first log dam and chute on the Ottawa River to circumvent Chaudiere Falls. When completed, the structure was able to pass whole cribs of timber around the falls and save considerable time, as workers did not have to completely reassemble the rafts from single timbers. The rafter-type splash dam shown in the sketch is a descendent of that first dam and chute.

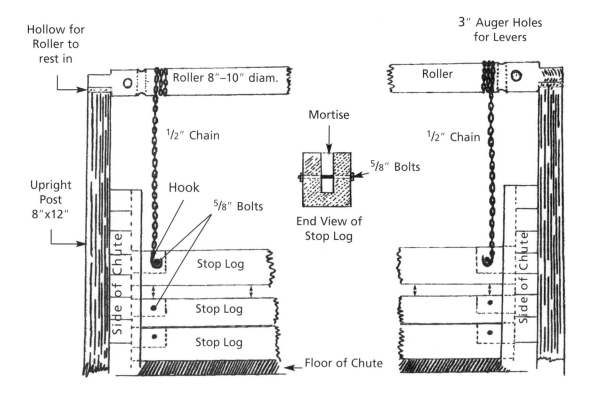

Figure 23 **Method of Raising Stop Logs in a Dam**
Source: R.L. Snow and C.W.R. Day, "Shier, Clintock Twp. Haliburton," *U of T Logging Report 59*, 1926.

Chapter 4: The Airy Mill Operations: 1922–1933

The logging dams of the 1840s to the 1940s were typical tools of the age: simple, durable and wonderfully functional. Rock-filled cribs formed the wings of the dam and framed the control mechanism. The roller-winch could raise or lower the stop logs that built up from the floor of the chute to give the desired height of water behind the dam. With a series of these dams, a lumber company could control the water level in a stream down which it was driving logs (Fig. 22).

Dams were simple yet ingenious structures. Although the dams leaked more as they aged, the basic structures lasted three or more decades. As shown in Figure 23, stop logs reg-ulated the water levels behind the dams as well as the flow down the log chute. To raise the stop logs and lower the level of water behind the dam, the chains would be lowered from the roller log by turning it with levers inserted into the holes in the roller log. The hooks would be attached to the bolts on the stop logs. The roller log would be turned in the opposite

Photo 23 Falls and Log Chute Between Pen Lake and Rock Lake
In this photo, the initial drop of the falls is to the left, while the log chute is to the right. There was a stop-log gate for each. The log structure in the middle is a deflection barrier to keep very high water from the falls from sweeping away the log chute.
Source: J.W. Ross. 1895. APM 83

direction. The raised stop logs would be stacked on the deck of the dam. To insert the stop logs and raise the level of water behind the dam, this procedure would be reversed (Fig. 7).

Photo 24 Narrow-Gauge Cart Railway between Rock and Galeairy Lakes

In this photo, members of the Barclay family are using the remains of the McRae track between Rock and Galeairy lakes. The Barclays, relatives of J.R. Booth, had a private estate on Rock Lake until 1956, when it was purchased and removed by the Department of Lands and Forests.

Source: APM 0196

The alligator, a steam-driven, flat-bottomed boat propelled by paddle wheels, replaced the cadge crib. It floated very high in the water and consequently could operate in shallow water. A powerful winch could move large booms of logs after the alligator was anchored. The winch could also pull the alligator across land. The McRae alligator operated as needed on Long (Galeairy), Hay, Rock, Whitefish and Pen lakes.

Why was such a strange name used for the boat? The reptilian alligator is an amphibious creature which is capable of living on land or in the water, hence the use of the term for this very useful boat. The alligator was invented in 1889 in Simcoe, Ontario by West & Peachey Ltd., who built over 200 of them (Photo 26).[78]

Moving the alligator from lake to lake across a portage was slow work and had to be done carefully. Two iron-clad 15-inch-thick timbers along the bottom of the alligator stretched from front to back and acted as runners. When moved across land, fresh peeled skid-logs were laid across the trail and the winch on the alligator pulled it along over the skid-logs. If going uphill, the steam engine had to be kept vertical. To do this, a jack screw attached to the engine was used to adjust the tilt of the engine. The men had to add water to the boiler to maintain steam as well as stoke the firebox with wood to keep the heat up.

Photo 25 Alligator, Pointer and hardwood Log Boom

The fact card at the Algonquin Park Museum (Visitor Centre) records that this alligator was delivered to McRae Lumber at Rock Lake station in 1928. It worked on Pen, Whitefish, Hay and Long (Galeairy) lakes. When shown the photo in June of 2005, Gordon Palbeski, who worked for the McRae Lumber Company, immediately said the boats were on Hay Lake and the pointer was being driven by Dave Bowers.

Source: Henry Taylor. APM 1881

Chapter 4: The Airy Mill Operations: 1922–1933

Photo 26 **Remains of McRae Alligator**
Henry Taylor is shown inspecting the alligator, which had been beached on Long (Galeairy) Lake.
The note on the photograph file at the Visitor Centre indicates that McRae Lumber Company bought the boat from Ferguson and Findlay Lumber Company of Renfrew, Ontario in 1928.
Source: Henry Taylor. APM 01292

Upon reaching Long (Galeairy) Lake, a tugboat was used to tow the cribs. Donald McRae said:

> Twelve men were required to move the hardwood from the dumps. One man was in charge; four men loaded the cribs; two men manned the tug; and there was one man in each launch. At the mill, five men unloaded the logs at the jack ladder.
>
> Probably 700 logs would keep the mill going for a shift. The steamboat burned wood as fuel. At the mill the crib would be pulled into position beside the jack ladder. The logs were then unwired and poled into the jack ladder.[79]

In the summer, the steamboat was a great favourite with the younger McRaes and their

Photo 27 The McRae Steam Tug
Source: M.A. Adamson, W.E. McGraw, D. McLaren and D.M. Parker, "McRae Lumber Company, Louisa Creek, Nightingale Township," *U of T Logging Report 57*, 1926.

friends. Marjorie McRae McGregor said it went up the lake twice a day,

> ...and brought down a boom of logs. They had big skidways up there and they put the logs in the water and they'd make booms, and every day they went up to get more logs. So it was a great thrill for us to be able to go up. Then, I think, they left at maybe six o'clock in the morning. I can remember they had a whistle and they'd blow the whistle and of course we'd have company: my cousins used to come up and we'd all go. It was run by steam, but they used wood to fire the boilers. And of course the sparks—you'd wear something and at the end of the day your clothes would be all full of holes from the sparks. But we'd go up there for the whole day, and then you'd have lunch there in the camp.[80]

Photo 28 Boomed Hardwood Logs, Man with Pike Pole and Pointer in Background
This photo was likely taken at the north end of Hay Lake about 1929–1930. From here the logs would have been floated down Hay Creek, then cribbed and towed to the mill at Airy. The logs in the photo are hardwood, mostly yellow birch. The workman's pike pole is resting on a softwood boom log—notice that this log is floating much higher in the water than the hardwoods. These logs show sweep or crook which is typical of hardwoods.
Source: McRae Lumber Company

Chapter 4: The Airy Mill Operations: 1922–1933

The Canisbay Licence and William Finlayson, Minister of the Department of Lands and Forests

In 1929, just before the Great Depression, J.S.L. decided to expand. J.S.L. McRae and his chief assistant, Sandy McGregor, made an estimate of the potential yield in Canisbay Township north of the railway line (Map 11). This estimate was used in the bidding process, which also involved the Department of Lands and Forests cruising the area as well. The lease negotiated was only for hardwood. There were few pines because the area had been logged before.

Obtaining this lease was not straightforward. In 1929, on hearing of the proposed timber sale to McRae, Cache Lake leaseholders, remembering the Munn episode in 1910, put pressure on the Lands and Forests Minister, William Finlayson, to deny the application. The Munn Company had enraged the tourists as well as Bartlett, the Park superintendent, by cutting down trees to a small size in plain view of the newly constructed Highland Inn. However, in 1929 the Depression was causing massive unemployment and the government was aware that McRae was the only functioning employer in Whitney. This fact and the government's need of stumpage fees persuaded Finlayson to grant the lease. But he made sure that there would be no repeat of the previous Munn situation. In a creative move, Finlayson placed a 100-foot reservation around the shoreline of Cache Lake before the licence was issued. This measure protected the scenic nature of Cache Lake and placated the leaseholders. It was the first time that the interests of tourists were put ahead of those of lumbermen in Park management.[81] Previous to this, timber could be cut to the shoreline. McRae Lumber stayed clear of the tourists by locating its operations some distance away from Cache Lake, at the north end of Cranberry (Canisbay) Lake.

The Depression was very hard on the Park area. In the winter of 1930–1931, McRae was the only mill operating in the area. Timber came from a camp located at Rock Lake, and a second that was operated by a jobber in Nightingale Township. Despite the Depression, McRae opened a third camp just north of Canisbay Lake after he purchased limits in Canisbay Township in 1930. Expansion of logging operations was unusual, as softwood prices were low because of the Depression. In the general vicinity, the Canoe Lake Lumber Company, and the Duff mill at Brule had closed down. These companies were cutting softwood. The J.R. Booth Company at the Egan Estate limit did, however, have 350 men working at three company camps as well as a jobber cutting pine only (Map 5). The Highland Inn was closed for the 1933 season. By the 1934–1935 cutting season, however, there was considerable improvement in employment within the Park, with 14 licence holders in operation and 1,500 men employed.[82]

By locating at Cranberry Lake, McRae only had to build a half-mile of tote road, as it could use the existing road to the then-closed Minnesing Lodge for much of the distance.

All supplies for the Cranberry Lake camp in Canisbay Township were shipped to Algonquin Park station on Cache Lake by rail. From there, they were toted four miles to the camp. There were two toting teams that moved about 2,700 pounds per load, making two trips per day if necessary. McRae was also able to run a spur telephone line from the one that ran along the road from the railway to the new camp and thereby connect to the office at Airy (Map 11).

Most new bush camps were built in September, and this gave plenty of time to complete them before severe and often very wet weather set in later in the fall. Occasionally, the building of bush camps was delayed and the inevitable happened. Rain and cold weather made for miserable conditions. Duncan MacGregor remembered one such time:

> I went into Cranberry [Canisbay] Lake in 1931 when they went in there. We built camps in October. We lived in tents. We didn't get cold because we went into the cookery to get warm. And they worked like the devil! It didn't take them long to build the camps because they wanted to stay in them.[83]

At the Cranberry Lake camp, about 80 per cent of the labour force was of Polish extraction, drawn from the local communities. The remainder were English, German and French-Canadian, again from nearby. The McRae camp at Rock Lake, however, was almost entirely French-Canadian.

A 2½ mile strip of forest was logged on either side of the bush camp. Since the operation was on one watershed, the logs could be readily hauled down Cranberry Lake, and only a short haul road from the lake outlet was necessary to reach Algonquin station on the CNR line. McRae was able to take advantage of a new method for loading the railway cars. At Algonquin station, logs were loaded on flat cars by a rigid-boom steam hoist, which had been purchased from the J.R. Booth Company. Seven cars a day were loaded with logs, which were then shipped by rail to the mill at Airy (Photos 30 and 31).

Photo 29 **Rigid-Boom Steam Hoist: Algonquin Park Station, Cache Lake**
The apparatus used here to load logs on the railway car was bought by McRae from J.R. Booth.
Source: APM 2462

Chapter 4: The Airy Mill Operations: 1922–1933

Photo 30 **Train Hauling Logs to Airy from Algonquin Park Station, Cache Lake**
Source: APM 3262

Accidents

Bush work has always been dangerous. Saws and axes are sharp. The behaviour of a falling tree is never entirely predictable. Accidents happened even to experienced workers. Two fatalities occurred during the days of the Airy operation.

In 1926, the *Eganville Leader* reported:

> A sad accident occurred in the Whitney woods on Friday morning last when Ovila St. Louis, a young man from Perrault [8 km south of Eganville], while at work in one of Mr. John McRae's camps received injuries which resulted in his death. He and a companion, Clayton Senior, of the Opeongo, were engaged in cutting down the standing timber. A tree falling, struck a stub birch which in turn, came with great force against the body of St. Louis and inflicted injuries of a fatal character to his back.[84]

Dr. McKay attended to him at the Red Cross hospital in Whitney where he died.

Five years later, the *Eganville Leader* reported another fatality. "While at work in the McRae lumber camp… Emil Zeeman of Golden Lake [10 km northwest of Eganville] was instantly killed on Tuesday, December 1, 1931, when a limb of a birch tree on its descent struck him down.[85]

Frank Shalla, a long-time McRae employee, vividly remembered the second fatality.

> It was in the thirties, and McRae said go up to the camp, the lads are going to come the next week. I was portaging the grub and that. I was goin' in there one night about four o'clock and I see another team comin' out. I stopped and said, "What's the matter?"
>
> And he said, "There was a lad got killed." They were takin' him out. They just made a rough box out of lumber and brought him out to where the train shipped him out. I was scared then! It was Zeeman; I think he was from Golden Lake.[86]

The Airy Mill: 1922–1932

The Airy Mill was located on the north shore of Long (Galeairy) Lake, a short distance to the north of the hamlet of Whitney. A short spur line connected the mill to the railway (then the Grand Trunk). As well as the mill, there was a sawmill office, a warehouse, a boarding house and stables. All were of undressed lumber covered with roofing paper. Three boilers, burning sawdust and small pieces of mill refuse, developed 300 HP and powered the mill. A burner disposed of the excess refuse. The mill operated from about the 24th of April until the end of October, or until all the logs went through the mill, the softwood generally being cut last.

Photo 31 **The Airy Mill Showing the Jack Ladder Passing Under the Railway**
An endless chain with hooks raised the logs to the second floor of the mill where they met the saws.
Source: McRae Lumber Company

Chapter 4: The Airy Mill Operations: 1922–1933

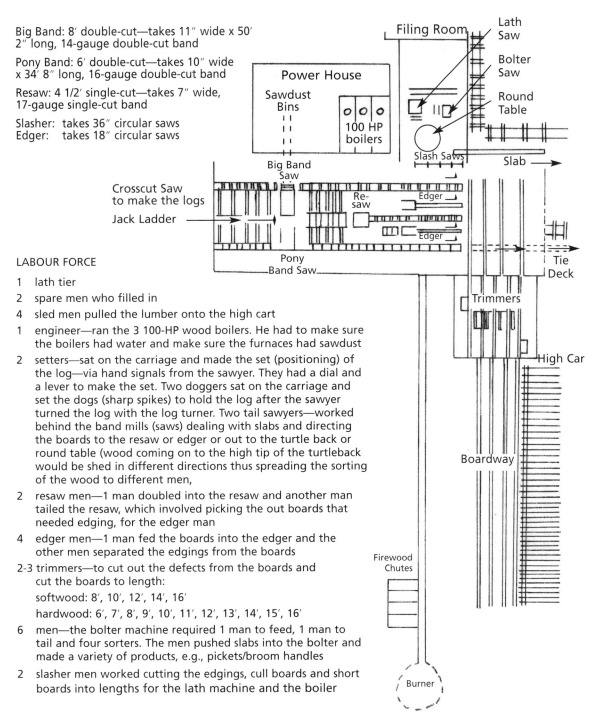

Big Band: 8′ double-cut—takes 11″ wide x 50′ 2″ long, 14-gauge double-cut band

Pony Band: 6′ double-cut—takes 10″ wide x 34′ 8″ long, 16-gauge double-cut band

Resaw: 4 1/2′ single-cut—takes 7″ wide, 17-gauge single-cut band

Slasher: takes 36″ circular saws
Edger: takes 18″ circular saws

LABOUR FORCE

1 lath tier
2 spare men who filled in
4 sled men pulled the lumber onto the high cart
1 engineer—ran the 3 100-HP wood boilers. He had to make sure the boilers had water and make sure the furnaces had sawdust
2 setters—sat on the carriage and made the set (positioning) of the log—via hand signals from the sawyer. They had a dial and a lever to make the set. Two doggers sat on the carriage and set the dogs (sharp spikes) to hold the log after the sawyer turned the log with the log turner. Two tail sawyers—worked behind the band mills (saws) dealing with slabs and directing the boards to the resaw or edger or out to the turtle back or round table (wood coming on to the high tip of the turtleback would be shed in different directions thus spreading the sorting of the wood to different men,
2 resaw men—1 man doubled into the resaw and another man tailed the resaw, which involved picking the out boards that needed edging, for the edger man
4 edger men—1 man fed the boards into the edger and the other men separated the edgings from the boards
2-3 trimmers—to cut out the defects from the boards and cut the boards to length:
   softwood: 8′, 10′, 12′, 14′, 16′
   hardwood: 6′, 7′, 8′, 9′, 10′, 11′, 12′, 13′, 14′, 15′, 16′
6 men—the bolter machine required 1 man to feed, 1 man to tail and four sorters. The men pushed slabs into the bolter and made a variety of products, e.g., pickets/broom handles
2 slasher men worked cutting the edgings, cull boards and short boards into lengths for the lath machine and the boiler

Figure 24 **Airy Mill Layout**
Source: Bob McRae

J.S.L. McRae would tell the mill foreman what size of boards he wanted cut. The foreman had some 48 to 50 men under him to operate the mill, many of whom also worked in the bush during fall and winter operations. The engineer looked after the steam system that powered the mill, along with other maintenance. A greaser lubricated and oiled all the equipment. The skill of the sawyer, together with that of the doggers, the saw-setters and the saw filer largely determined the quality and quantity of the wood produced by the mill, and therefore the mill's profitability. The Airy mill was double-sided, meaning that it had two carriages, two band saws and two sets of edgers.

Jack-ladder men met the cribs of hardwood or booms of softwood brought down the lake. The hardwood logs were brought directly from the transportation cribs into the mill, while the softwood logs would be held in booms until later in the summer when they would be milled after all the hardwood. The jack-ladder men, using pike poles, pushed the logs into the mouth of the jack ladder, where an endless chain of moving hooks caught the back ends of the logs.

The jack ladder then carried the logs to the second floor of the mill, where they were landed or decked. Each log was then rolled on to one of the log carriages. At this point the sawyer took charge. A near instantaneous decision on how to set, or position, the log was required of the sawyer in order to get the best value out of the log. The act of reading the log was important, for it determined how the grain of the log was to be cut and how many boards of what thickness would be made from it. It was important to get the best value from a log while at the same time following the production instructions given to him by the mill foreman. The sawyer sat beside the saw and used hand-finger signals to tell the first dogger how to adjust the dog or clamp that could turn and then hold the log to get the grade he wanted,

Table 8 **Airy Mill Labour Force, 1926**

1	foreman				
1	millwright				
1	engineer	1	greaser		
1	filer	1	assistant filer		
2	sawyers	2	doggers	2	setters
2	resawers	2	tail sawyers	4	yard men
2	edgermen	1	lath machine feeder	2	tailing edgers
2	trimmers	1	lath tier	2	slashers
4	sled men	4	sorters	1	deckman
1	tail bolter	4-6	jack ladder men	1	bolt feeder
2	spare men				

Source: M.A. Adamson, W.E. McGraw, D. McLaren and D.M. Parker, "McRae Lumber Company, Airy Mill," *U of T Logging Report 57*, 1926.

Chapter 4: The Airy Mill Operations: 1922–1933

and the setter made adjustments to the bunks that controlled the eventual thickness of the boards to be cut from the log (Fig. 25).

In an interview, Bob McRae described the process:

> The carriage then moved forward carrying the log into the saw and then back through the saw on the return trip producing a slab for the waste chute and a board for the edger. The sawyer would then use the steam log-turner to turn the log on the carriage and the process would be repeated for each side of the log until the sawyer had a square cant on the carriage. Then the sawyer would cut double boards off the outside of the cant for the resaw, trying to get the best grade possible by always turning the cant and boxing the heart of the log into a 6"x6" or 7"x9" railroad tie. At the resaw, the doubles would be split, with any waney boards [boards with one rounded side] sent to the edger. At the edger, the waney rounded edges of a board were removed. Boards were the high value product of the mill from the outside of the hardwood log and ties were the product of the lower-valued heartwood of the log. At the trimmers, the ends of the boards were squared and cut to length.[87]

Jamie McRae made the following comments when asked about measurement in lumber.

> In hardwood and pine a 2"x4" or 3"x3" is pretty much a true dimension (i.e., not planed). However, in other softwood used in construction and industrial purposes, dimensional wood, for example 4"x4"s for wall studs, is more likely 3.5"x3.5", with a standard length in a package of, say, 8 or 10 feet. As well, the wood has likely been put through the planer and dressed, so that takes off a bit as well.[88]

One side of the Airy mill had a 40-foot double-cut band saw, and the other side, a 34-foot double-cut band saw. The saws were sharpened and changed twice daily. Both sides of this mill had edger saws.

1 finger up = 4/4
2 fingers up = 8/4
3 fingers up = 12/4
1 finger down = turn the cant

Figure 25 **Hand Signals Used by Hardwood Sawyers**
One finger up would signal 4/4, or a 1" board; 8/4 (two fingers), a 2" board; while 12/4 (three fingers) would be a 3" board. These signals were used amidst the bedlam of the screaming band saw, with other machinery simply adding to the din.
Source: McRae Lumber Company

After cutting, the boards still "green," or heavy with moisture, would be moved to the boardway or loading dock, which was even, or level, with the high level of the carts or wagons (Photo 5). Posts at each corner joined by crossing timbers created an upper level on each wagon.

The lumber was sorted by grade, size and species onto the four-wheeled cars that ran on standard-gauge tracks. Yardmen, with one horse pulling the cart on rail tracks, then took the freshly cut planks to the drying yard. Because the carts had the lumber at a high level, the off-loading of the planks was downward onto the drying piles, which was much easier than lifting the boards up from near ground level would have been.

> In the mill, the wood sent to the basement consisted of outside slabs from the resaw, edgings from the edgers and long trim ends from the trimmers. Considerable manpower was used to extract any remaining value from this wood. The two slasher men cut this wood into 48-inch lengths in both softwood and hardwood—the standard length of lath. The slasher men would pick the rough 48-inch material putting the useable material on the sorting table. The scrap material would drop onto the chain conveyor going to the burner. The bolterman used the circular bolt saw to process slabs into useable boards. The lath machine was used to convert softwood boards and edgings into $1/4"$ x $1"$ x $48"$ lath and the hardwood boards into a variety of products.
>
> In softwood, a major product was lath, with the standard dimensions of $1/4"$ x $1"$ x $48"$. Before the development of drywall panels, walls were plastered. In this procedure, $2"$ x $4"$ wall studs were latticed horizontally with strips of wood called lath. A rough coat of plaster was applied and, when dry, a finish coat was added, completing the walls in houses and offices. In hardwood, the products were more varied, consisting of wood for pickets, brush and broom handles, clothes peg stock, high heel stock and toy stock. In hardwood, slabwood for furnaces was also put out. The wood salvaged and sold from the basement helped greatly to offset the operating costs of the mill.[89]

At the bottom of the jack ladder, low-grade hemlock that had been cut in the bush to a length of 17 feet, 8 inches was sawn in half by a circular job saw. This began the process of producing railway ties that finished at the band saw when two planks on opposite sides were cut off with the wane and sent to the basement as waste wood for the burner. A deckman moved the heavy two-sided ties on rollers to a railway car. When filled, the load of railway ties was sold and shipped out. The railway companies liked hemlock ties because spikes driven into them did not pull out.

The two sides of the mill cut about 120,000 logs a season. The daily output was 50,000 board feet; the annual total was between 6 and 7 million board feet per year. In 1926, the cost for sawmill work was $4 to $5 per thousand board feet for softwoods, with costs being somewhat higher for hardwoods.

The drying, or conditioning, of the lumber, was done by weather seasoning it in the yard. The yard was about one mile in length by one

Chapter 4: The Airy Mill Operations: 1922–1933

Photo 32 **Interior of Airy Mill Showing Conveyors**
Areas around the conveyors had to be constantly cleaned of sawdust. It was also important to clean the belts that took power from the driveshaft in the basement to the machines on the floor above.
Source: M.A. Adamson, W.E. McGraw, D. McLaren and D.M. Parker, "McRae Lumber Company, Airy Mill," *U of T Logging Report 57*, 1926.

hundred yards wide and extended up a narrow protected valley. There were two narrow-gauge or dolly railways on which both mill wagons and freight cars could run. The lumber was piled on eight-foot-by-eight-foot sills, or platforms, by grade, size and species. When a shipment was prepared, a freight car was moved into the yard, loaded and then taken back to the spur line to await attachment to the appropriate train. The total capacity of the yard was about 6 million board feet per year.

While it was considerably smaller than the St. Anthony/Dennis Canadian mill, the Airy mill, since it had been built to cut softwood, was really too large a mill for McRae's hardwood operation. Only one side was needed, not two—yet both sides were used, which required two sets of saws and twice the labour. Duncan MacGregor commented that "when he had the big mill at Airy, I don't think John made

Photo 33 **Disposal of Waste by Burning, Airy Mill, 1903**

Note the slides next to the conveyor belt leading to burner. Here, a final sorting or "picking" of wood took place. Later a small lath plant was added in the basement and the wood was picked there. Despite the high labour cost of picking it, the wood picked at the end of the milling process often helped to determine whether or not a mill was profitable.

Source: M.A. Adamson, W.E. McGraw, D. McLaren and D.M. Parker, "McRae Lumber Company, Airy Mill," *U of T Logging Report 57*, 1926.

too much money. It was too expensive to operate, too big a mill for the cut."⁹⁰ When McRae built his own mills, they were smaller and more efficient. Only one side would be used.

The McRae Lumber Company was mainly in the business of cutting and selling hardwood. Muskoka Wood in Huntsville took hardwood for many years, which they made into flooring. Sales were also made to the Pembroke Box Company and to a slabwood dealer in Ottawa. Duncan MacGregor said that the market for hardwood then expanded. "John told me that when they started to build a lot of cars down at Detroit, that it was a great boost to the hardwood business. All the frames of the cars were made of wood, you know. There was such a demand for it, they'd ship it green."⁹¹

Government Regulations of the Day

Working conditions in the developing lumber industry on the Ottawa River were very tough. By the 1920s, when J.S.L. McRae was starting his business, conditions for workers had improved and the beginnings of health regulations and worker insurance for employment injuries were in place.

In 1882, the Provincial Board of Health began the systematic collection of data on health and its relationship to the physical work environment. Included were such items as the location of cesspools and the quality of ventilation. Results were published weekly and included the incidence of disease.

The employer was required to provide to the Board of Health the following information:

- A list of camps with location of each camp
- Name of foreman of each camp
- Means of access to each camp
- Average number of men in each camp
- Name and residence of physician contracted with, and whether located in camp or not
- Date of contract and duration
- Presence or absence of a permanent hospital in camp

By regulation, in 1901 all employers of labour were required to have a contract with a physician who would inspect all camps and give medical and surgical services. The employer had to provide hospitalization in case of injury but could deduct 50 cents to one dollar per month from wages to cover costs for this. This system proved to be unsatisfactory to both workers and physicians, and, in 1906, that system was replaced with new regulations whereby the employer assumed full responsibility for disease and injury originating in the workplace.

While these measures were being developed in their interest, the workers sometimes didn't readily accept what was being offered. The winter of 1901–1902 saw a high incidence of smallpox. Of the lumber companies operating in Algonquin Park, Booth, Mickle, Dyment and McLachlin Brothers reported cases. Despite the danger of the disease, however, some men displayed reluctance to getting vaccinated. Fortunately, no deaths were recorded as a result of this outbreak.⁹²

A 1912 government regulation required the

Chapter 4: The Airy Mill Operations: 1922–1933

providing of single bunks, a clean blanket and a mattress or tick, along with a public laundry room containing a boiler and laundry tubs as well as adequate ventilation and windows. McRae applied these regulations at Canisbay. To the lumbermen, the necessity of providing a single bunk for each worker was the most irritating change demanded. The regulation required twice as many bunks and sleeping quarters of almost double the size. Previously, men slept two to a bunk and entered them from the head of the bed. Such bunks were "muzzle-loading" bunks.

More regulations were introduced under the *Public Health Act* of 1921: " Comfort and Health of Men in Lumber Camps."

The 1921 *Act* stipulated that each camp must conform to the standards for Class "A," "B" and "C" camps. The group classification depended upon the size of a particular camp. Class "A" required a camp capacity of 100 or more men, Class "B," 50 men and Class "C" a capacity of 26 men. The new regulations stated that bunkhouses were to be nine feet high and that each man should have 400 cubic feet of air space. Regulations required that a Class "A" bunkhouse be fitted with three roof ventilators and the requisite number of windows.

Photo 34 Interior of Bunkhouse, Canisbay Township, 1930–1931
Note the overhead poles on which wet clothing was hung. Heat rising from the stove would help to dry them. The odour coming from the clothes was distinctive!
Source: J.E. Bier and A. Crelock, "McRae Lumber Company, Cranberry Lake, Canisbay Township," *U of T Logging Report 75*, 1930–1931.

The *Act* also stated,

> If a public laundry satisfactory to the Provincial Board is not available, a laundry building is to be provided. ... The laundry shall contain a built-in or iron boiler, together with laundry tubs.[93]

The Department of Health found it necessary to follow up the above instructions with the following letter to all lumber camp operators.

This was implemented when the mill moved to Lake of Two Rivers.

Ontario Department of Health
16 Bye Street
North Bay, Ont.
Aug. 1937

TO LUMBER OPERATORS

Dear Sirs,-

Owing to a number of complaints in regard to the construction of lumber camps and the tendency to overcrowding same during last season's operations, I would call your attention that in the building of camps four hundred cubic feet of air space must be provided for each person. Camps must, therefore, be built accordingly. Single double deck bunks of iron or wood, with either mattresses or ticks must be supplied; an eighteen inch space between each set of bunks is required.
Ventilation
To consist of fresh air inlet and roof vents. Windows: four in each gable end, and roof lights according to the size of the sleep camp. Cook Quarters to be partitioned off from the kitchen. Cesspools for the disposal of slop and wash water must also be provided. Garbage and other waste matter must be collected at regular intervals and disposed of by burning or deposited in a covered pit. Particular attention should be paid to the locating of the water supply in order to prevent any danger of pollution.
Camps must be at least 100 feet from any lake, river, stream or other waters. Stables to be located at not less than 175 feet from the water supply and camp buildings. Manure to be collected at intervals and spread on the ground in thin layers at some considerable distance from the camp and water supply.
The Regulations of the Department of Health will be strictly enforced.

Lumber Companies are urged to make known to all contractors, sub-contractors, jobbers and sub-jobbers what is required in regard to camp construction and sanitation before commencing operations.

Your co-operation is earnestly requested.

Yours very truly,
John Richardson
Provincial Sanitary Inspector

Figure 26 **Letter from the Ontario Department of Health, 1937**
Source: Ontario Department of Health

Chapter 4: The Airy Mill Operations: 1922–1933

Photo 35 Interior of Washhouse, Canisbay Township, 1930–1931
Note the axe-sharpening stone against the wall. Space was not wasted. While all saws were sharpened by a skilled man in the camp, the sharpening of his axe was the responsibility of each man.
Source: J.E. Bier and A. Crelock, "McRae Lumber Company, Cranberry Lake, Canisbay Township," *U of T Logging Report 75*, 1930–1931.

Prior to 1915, bush camp employees could receive some sort of compensation for injuries, as most companies carried a limited amount of liability insurance. The cost of this insurance was met by a deduction from the men's monthly wages. This type of employee protection was often inadequate so the company itself had to reach a settlement with the employee. As a result of counterclaims by management and the employee involved, the settlement was often delayed by long and costly litigation. Crossen, in his work *A Study of Lumbering in North Central Ontario 1885-1930 with Special Reference to the Austin-Nicholson Company*, commented that while the *Act* was

> denounced by many companies as being too expensive, the *Workmen's Compensation Act* of 1915 guaranteed the bush worker a degree of protection he had never before enjoyed. Although meagre by today's standards the amounts awarded by the Board for an injury were no doubt fair at the time.[94]

While preoccupation with World War I slowed the cause of improving working conditions and wages, worker reform gained force again in the Depression of the 1930s. Workmen's Compensation applied to both bush and mill workers. In *Mississagi Country: A Study in Logging History*, G.A. MacDonald said,

> It would be far from accurate to say that life in the pre-Depression camps was easy or financially rewarding. Men worked harder and longer than they should, for

too little; but the spirit of camaraderie and the benefits of fresh air and sufficient food were real and the work honourable and skilled. That the economic circumstances made the men pawns of enterprise is undeniable, but that the lumber companies took a certain paternalistic concern for the employment and feeding of any who wished to work is also true.[95]

The End of the Airy Mill

Because of the Depression, the Airy mill was closed temporarily in 1932, and in 1933 it burned down while it was idle. Janet McRae recalled:

There was a big room with leather in it. Men used to come in to get pieces for the bottoms of their shoes. And that's where the fire started. Somebody came in there to get a piece of leather to fix their shoes and must have dropped a match or something. I was the first to see the thing burning. I was upstairs, and I shouted down to Dad. And the Forestry Fire Patrol came in from Whitney, but, oh, you couldn't stop it.[96]

Yet despite the Depression and the losses caused by the fire, J.S.L. forged ahead and soon was back in business.

Lumbermen and all workers in wood, like agriculturists and the miners and manufacturers of the metals, are the world's real benefactors. They, like agriculturists, contribute more to the world's wealth than the followers of any other pursuit. The products they utilize are natures' gifts, whether it is the food men subsist upon or the clothes they wear, the tools they work with or the houses they dwell in. The produce of their labour is clear gain, and all the occupations are dependent upon them. They are no more useful members of the community than the men who fell the forest trees and fashion the wood into articles of utility.

Musson's Improved Lumber and Log Book
New and Revised—1905
—Illustrated Edition

Chapter 5
The Lake of Two Rivers Operations: 1933–1942

Donald McRae remembered the difficult times: "We had closed down in the Depression. I think the last year they run that mill at Whitney was in '30 or '31. It was down a year before it burned. The bottom just dropped out of the market."[97]

When the Airy mill burned down there was essentially nothing to salvage. J.S.L. put together enough money from insurance, from the sale of lumber that was saved from the fire, and from a loan from the Bank of Montreal to build the mill at Lake of Two Rivers. Donald McRae recalled:

> There was no money then. It was pretty rough. We borrowed money when we started at Lake of Two Rivers and owed the bank for a long time. We dealt with the Bank of Montreal, Barry's Bay. There was nothing wrong with the bank: that was the way business was done. As far as I am concerned, the only way to do business with the banks was when you owed them nothing, and they owed you![98]

Ten years were spent at the Two Rivers mill. The Depression years were endured; indeed, the company prospered. Inevitably changes took place. During this time, trucks were used for the first time to tow log-laden sleighs. Trucks greatly increased the distance from the mill that trees could be harvested and hauled.

Operations in the Woods

Bush camps remained essentially the same as those that served the Airy mill. They consisted of a sleep camp, a cookery and an office, as well as a saw filer's shack and blacksmith's shop.

Photo 36 Barrienger Brake

The Barrienger brake consisted of a low steel frame and a number of in-line steel cable wheels through which was woven a steel cable. The brake, firmly anchored at the top of a hill to a tree or stump, was operated with two levers that controlled the rate at which the cable was let out. The cable, attached to the back of a loaded sleigh, gently let the load down the hill. The weaving of the cable was quite complicated. To while away the time, loggers sometimes attempted to whittle a model of the machine. The intricacies of weaving the thread to imitate the cable on the machine caused the loggers to call it the "crazy wheel."
Source: G.J. Paul, C.R. Greaves and J.S. Hazel, "J.J. McFadden Lumber Company," *U of T Logging Report 39*, 1949.

The mill was located in the middle of the company's cutting limits. For a time in the 1930s, there was a train a day into the mill, where goods were unloaded and toted by horse and wagon eight or ten miles into the bush camps.

The company first cut south Canisbay Township from Head Lake into the valley of Mosquito Creek. Hardwood from south of the mill was brought directly to the mill. Coming down into the Madawaska Valley, McRae used a Barrienger brake to prevent the heavily loaded sleighs from speeding out of control.

McRae then cut the north side of Canisbay Township. Following this the company bought licences in McLaughlin Township from the McLachlin Brothers estate (Appendix IV). These limits east and south of Burnt Island Lake were cut after trucks proved to be capable of the long haul from as far as Crossbill Lake.

The bush camp at this location is marked as a historic site on the MNR Canoe Routes map. This McRae camp was sufficient for 80 to 100 men. Active in 1940–1941, the camp was run by Alex Luckasavitch.[99] In 1980, the two-acre site was slightly overgrown with one building still intact, while 12 other buildings were in ruins. An old road ran to the south, following the valley of the North Madawaska River. The bottom part of the road is still active today from Highway 60 as far as the Wildlife Station on Lake Sasajewan. Researchers there refer to this road as the Old Tote Road. There was a McRae bush camp at the north end of Sasajewan from 1938 to 1940 (Map 11).

The Crossbill Lake camp was somewhat awkward to reach. During open water, the supply truck went in to Burnt Island Lake along the then-abandoned road to Minnesing Lodge.[100] From there, goods were toted up to the northeast corner of the lake by a ten-horsepower motor on a pointer. A scow was used for transporting hay. There was a warehouse at the end of Burnt Island, and the tote road ran a mile or so back into the camp. The haul road from Crossbill Lake, much of which shows on the NTS (National Topographic Sheet) of 1934 for Algonquin, basically followed the valley of the North Branch Madawaska River southward to the mill at Lake of Two Rivers.

Hardwood logs brought out of the bush from north of the mill were piled up where the present Lake of Two Rivers campground is now located. The logs were brought to the mill as needed. The softwood was piled up on the ice of Lake of Two Rivers during the winter, boomed in the spring and towed to the mill by launch near the end of summer.

The big pine in the area had likely been cut by Perley and Pattee in the 1880s, while St. Anthony had cut most of the smaller sawlog-sized pine after the turn of the century. Occasionally, the cutters would run into an exceptionally large tree that had been overlooked. In 1940, a huge pine was cut and reported in the papers.

> One of the biggest white pine trees ever cut in Algonquin Park was felled recently on J.S.L. McRae's limit on the north branch of the Petawawa River [North Branch Madawaska River], reports the *Pembroke Standard*. The tree stood

Chapter 5: The Lake of Two Rivers Operations: 1933–1942

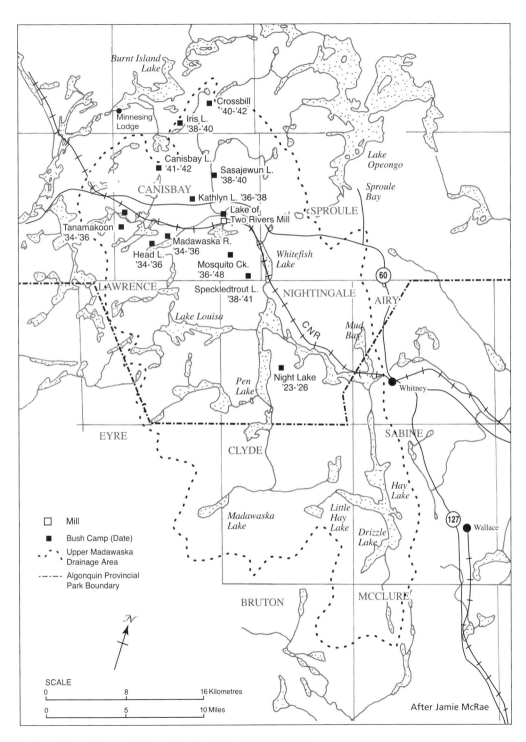

Map 11 **Lake of Two Rivers Mill and Bush Camps: 1933–1942**

116 feet in height and contained 7,787 feet of timber. The huge pine produced six sixteen-foot logs. The first had a diameter of fifty inches and contained 2,116 feet.[101]

Donald McRae and Duncan MacGregor talked about big trees to Ian Radforth in 1986.

Duncan: The biggest birch I've seen was 36 inches; I've seen quite a few 30 inches, which is a tremendous size for a birch tree.
Donald: I've seen bigger that that: 36 inches easily.
Duncan: The biggest pine we ever had was at Two Rivers.
Donald: I was there and I wish to hell it had stayed where it was.
Duncan: We should have left that for posterity. A big tree like that you should never had cut it down.
Donald: It cost us more than it was worth. You couldn't get it up the jack ladder.
Duncan: It shattered the lumber. It was a disgrace. That was the biggest pine I ever seen in my life; it was 60 inches. I don't know how many logs were made.
Donald: The first two logs were no damn good; I mean, we couldn't get it in.
Duncan: The Park should have left a nice area—say four or five miles each way—of those old pines, just for people to look at.[102]

When the mill was at Lake of Two Rivers, two bush camps were operated each winter with a hundred men in each. At various times, bush camps were located at Head Lake, Tanamakoon Lake, and at points along the North Madawaska Valley such as Crossbill Lake east of Burnt Island (Map 11).

Sometimes the weather didn't cooperate in the building of the camps. Duncan MacGregor recalled: "We stayed another time in tents—at Mosquito Creek. We went down in November some time. It rained so darn hard. It was all mud. Oh God! I was glad when we got into the camp, I can tell you that."[103]

McRae also had a camp on the Little Madawaska River about one km east of Found Lake, just south of Highway 60. The site was later used by the Park as a small picnic area. Remains of a root house can still be seen. In 1945, the abandoned camp was dismantled and the fine spruce logs supplied logs for two cottages and a number of staff cabins at Killarney Lodge on Lake of Two Rivers.[104]

Photo 37 Abandoned McRae Bush Camp on Little Madawaska River
Source: G. Garland and R. Cond

A cutting gang was comprised of two cutters with a crosscut saw, a trail cutter, a teamster with his horse, and a piler. They worked as a team, as was the case during the Airy mill days. Donald McRae, who began with the company as a clerk out of Lake of Two Rivers, said of those days:

> They only cut the best of the trees. There was no marking or anything. In the early days you knew that if you couldn't sell it you didn't take it. If your maple was tough to sell, you went after the birch. Nothing like it is today. There was lots of timber around. You had to take a class of timber you could make a dollar off. If you took poor-grade logs you'd be in the hole before long.
>
> The cutters knew what trees to cut out of experience. And your foremen went around all the time and they knew what you wanted. The foreman had 10 or 12 cutting gangs in the bush. He'd be out with them every day, keeping them moving into the cuts and so forth. In those days you wouldn't touch a birch unless it was, I forget, 14 or 16 inches on the stump. One gang, five men, would put up about 40 logs a day. Most of the logs were fourteens and sixteens to get a fairly good price.[105]

Duncan MacGregor commented: "I'd say he was cuttin' veneer [logs] that didn't cost him any more than $50 to cut it and take it to Pembroke, where they got $300 for it. They had a million feet—which is a lot of money!"[106]

To make the thin veneer sheets, high quality logs are peeled at a specialized mill. Then the thin wood sheets are layered, as in a sandwich, with the grain of each running at right angles to that of the sheet below. These layers are then glued together to make panels, commonly four feet by eight feet in size. The strength of these panels can be increased by increasing the number of glued panels. The value of each panel can be increased by using valuable furniture wood such as birch for one outer layer.

The Saw Filer and the Blacksmith

The tree fellers, of hardwood especially, depended on the saw filer to keep their saws sharp. The saw filer's tools consisted of a wooden clamp to hold the saw blade, a tooth gauge, or "spider," to set the rakers and joint teeth to proper height, a swage, a set hammer, flat files and a hand anvil.

Donald McRae commented:

> It was a little bit harder to cut a hardwood tree down, but in the thirties we never had trouble finding men. The only thing was, you had to have good filers to file the saw. If you had a bum filer, that's when the shoutin' would happen—if the saws weren't as sharp as they shoulda been. Frank Bleskie was one of the filers. Filers, that's about all they ever did. They had a little bit of a shack of their own where they filed the saw, right at the camp. Each gang had two crosscuts, and each night they'd bring one in and he'd file. And they'd take the one that was already filed. He could get through all the saws in one day.[107]

The blacksmith was very important in the operation of a bush camp also.

> The leather-aproned blacksmith and his handyman, pumping forge coal with a hand-blower in their clanging, smoky shop, shoed horses, built sleighs, made cant hooks, peaveys, and pike poles and chains, and fashioned bunkhouse stoves from 45-gallon oil drums.

"I worked seven days a week from six in the morning sometimes to nearly midnight," said Tony Shushack at Mattawa on the upper Ottawa. "I worked 57 years and I was never fired off a job. My family was Polish and I worked on their farm at Wilno near Barry's Bay and went to school only about a year. I was working as a handyman for Jack McRae in Algonquin Park when he said, 'I want you to go and be a blacksmith.' I learned to use the fire.

"Horseshoes came in barrels, ready-made, with eight nail holes in the sides but you shaped each shoe in the forge and anvil because horses are like people, their feet are different sizes. If you put the wrong shoe on, it would bother his feet. You had to weld the caulks on, blunt for skidding in autumn and sharp for the ice roads. When you picked up a foot to put a shoe on you'd trim the hoof and rasp it off and if it was an inch too narrow for the shoe you'd shape the shoe to fit. Some blacksmiths left the shoe sticking out too far and the horse would catch his front foot with his hind foot and pull the shoe off."[108]

It took a while to change over from cutting to hauling because the blacksmith had to change the way the horses were shod. Don McRae found the weather didn't always co-operate.

> We got all the roads ready and, oh, did ever a thaw come … I put all the gangs back cuttin'. We had enough logs back in the bush … all we wanted. We were just prayin' it'd turn cold. Then we took two gangs off and started to build the roads again … and it started to freeze. It was about four days before we got everything ready to go again. The worst was you had to take your shoes off the horses and change them from the sharp. I think Steve Wazinsky was blacksmith that year. And oh, gee, he was ready to kill me![109]

Felix Voldock remembered that

> all the men would know the jobs they were doing … You had to be at work at seven o'clock in the morning, but on the sleigh haul you might leave camp before six, sometimes in 40-below weather. At times, the teamsters were at work until eight o'clock at night, feeding and bedding down their horses before they could rest themselves.[110]

Big horses, over seventeen hundred pounds apiece, were needed to pull the heavy loads of hardwood logs that would weigh around 16 tons. From the skid piles to the dumps was a three- or four-mile pull on average. At the beginning of January, daylight lasted only a little over eight hours, so the workday started before sunrise. Teamsters were out to the

Chapter 5: The Lake of Two Rivers Operations: 1933–1942

stables by five o'clock to feed and harness up their teams. Then it was breakfast for themselves and then out to the skidways on the haul road. Torches would be used to see to the loading of the first log sleighs and it would be dark before their day ended. The lead teamsters wouldn't see the camps in daylight until Sunday—the day of rest.

The horses were bought in the Ottawa Valley from around Renfrew, Carleton Place and Smith Falls. A veterinarian, Dr. Foster, knew the farms where the most suitable horses were.

Donald McRae said, "The best bunch of horses I ever bought was from a dairy in Ottawa. They got rid of their horses and went for the trucks before the war."[111]

Even though they were well-tended, the haul was hard on the horses according to Frank Shalla.

> On the sleigh haul, the horses'd be limpin' or hurt. Jim Taylor'd say, "So long as they're on three legs, they can go on the sleigh haul." Them horses'd be so tired! One time this horse was so tired, he was on the lake and just fell down there. And they got another team and a sleigh and just rolled him on the sleigh and brought him to the stable. He got better; he was just tired, you know.
>
> I used to pick my own horses out. I drove one team six years steady. One time they put 40 big birch logs on my sleigh. I thought I'd never get down that hill, but I did. Some of them hills there was so steep and long, when you'd come to the top of the hill there, they put a rope around a big stump, and they had sand on it [the hill], and they'd let you down easy, just like that.
>
> One time they let me down too quick and I went right down between the whippletree and the two horses. I went right over. I used to be afraid I'd kill my horses on a big hill like that. We used to call that a run. I only jumped off a sleigh once. A fella never put any straw on the hill. He said, "All right, the hill is good." I darn near killed the horses. There was lots of horses that got killed. Some of the men couldn't drive at all![112]

The top loader supervised the loading of the heavy logs, lifted by a decking line or jammer. The logs had to be placed so that they would not shift. Then the teamster would centre himself atop the logs seated on his "dry-ass," which was a gunny sack stuffed with hay that he used as a seat.

They had torches on the hills and the loaders had torches, too. It was pretty miserable coming down the haul roads at 30 to 35 below zero. Frank Shalla remembered:

> You'd be comin' down the big sand hill, and them sleigh's 'd be jumpin' like that! You'd have to stop, because if you wouldn't, the brake'd get too hot. And when they cooled off, you'd go down again, and pretty soon them sleigh's be bang, bang, bang, again.
>
> First time I was on the sleighs, I didn't know how to go down the sand hill. So Jim Taylor, he said, "Frank did you ever go down a sand hill?" I said, "No." He said, "Give me the line, I'll show you

how to go down." He took me down once, and after that I was all right. I must have been about 15 or 16. I used to steal lots of brown sugar on the cook. He'd have it on the table, and I'd take a handful and put it in my pocket. I used to give it to the horses. They used to like brown sugar.[113]

The foreman could be hard on the men as well. Frank Shalla recalled:

> Albert Perry used to work with me up at Cranberry [Canisbay] Lake in about 1930. Taylor used to call him "Jimmie." Anyways, Albert was runnin' the hill and he broke the neck piece, or collar, on the horses. Taylor said, "You have to gain that trip somewhere, even if you have to stay out all night," Albert said, "By Jesus, Taylor said I have to make that trip or I have to go home." I don't know what time he got in that night! And breakin' the collar wasn't his fault.[114]

In summer, the animals would be taken back to their original farms or put out to pasture around Whitney. One summer they were even taken up Lake Opeongo and pastured on the abandoned Dennison farm. Janet McRae remembered:

> There was a big farm [Dennison Farm] on Lake Opeongo there, way down the lake. And they'd drive the horses down there for the summer. There was lots of feed there for them. They'd have to get about six horses, old Paul Shalla said, and drive them down about 14 [miles]. This was when the mill was at Two Rivers. I remember one time, oh, they were so mad that they had taken a bunch of kids to help. The horses got frightened, and Paul was so mad at those kids he chased them all home. The horses would follow a good team, but if you started throwing stuff at them and shouting and hollering, they'd start to run into the bush and you'd have to go and get them. Sometimes they'd put the horses on a scow and they'd tow it behind the boat with a big rope.[115]

The Beginnings of Mechanization

In the early years of the Lake of Two Rivers mill, the hauling of logs was carried out just as it had been during the Airy operation, but it underwent a significant change in the mid 1930s with the introduction of trucks. Work in the bush was never quite the same after that. The beginning of mechanization, when trucks replaced horses, took place on the long haul from Crossbill Lake. The company first tried hauling, with trucks pulling sleighs, in 1935, but there was a lot of trouble with chains on the trucks and with hitches for the sleighs.

Gary Cannon said,

> I remember when McRae bought their first trucks. They bought two Mack trucks in about '35. The first trucks we used were for pulling sleighs. They were 2.5 and three tons. They'd take the dump box off the back of the truck (you couldn't see over it very well), and put a little box on, about eight-foot square, and fill it full of sandbags—maybe two tons of sand and put ice chains on it. And that was your hauling for the winter, pulling the sleighs over iced roads.[116]

Chapter 5: The Lake of Two Rivers Operations: 1933–1942

According to Bernie Stubbs Sr., the system that eventually succeeded involved using long and short tongues—the long tongue for horses hauling a sleigh and the short tongue for trucks. A tongue was the connecting shaft that joined the sleigh to the truck.[117]

J.S.L.'s wife, Janet, remembered those times vividly.

> See, the foremen didn't know anything about trucks—to fix a truck—that was one awful thing I'll never forget, because they were breaking the trucks and the trucks weren't going, and they were drowning the trucks. They'd miscalculate the ice. You couldn't build your ice up for trucks. You can for horses and sleighs because the weight is more distributed.

> Up here on the Opeongo … two men drowned and a whole load of logs went down. And that lake is very, very deep. Never got the men. Omanique was there. And machinery was dear then. You couldn't insure it, if you were going to put machinery on ice.[118]

Photo 38 Early Truck and Sleigh
Source: McRae Lumber Company, Marjorie McRae

Lorne Boldt, a long-time McRae employee and saw filer in the mill, had some interesting experiences driving the early trucks.

> I drove trucks in the winter. I used to like that. Broke through the ice a couple of times. It'd scare ya! One time the cab was just stickin' up. Another time I was goin' like hell down the lake. She broke and I jumped. Always had my hand on the door. Tore the whole rear end off. I lost

Photo 39 Lake of Two Rivers Mill and Emergency Airfield, 1938

Note the main line of the CNR (originally the OA&PS) railway and the land cleared for an emergency aircraft landing field.

Source: Henry Taylor. APM 1283

the rear end and the rack just came and never went down, just went back up on solid ice. But the rear end is still in the lake, tires and all! Ha, ha! It was gettin' pretty well into the spring.

We used to draw with sleighs. That was good. You never broke in them. It was better than the weight bein' all on your truck. We drove GMs—three tons. One time I drove a five ton and I brought seven sleigh-loads of logs out of the bush one trip. Usually you take two. I brought out the big load just to see if I could bring them out. Ha, ha. The roads weren't rutted—just glare, the glarer the better.[119]

Photo 40 **Marjorie McRae at Lake of Two Rivers**
Source: McRae Family

An obvious attraction of the mill site at Lake of Two Rivers was its proximity to the CNR railway. A siding and tracks were laid into the yard so lumber could be conveniently shipped out and supplies brought into the mill (Photo 39). This was a very efficient operation. Although the CNR discontinued through service on the line after trestles near Cache Lake were condemned in 1933, lumber and supplies were still shipped in and out of the mill east to Whitney. To move men, the railroad operated a gas car on the line from Whitney, and the company had the use of a smaller jigger.

The Mill

This mill was smaller than the Airy mill, with only one side, which had a band head rig and log carriage, a flatbed resaw and edger. Paul Wolff of Arnprior supervised construction and Isaac Buckman was the millwright. Isaac Buckman had helped to build the mill at Airy and he stayed on at Lake of Two Rivers as the mill manager/supervisor from 1932 to 1941. Isaac Buckman was quite talented technically. Auguste Bissonette was the sawyer at the mill until 1936, when Matt LaHaie succeeded him.[120]

At Lake of Two Rivers, 60 men were employed, with 30 to 35 being in the yard. Besides the mill, there were three bunkhouses, a cookery and an office, as well as a garage and stables, all made out of logs. Some 50 or 60 pigs were kept, including a boar and 8 to 10 sows plus their young. This recycling unit was fed garbage from the camp kitchen.[121]

Two Watrous horizontal boilers that came from North Bay powered the Two Rivers mill. The fireboxes were Dutch ovens, and were

Chapter 5: The Lake of Two Rivers Operations: 1933–1942

Photo 41 **Lake of Two Rivers Mill Showing Jack Ladder**
Source: W.L. Brice. APM 01492

fuelled by sawdust and some slabwood. Donald McRae explained that this engine is now in Ottawa; it was donated to the Museum of Science and Technology when the Rock Lake mill closed. "The old Coreless steam engine come out of Pembroke Electric. It's now down at the museum in Ottawa. It had lots of power, with a great big wheel, 14 feet, I guess. A big one. And we had an electric generator too, powered by the same steam."[122]

The Lake of Two Rivers mill was the first company mill to have a hot pond. Steam from the mill heated a pond of water. Frozen logs were put into the pond to be thawed out and then sawn. This allowed the mill to operate year round. In winter, logs were brought directly from the dump to the hot pond, from where they would go up the jack ladder into the mill to be sawn.

The Lake of Two Rivers mill cut between 30,000 and 35,000 board feet of hardwood per shift or 40,000 to 45,000 board feet of softwood per shift, for a total of 6 to 7 million board feet per year. In addition, the mill would cut 30,000 to 35,000 board feet of pine. In terms of revenue, 70 to 80 per cent came from hardwood, and the balance from the softwood. In 1939 and 1940, there was an acceleration in harvesting, in order to provide veneer for Mosquito aircraft for the RCAF and RAF, birch lumber for corvettes for the Canadian navy and spruce decking for the

Queen Mary. Birch gun-stock blanks were also produced.

At the mill, the lumber was piled on high wagons that were pulled out into the yard. With rollers, the lumber could be piled at least three piles away from the tracks in the lumberyard. However, the wood cut for use in building corvettes was so heavy that a decking line had to be employed on each piece of timber.

Besides dimensional stock, the mill turned out lath, 1″ x 1″s, 1½″ x 1½″ s and 2″ x 2″s, which were used for railings and chairs. All the lumber was used, down to the small pieces. As with the Airy mill, a small picket line employed six men. They picked out the short slabs and edgings. These pieces were shipped and manufactured into articles such as broom handles. The poorer pieces went for pickets for fences.

Slabs were picked out for fuel wood and piled along the tracks. There is a story that during Christmas week of 1942 all the sawmills in the Ottawa Valley were required to forcefully conscript all the men to load boxcars with slabwood for heating in the cities. C.D. Howe, the (very powerful) Minister of Supply during World War II, had sent all the coal to the steel mills.[123]

Donald McRae commented on the importance of sawing to the product.

> The pine business and the hardwood business: there's not a hell of a lot of difference, I think. In those days you were sawing for grade. With the spruce and hemlock you can ram it through; there's only one grade. If it's not a standard thickness or width, it's fixed at the planer. Your hardwood isn't dressed much. Most has to be a standard thickness, an inch-and-an eighth, from one end to the other. And you've got to get your grade out. Always there's been a few hundred dollars difference between the selects and your 2 and 3 commons. And it is the same in piling. You got to pile pine so it don't stain—the same with hardwood—so it gets the air. It's a bit trickier from that point of view; you've got to watch all the time. Make damn sure your men are doing what they should be doing.[124]

Veneer birch was the big money-maker according to Donald McRae:

> The birch was the only one you could get anything for. They shipped it down to Saint John, New Brunswick or direct to the veneer mills in England. That was just before the war. They made Mosquito aircraft out of it. A log buyer used to come up from Saint John, N.B., and he picked out the absolute best. The only veneer plant in Canada in those days was down in Saint John. When I was overseas, in Britain, I saw Canadian birch there goin' to a veneer plant. The time we were at Fort Call, not too far from Bristol. I was in the port of Bristol, and I saw them loading Canadian birch on flat cars.[125]

Once J.S.L. McRae got to Two Rivers he started to make money. It was a smaller and more efficient mill, and the haul was shorter for the horses.

Although the mill was in a building, inclement weather created difficult working conditions that the men simply endured. Frank Shalla had vivid memories of it.

That goddamn sawmill had no heat in it either, you know. It'd be 40 to 50 degrees below zero. And nobody'd complain. I worked pilin' that lumber by a little gas lamp. And snow and blowin'. Jesus, I worked hard pilin' that lumber. Just a little wee gas lantern, and pilin' all day, all night.[126]

McRae was always on the lookout for steady workers, especially ones with talent. Ted Kuiack, for example, began working in the yard. When it was learned that he had some experience working with steam, he became a night watchman caring for the boilers. Later he was put on days on the log deck.

At the log deck, there was a man at the lake and a conveyor belt with a chain. You put the log on the chain. I started the chain and the mill brought the log up. I stopped the chain and rolled the log off and towards the carriage. I had to keep track of how many logs a day. Then start the chain again … There was nothing to it. It was good work.

When I was there, the filing room was just across from the carriage, and the filer always needed help to turn the blades over—they were about five feet long—he needed help to turn them over on his grinding machine. Since it took a long time for the logs to come up, I would go over and help the filer to turn the blades over. That was really something for me. I helped to turn them over and then I watched how he was hitting the thing, and I helped to set it after a while.

Afterwards, I worked as a filer. That filer, he used to booze quite a bit, but he was a good filer too. By Monday he was so sick that he could do nothing. When I was there after a certain length of time, on Monday when he didn't come to work and he wasn't there when they changed the saws, I would call someone to come and help me to put the saw on the grinder and I started to file. From where I was working, from the log deck, I could see the grinder and the way it worked—it was automatic. So I started the machine myself one Monday. But all I did was fit them; didn't do any more than that. So on Tuesday they put the saw on and it would saw like nobody's business.[127]

Donald McRae as Clerk

Donald McRae, son of J.S.L., started as a clerk based at the Lake of Two Rivers mill and ended as the superintendent. He worked with a foreman.

> The foreman planned and inspected the roads, and looked after directing the cutting gangs. The clerk counted and measured the logs cut by each gang. He spent considerable time with the foreman. If the clerk saw something wrong, he talked it over with the foreman, who gave the orders. At night, the clerk sold goods out of the van—tobacco and small articles of clothing.[128]

The foreman handled the men directly. The bush superintendent, or walking boss, Sandy McGregor, a relative of Donald's on his mother's side of the family, continued to plan and supervise the cuts, the haul and the camps.

To recruit men, McRae never had to use employment agencies, unlike many mills across the north. Most of his workers came from the farms of the surrounding area. Donald McRae said,

> They got paid the same throughout the season. Sometimes there was a bonus at the end of the season; it depended on the man, just like today. They bargained all the time. Whatever the goin' rate was you paid it, or a little more. I never had too much labour trouble, only once or twice—none in the thirties. In those days you could change a camp overnight. You pretty well had the same gang workin' for you year in and year out, as I remember it. When they went into the bush in '33, I know, when the foreman hired his gang there must have been another 50 come up on the train and went back. He picked the ones he knew, the ones who'd worked for him before. The foreman hired their own men. We didn't have anything to do with that; it was up to the foreman. If the Polish foreman, Chippior, wanted another gang he'd phone down to Wilno; he'd have names of people who were lookin' for work. And the same with Taylor, who always brought his own men from down around Boulter.[129]

The importance of the Polish-Canadian labour force to the McRae Lumber Company is made very clear in the following exchange between Duncan MacGregor and Frank Shalla.

Frank: You made lots of money for McRae.
Duncan: No I never made money for McRae; the Polish and the Germans made money for McRae. I'd say at the time [1930s] it was 95 per cent Polish.[130]

Photo 42 Donald McRae at the Lake of Two Rivers Lumber Yard

Donald McRae started his formal apprenticeship into the business at this mill.
Source: McRae Lumber Company

J.S.L. McRae and Business Opportunities

The Depression of the 1930s was very difficult across the province, indeed throughout the country. In 1935, the Province of Ontario responded to the needs of the forest industry and reduced stumpage fees (Appendix III).

It may have been Depression times but J.S.L. McRae aggressively sought out business opportunities. In doing so, he prospered himself and provided employment to others at a time when jobs were in very short supply.

McRae had a small operation in Wicklow Township, north and east of Maynooth. The licence had been taken over from E.T. & J. Lumb. In 1936, McRae was having trouble with the local owner of Lot 3, Concession 2. In a letter from the Department of Lands and Forests dated November 23, 1936, Renfrew, Ontario, the Crown timber agent and assistant forester advised J.S.L McRae:

> The Locatee has the privilege to cut up to 40,000 feet of the pine thereon for his own use as building timber, fencing, fuelwood etc. but has no authority to reserve any of the standing pine or to prevent the Licensee from cutting and removing it. Therefore all you have to do is go ahead and cut it; and if the Locatee attempts to stop you, have him arrested if no other way out. Of course if he started to cut the pine for his own use, you could not stop him either.[131]

Tired of the hassle and not impressed with the quality of timber, McRae sold the limit. One person who did get a good deal out of the situation was Duncan MacGregor, who was at the location grading the lumber. "I was down at that little mill below Maynooth. I was getting $5 a day and fellas was getting a dollar and a half and boardin' at home. My board was paid. It was a jobber's mill who been cleaning up a bit of timber down there for us."[132]

Accidents Remained a Part of Work in the Bush

Mention has been made of two fatalities that took place during the time of the Airy mill. Unfortunately, another fatality took place during the Lake of Two Rivers operation. Donald McRae commented:

> We had some serious accidents. I think there was one lad killed at Lake of Two Rivers: Leonard Fuller. Down at the dump, the log come off the top and hit him. I was there at the camp. He was killed outright. That's the only serious accident I remember. We never had any drownings on the drive. Of course, there were always lots of cuts and so forth. They were always cuttin' themselves with a saw or an axe—but not too many. You always had a few compensation cases, but not too many. You had a few bad cuts sure; you were 10 or 15 miles in there. The first doctor that was around was Dr. MacKay, and then Post. He'd come in once a month or else you'd call him if there was a bad accident or if somebody was pretty sick. The men were deducted a dollar a month for the doctor.[133]

Dr. Post gave Donald McRae some first aid training. It came in handy later during the war.

Photo 43 **Dr. Post Coming up the Jack Ladder at the Lake of Two Rivers Mill**
Source: McRae Lumber Company

Lighter Times

After a long day of hard physical work and a big meal at supper, bed time came shortly after. Nonetheless, if a man could play a few tunes on a harmonica, it was welcomed. Saturday night might see a square dance if someone could play a jig on a violin. Sunday was the rest day.

Work, yes, lots of it in the bushcamps, but there were distractions. Duncan MacGregor remembered some:

> The priest used to come up. We Protestants'd always go too. The priest'd take up a collection. Then these ministers'd come in. The boys who were Catholics, they'd all give to … that old fella, Simpson, who used to come and rant and roar. Then he went to Madawaska and got drunk! Ha, ha. All those holy fellas'd listen to him, because he'd preach hell and high water, roarin' and shoutin'. They'd give him money. Then he went to Madawaska and be drunk for a week.
>
> One time I saw him up at Mink Lake with this swell lookin' girl. And I said, "That's a nice lookin' daughter, you've got." And he said, "That's not my daughter, that's my wife!"
>
> Ha, ha. She was at least 25 years younger than he was. He was quite prosperous, you see. Preachin' must've been a payin' proposition.[134]

Frank Shalla added, "Remember those girls sellin' magazines? Some of those fellas couldn't read, you know, but they bought magazines. These was good lookin' girls, oh my gosh. All those fellas buyin' magazines and they couldn't read them at all."[135]

J.S.L. and Duncan kept an eye on things in their own way! Duncan remembered:

> There was very little drinkin' done at the camps. Only when ya come back after Christmas. John'd be on the train with the lads comin' back. They'd all want to give him a drink out of their bottles. Well, he was the boss; he'd given them the jobs. Instead of given' the bottle back to the man, he put it in his pocket. The lad come to me and said, "Christ, that old bitch took my bottle and never gave it back!" I said, "You just shut up. You either let him keep that bottle or you'll have no goddamn job." So he let it go.[136]

Oh yes! There was also poaching in the Park. Henry Taylor tells of two men who used working in a lumber camp as cover for their trapping.

> The two lads are dead now so I can tell the story. They come in there, to Head Lake camp, for no other reason. Their excuse was working in the camp. They had traps with them and a sawed-off .22. Every Sunday, I guess, they spent the whole Sunday setting traps and so on.
>
> We had a network of roads running south to Head Lake way, away up to Harris Lake [perhaps Harness Lake] and Source Lake. Up on the east side of Harris Lake, this big gang was cutting there. This big tree, they didn't need to bother it, but they did anyway. And they hit it with an axe, and there was a big old bear denned up in it in a hole between two roots, just like a dog house. She stuck her head out and one of the lads drove his axe into her head and pulled her out. By and by a cub stuck his head out and he got the same. Another one stuck his head out and he got the same.
> Of course, they weren't going to let the Park rangers know anything about that. They piled up brush on top of the bears. But it was too big of a story to keep from talking about, so one of the Park rangers got a hold of it. They sent in orders for those bears to be turned over to the Park rangers, to skin them and take them out. Then they could report that they were nuisance bears.
> Of course they weren't, but some report had to be made. So me and another fellow was sent in to skin the bears. Well, it was near Christmas-time and the bears were froze hard as stone. We made a fire and got the old she-bear up to the fire and got the feet thawed so we could start. We tied her to a tree on the side of the hill and just tore the hide off. There was lots of fat and meat went with it. I took one cub and rolled it up in the mother's hide and the other lad took the other cub. We took them out and let the Park rangers do whatever they wanted to do with them.
>
> The first Sunday after that, these two gentlemen doing the trapping, they came along and wanted to know what I had done with the bears. I said the carcass was just where I skinned it. Of course they wanted it for bait. A couple of other lads came along about 15 minutes afterward and wanted to know what I had done with the bear. They wanted some of the fat off it to grease the leather tops of their boots.
>
> A fellow from Barry's Bay was buying furs. One night he got a message from Whitney, for him to come to Whitney and meet this lad. And he had about a thousand dollars of furs in his packsack, mostly fishers.[137]

Henry Taylor also described catching fisher.

> Fishers were the prized fur. Just before Christmas, a gang of men had two fishers up a tree. They all had clubs, and they were standing in a circle, and one fellow in the middle was going to fall the tree. They managed to get one fisher and the other lad got away.

Well. What to do with the fisher when they got him skinned? They decided they would cast lots to see who would go out and trade that fur for whiskey for Christmas.

The lot fell to Joe to go. They threw the fisher skin in a packsack with dirty socks on top of it, and let on that he was going to get his washing done. He walked out to Rock Lake station.

Old McIntyre, the Park ranger, had a little spaniel dog. He had that dog trained to sniff stuff in packsacks. He was there at the station and the good luck that Joe had, he was talking to somebody and he didn't notice the performance of the little dog. The blamed dog had his front paws right up on the calves of his legs and his nose in right up on the bottom of the packsack.

Joe backed around the corner of the station and he put the rubbers to that fellow, discouraging him from sniffing his packsack. When the train pulled into the station he just booted it real fast and he got away, and traded the fisher skin in Barry's Bay for a packsack full of whiskey.[138]

Those were Depression days. Men worked for $50 a month and were grateful for employment. Most men in the area didn't have jobs. A fisher skin could be worth $300! The temptation was obvious.

Coming into the forties, the cut was nearing an end at the Lake of Two Rivers site. The

Photo 44 **Jim and Henry Taylor**
Source: APM 03299

time spent there had been quite successful. The Depression had been survived—indeed it had been prosperous. Significant changes had been made, particularly in bush operations, where trucks had been successfully used to haul sleigh loads of logs to the mill over distances that would have been impossible for horses to manage.

In 1942, the mill at Lake of Two Rivers site was dismantled and moved to Hay Lake, south of Whitney, off Highway 127. Transportation by railway was available by CNR from Whitney and from Wallace on the former Central Ontario Railway.

However, the establishing of the new mill on Hay Lake did not go smoothly.

Chapter 6
The Hay Lake Mill Operations: 1942–1952

Photo 45 **Hay Lake Mill from the water.**

Establishing this mill was difficult because of shortages of men and materials caused by restrictions due to World War II. "Making do" was the standard procedure. Also, all the younger men were in the services. In addition, there did seem to be a problem with the head millwright, Isaac Buckman.

Problems in Starting Up

The men knew the mill was flawed. Pete Kuiack said,

> They had trouble getting the mill going right, here at Hay Lake. Instead of putting conveyors in the mill, Buckman was goin' to blow the sawdust out—I don't

know where. But, my God! Everything was plugged up with sawdust for a few acres around. To fire your boilers, you had to first blow everything out. They had to stop and put conveyors in. It wasn't worth the cost. He really messed up there. Could be he messed up on purpose. It was something of a mess. He should [have] know[n], because he was a really good mill man.[139]

Ted Kuiack thought he knew why the mill didn't work.

> When we were finishing the last year at Two Rivers, then McRae had started to build a mill at Hay Lake, and he got this old Isaac Buckman's brother, John, to build this one. McRae had fired that John Buckman over something years ago when they were near here. And old Mr. Buckman said,"I'll get it over McRae yet." So he did that mill at Hay Lake, and he did put it over. That was the worst thing that ever was built.[140]

Alex Cenzura worked on the mill and straightened it out. He was a local man given the opportunity to advance by J.S.L. He became the millwright and foreman at three mills. The two mills that he built later, Mink and Rock Lake, worked well. Commenting on the Hay Lake mill almost 30 years after it had been built, Cenzura said,

> At Hay Lake we had lots of trouble with the mill because Buckman built it too small and too narrow in the first place. There was no room at the back of the carriage. Then he set the resaw right up by the edgers instead of being further back, so that was too tight. So we put a 15-foot addition on the end and moved the resaw to the far end of the mill—to make more room for lumber. There was lots of flukes. I was changing this and that; some of the shafts didn't work right and didn't move. It took me about a year and a half before the mill was going good.
>
> Buckman used to build mills in Africa and all over when he was younger. He built the mill at Airy, the original one. Then Buckman built the mill at Two Rivers for McRae, so he had lots of experience. But what happened at Hay Lake? I don't know. Maybe he was just too old a man then. He didn't have it figured out right.
>
> Some people say he was mad at McRae and he was trying to get back at him, making a mess of it. But what trouble they had I don't know. There was something like that. I guess he certainly did get McRae because McRae, he was worrying about the mill, around six, seven hours a day at the mill, poking around, looking. Maybe every half an hour something happen[ed]. McRae was in bad shape.
>
> Then I kept working on it, fixing this and fixing that. I got the mill running about nine hours a day. McRae was glad; that's why he put me to build the mill at Mink Lake and then put me back at Rock Lake.[141]

The Hay Lake mill had a capacity of some 5 to 7 million board feet per year. The carriage had an automatic control system, worked by air pressure. The carriage was a movable platform that carried the log to be cut, which was held with clamps, called dogs, into the band saw. The main cut was by one double-edged band saw and one resaw, the same system used at Lake of Two Rivers. For power, the mill used steam provided by one central engine fired by lumber scraps and sawdust. All exhaust was piped directly into a concrete hot pond that could hold one day's cut of logs. This enabled the mill to operate year round on frozen hardwood.

The Product and Markets

The average yield per log was 125 board feet. The lumber from the mill was inspected and sorted into one of 12 different grades, with most of the lumber going by truck to Muskoka Wood in Huntsville for flooring. The better grades were kiln-dried while a small amount was air-dried in a yard located a short distance from the mill.

Good logs, 20 inches and over in diameter and quite clear or free of knots, were picked for veneer by the scalers at the dumps and sold separately from mill lumber products. A change in mode of transportation was noted by Donald McRae:

> We used to truck it out of Hay Lake to the siding at Sabine or to Whitney ... It depended on the freight rate ... if it was going to Montreal or the eastern United States, it was better to ship from Whitney. If it was going to Detroit, Toronto or southwestern Ontario, it was better to go from Sabine [Wallace]. And then about 1950 we started trucking out lumber. And it was all trucking after that.[142]

Photo 46 **Gin-Pole Loading Veneer Logs at Whitney Station, 1949**

High quality veneer logs were sold separately from mill products.

Source: I. Cahill, D.W. MacGregor and J.W. Likinuk, "McRae Lumber Company, Hay Lake, Sabine Township," *U of T Logging Report 79*, 1949.

Government Regulations and Forest Regeneration

Despite the war, the Ontario Department of Health was on the job. In 1945, a report ordered the end of placing sleeping bunks in the halls, along with overcrowding in bedrooms in bunkhouses (Appendix 6).

In the late 1940s, the most important cutting regulation was that all white pine was to be 13 inches on the stump before being cut. Other diameter limits were determined by economic rather than by forestry principles. The cutting practice was essentially one of selective high-grading. By cutting white pine 13 inches and over, and only those hardwoods that would give at least one 16-foot log to a 10-inch top, (meaning that the log on the small end was 10 inches in diameter), a stand of younger and smaller trees was left, which resulted in good natural hardwood regeneration in this area. The fact that the stand was not too dense meant that the smaller trees were not badly damaged by skidding. The fairly open, unevenly aged stand also gave an opportunity for seed trees to keep the natural regeneration ahead of dense underbrush growth. The stand could be ready for cutting again in approximately 25 years. McRae, in effect, was practicing selection cutting some 20 years before it became mandatory to do so.

The leasing procedure for a timber licence was essentially unchanged from when McRae took up leases in McLaughlin Township in 1937. Government cruisers placed an estimate on the volume and value of the species, then the government put the area up for auction and operators submitted bids based on this estimate, plus a bonus. The contract didn't necessarily go to the highest bidder, as the past record of the prospective operator was also considered.

The two government scalers then active on the cutting operations of the bidding company reported on the activities of the company relative to adherence to government regulations. These regulations covered diameter limits, stump height and the size of the tops left in the bush. This report was turned in to the government and served as a basis for government opinion on any particular operator. All log piles were government-scaled on the skidways before the haul started. Regulations required the removal of all felled trees down to a nine-inch top, and the government also suggested the removal of more birch and hemlock. For economic reasons, the company avoided all but the best birch and hemlock and cut to a 12-inch top.

This period saw the beginning of the movement to a sustained yield from the forest. The company was required by the government to prepare a 20-year cutting plan. McRae Lumber Company hired a professional forester, Felix Tomaszewski, to set up the company's forestry plan on a sustained yield basis. This was a very significant move because Tomaszewski educated the management of the company in modern forestry techniques and conditioned them to the ideas behind the silvicultural procedures and harvesting techniques that the government came to insist upon in the next 25 years.

Bush Camps and Woods Operations

To supply the mill at Hay Lake, McRae harvested leased timber limits in Clyde, Sproule, Bruton and Bower townships in Algonquin Park. The company also had leased timber limits and privately owned lands adjacent to the Park in Airy, McClure and Sabine townships, as well as the privately owned English lands in Eyre Township.

The cutting plan, as before, was based on a cruise by the woods supervisor, Sandy McGregor. From this survey, the road network was planned, and an estimation of the yield of the number of logs per mile of road was made. A check on this estimate was always maintained during cutting. This was done not only to have eight miles of cleared strip road ahead of the cutters, but also to make certain that acceptable progress on the year's cut was being made.

Ten per cent of the total cut was made up of spruce, white pine, beech and hemlock. The latter two species were avoided as much as possible, because beech brought a poor price, and hemlock usually produced a very low grade of lumber due to butt-shake, a separation between the growth rings near the base of the stem. The main cut was divided about evenly between sugar maple and yellow birch.

Log cutting began when the sawmill was shut down, usually about the middle of October. The cutting gang consisted of two cutters and one skidder, who helped each other. The swamper, or trail cutter, employed at Lake of Two Rivers, had been dropped. Only one supervising foreman was needed per camp since the work force was experienced. The most common saw in use was the six-foot crosscut. To service these, there was an experienced saw filer in each camp.

Skidding logs to the main haul road took place at the same time as the cutting was done. Skidding crews were paid $9 per thousand board feet for hardwood. A cross haul was used to deck the logs on skidways that contained between 150 and 400 logs each. Softwood logs were kept to the back of the skidway and were hauled separately.

As was the usual practice, the camp foreman laid out the haul road network in mid-summer. Stumps were cut flush with the ground and overhead trees were removed to permit the passage of the skyline jammer that was used to pile logs at the dump (Photo 54). The farthest distance between side roads was some 30 chains (2,000 feet), but most were much closer. Men worked in clearing crews of three or four and were paid at a piecework rate dependent upon the length of road cleared daily. The job of levelling or filling in the road was done by one of the bulldozers.

The main haul road used by trucks was cut in the fall so that, when the snow came, it could be packed with a roller and watered to make a smooth ice road, which usually occurred after Christmas. Roads used mainly by horses were cleared of heavy snow by a bulldozer, but were not watered.

Because the haul was mainly downhill, snow acted as a brake for the sleighs where the road was not iced. On the steeper grades, straw was placed in the sleigh tracks to increase friction and thus prevent sleighs from

Photo 47 Tanking Sleigh and Team of Horses
Most of the icing was done at night but if the conditions were favourable, a crew was put on during the day.
Source: McRae Lumber Company

speeding out of control. Truck roads were both iced to cut down on friction and banked to prevent the sleighs and trucks from jumping off the road.

In the bush, the minimum ice thickness for the ice roads was between five and six inches. On lakes, it was necessary to have the ice approximately 20 inches thick before any amount of heavy hauling was done. The ice roads on lakes were not built through narrows where currents could make for thin ice. Where the juncture between lake ice and land occurred, brush, saplings and snow were built up and iced to reinforce this weak point.

For icing purposes, a homemade tank mounted on two sleigh-bobs was used. The tank was filled in several ways: by the use of an ordinary fire pump fixed to the back of a Caterpillar D-4 tractor or an army three-ton truck. In the event of failure of the pumps, there was a barrel with the necessary block and decking line attached to all tanks.

Cold weather was critical for the building and use of the haul road, and winters usually cooperated. But January thaws did occur. Once when McRae was hauling from Sproule Township, it rained heavily and the haul road deteriorated rapidly. Don George, the MNR management unit supervisor out of Whitney, tells the following story:

> J.S.L. had an old Philco radio powered by a battery hookup. Up early in the morning, he was bent over listening to the weather report that was coming in faint but clearly. When the report indi-

cated five more days of rain, J.S.L. drove his fist through the radio and roared, "That's the last lousy goddammed weather report you're going to give!"[143]

Snowplowing involved the use of two different systems, one for bush roads and the other for roads that crossed lakes. For use in the bush, the Otaco plow was used (Fig. 15). A team of horses could pull it, but a D-4 was usually employed. On the lakes, where there were long straight stretches of road, speed was essential in snowplowing and a plow fixed to the front of a truck was used. The big difficulty with lake roads was drifting snow. Where drifting snow filled the sides of a road, there was a tendency for the road to build up on that side and it was the road monkey's job to keep this from happening.

Mechanization: Tractors and Trucks

Despite wartime constraints, mechanization increased at Hay Lake. In 1944, four new trucks were purchased by McRae. In addition, a 25-passenger bus was obtained to move the men back and forth from Whitney and, in some cases, from the mill site to bush camps. Two boats moved boomed softwood logs on Hay Lake. A steel-hulled, diesel-engine boat along with another hull and outboard motor were purchased in 1946. Traditional equipment was also replaced, as five sets of sleighs, two jammers and one Otaco plow were purchased. Previously, this equipment had been made in the camps by the blacksmith and his helper.

Truck use was not trouble-free, as replacement parts were needed. In 1948, 12 replacement sets of dual rear ends were ordered along with tires and tubes and three new GM three-ton trucks. All trucks and tractors were company owned. Although this necessitated having maintenance and servicing facilities on the site, it enabled the company to get a truck back on the job faster after a breakdown than if the repairs were contracted.

To increase the efficiency in building large log dumps, two power jammers were purchased—one for bush operations and the other for the mill. The first purchase of a power saw was recorded in 1950, when a Super Irwin chainsaw was obtained. In the same year, a Caterpillar D-4 with an HT4 Traxcavator forklift and blade came into use for moving logs from the dump to the mill.

Eventually the company had five Caterpillar tractors, two D-6s and three D-4s. The D-6s were used as bulldozers, on road-making in the summer and on snow clearing around the dumps in winter. The D-4s were used

Photo 48 Truck Pulling Sleigh of Logs, 1946
The box of the truck was filled with sand bags and chains were on the drive tires. Ambrose Pordonick was the truck driver in this photograph.
Source: Ambrose Pordonick

Photo 49 Empty Sleighs Showing Long Horse Tongue at Pick-up Point

The long tongue was used when the sleigh was being pulled by a horse. The short tongue was used when a truck was pulling the sleigh because the short tongue led to less swaying by the sleigh.
Source: A.J. Herridge, J.A. Hawtin and J.H. Jamieson, "McRae Lumber Company, Hay Lake, Sabine Township," *U of T Logging Report 77*, 1948.

mainly for towing the watering tanks, but they were also used in breaking and clearing new branch roads. D-4s were also used to load logs in the bush from 1950 to 1958.

The Haul

The haul, begun by mid-January, moved the logs from the skidways to the mill. At Hay Lake, where the use of trucks was being extended, yarding points were used. These were flat areas, usually in a swampy area. Here, bulldozers cleared off the alder and other brush, packed down the snow and iced the area to support the trucks. Horses were used to bring the loaded sleighs to this point. Trucks were then used, as they were more efficient than horses in taking loaded sleighs from this yard to the log dumps. The procedure of using trucks to tow sleighs had been worked out in the last years the mill was operating at Lake of Two Rivers.

The sleigh loads varied in size depending on the type of road the load was to be taken over and the size of the logs. The average load was 15 to 20 logs, making a load of about ten tons.

The logs were held on by a chain that held the two outer logs in place. Two chains circled the bottom log on each side, one from each side. A bitch hook tightened the chain from the inside down. Then another log was placed on top of the first log and the chain was swung over this log, across the load and fastened leaving

Chapter 6: The Hay Lake Operations: 1942–1952

Photo 50 **Loaded Sleighs Showing Binding Chains**

The method of securing logs to a sleigh was to first place a large log on each side at the outer ends of the bunks. Four corner-bind chains securely fastened these logs. The space between was then filled in and the load was built up. Two wrapper chains tightened by pinch bars were passed around the load for added security. Loose logs topped off the load and these tightened the wrapper chains with their weight.

Source: A.J. Herridge, J.A. Hawtin and J.H. Jamieson, "McRae Lumber Company, Hay Lake, Sabine Township," *U of T Logging Report 77*, 1948.

Photo 51 **Loading Sleighs with Gin-Pole**

Safety was always a concern when using this method of loading logs. If the hooks, driven into the ends of the logs being lifted, came out and the log fell, a serious accident could result.

Source: L Cahill, D.W. MacGregor and S.W. Lukinuk, "McRae Lumber Company, Hay Lake, Sabine Township," *U of T Logging Report 79*, 1949.

Photo 52 Loading Veneer Logs on Truck with Jammer
Veneer logs paid a premium price and didn't have to be put through the mill.
Source: L Cahill, D.W. MacGregor and S.W. Lukinuk, "McRae Lumber Company, Hay Lake, Sabine Township," *U of T Logging Report 121*, 1949.

slack enough that a log placed on the chain in the centre would tighten the chain, thus holding the load in position. The centre area was then filled in and the load completed (Photo 50).

Horses were still very important to the operation. The loaded sleighs were drawn from the skidways by teams of horses to the pick-up or yarding point on the main road. Here the sleighs were lined up, ready for hauling to the dumps by trucks. The trucks travelled together to the dump in order to prevent traffic tie-ups with trucks from other branch roads. At the dump, the trucks unhooked and picked up an empty sleigh and drove back to the yarding point.

The empty sleighs were drawn back to the skidway by a team of horses where the loading gang was working in the bush. A crew of four men, a team of horses and a portable gin-pole made up the loading gang (Photo 51). There was one man on the load placing the logs, and a top-loader. Two men drove hooks called pig's feet into the ends of the log and then guided the log while it was being lifted by the gin-pole. Any extra teamsters helped in rolling.

At the end of March 1951, McRae Lumber had 14 teams of horses. Unfortunately, during the winter season three horses had to be destroyed because of accident or illness.

The average truck haul to a log dump was about seven miles; about three-quarters of the haul road was on lake and marsh and was therefore quite level. But there were several steep grades on the haul road in the bush section and this limited each truck to towing one sleigh. From practical experience, it was found that one loaded sleigh per truck, with sand bag

Chapter 6: The Hay Lake Operations: 1942–1952

Photo 53 **Gin-Pole Piling Logs at Dump**

With the coming of open water, the bush dumps were taken down with gin-poles. The hardwood logs were then cribbed and towed to the mill.

The crew at the dump consisted of a team of six men: a top loader, two men on the pig's feet to drive these hooks into the ends of the logs to be lifted, two men unloading and rolling the logs, and a foreman. A high gin-pole decked the logs. The logs were unloaded from the sleighs by two men and rolled to the base of the dump. Here the logs were lifted using pig's feet and pulley block on the gin-pole. They were taken up to the spot designated by the top-decker. A man at each end of the log kept it straight with ropes attached to the pig's feet. The piles could be almost any size, but if the pile was made too high, the danger from rolling logs increased, and if the pile was made too long, it meant extra rolling at cribbing time

Source: L Cahill, D.W. MacGregor and S.W. Lukinuk, "McRae Lumber Company, Hay Lake, Sabine Township," *U of T Logging Report 121*, 1949.

ballast on the truck, was most efficient and gave the least trouble.

Gary Cannon remembered that "in 1948, they got rid of the sleighs and bought three three-ton GMC trucks, a little heavier than what they pulled sleighs with. They had a tag-along axle and we called them pup trucks. We used them from 1949 until 1954."[144]

There were four different bush dumps that fed the mill. The logs from the nearest camp were dumped at the mill, ready for early spring sawing prior to the breakup of ice on the lakes and streams. The company planned on dumping 10,000 logs right alongside the mill. There were two other dumps, one along the edge of the marsh stream for hardwoods and the other right on the lake for softwoods. The fourth was at Opeongo, for both hardwood and softwood.

Log Dumps

The Opeongo dump contained both hardwood and softwood from one camp in Sproule Township. The logs were trucked across the ice of Sproule Bay and then down the Opeongo Road. The softwood logs were taken to the dump at West Smith Lake. Veneer logs were trucked to Whitney, where a representative of Hay & Company re-inspected the logs chosen to be shipped to Woodstock (Photo 59). Rejected hardwood logs were returned for sawing to the West Smith Lake mill in the summer (Map 12).

The West Smith Lake dump and the mill of the Whitney Lumber Company received spruce, balsam and hemlock logs trucked from the Opeongo dump. The purpose of this dump was to allow this small mill to operate on softwood when half-loading regulations, which effectively cut off the supply of hardwood, were enforced in the spring along Highway 60. A small amount of hardwood was also cut here (Map 12).

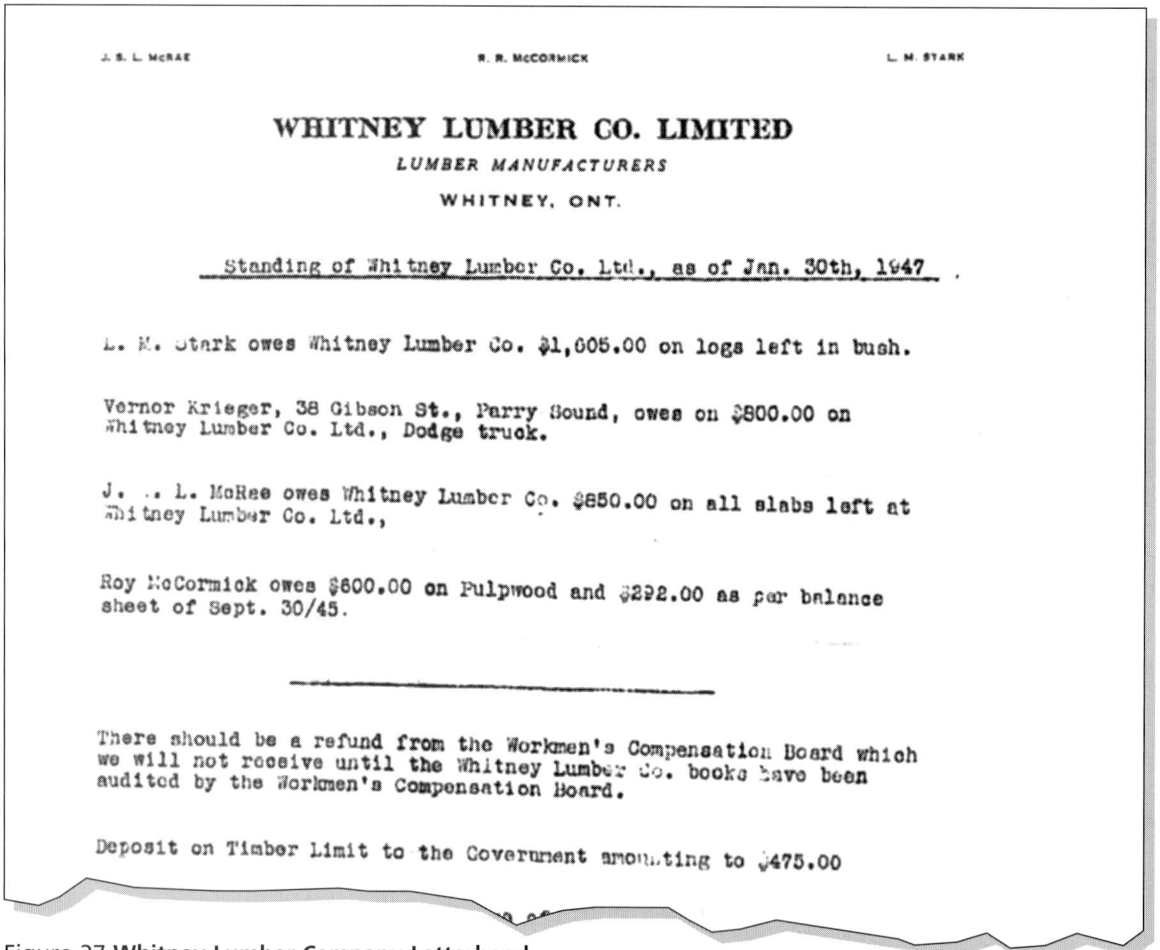

```
J. S. L. McRAE                R. R. McCORMICK                L. M. STARK

                    WHITNEY LUMBER CO. LIMITED
                         LUMBER MANUFACTURERS
                              WHITNEY, ONT.

              Standing of Whitney Lumber Co. Ltd., as of Jan. 30th, 1947.

L. M. Stark owes Whitney Lumber Co. $1,605.00 on logs left in bush.

Vernor Krieger, 38 Gibson St., Parry Sound, owes on $800.00 on
Whitney Lumber Co. Ltd., Dodge truck.

J. S. L. McRae owes Whitney Lumber Co. $850.00 on all slabs left at
Whitney Lumber Co. Ltd.,

Roy McCormick owes $600.00 on Pulpwood and $292.00 as per balance
sheet of Sept. 30/45.

          ─────────────────────────────

There should be a refund from the Workmen's Compensation Board which
we will not receive until the Whitney Lumber Co. books have been
audited by the Workmen's Compensation Board.

Deposit on Timber Limit to the Government amounting to $475.00
```

Figure 27 **Whitney Lumber Company Letterhead, 1947**

The location of this mill is marked on the Friends of Algonquin canoe route map as an historic site. West Smith Lake is signed on Highway 60, but the mill was located on the east side of the lake. Access to the mill was via the old "Automobile Road," which intersected Highway 60 to the south. All of this seems strange because in 1947 Highway 60 was in use and the "Automobile Road" had been abandoned for up to ten years. There is little solid evidence concerning the history of the Whitney Lumber Company Limited.

Source: McRae Lumber Company

Chapter 6: The Hay Lake Operations: 1942–1952

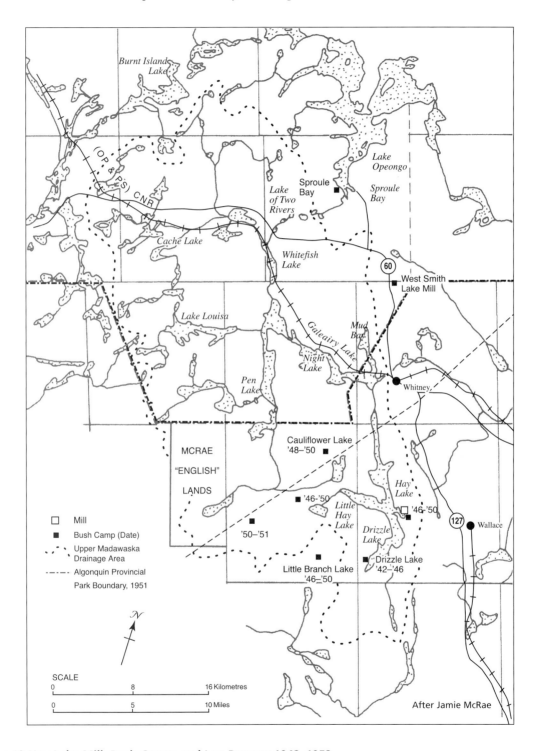

Map 12 **Hay Lake Mill, Bush Camps and Log Dumps: 1942–1952**

Photo 54 **Skyline Power Jammer**
Source: S. Brackenbury, W. Cleaverley and F.T. Collict, "McRae Lumber Company, Hay Lake, Sabine Township," *U of T Logging Report 125*, 1950.

Some confusion surrounds this mill site. The letterhead of the Whitney Lumber Company Limited of 1947 shows J.S.L. McRae, R.R. McCormick and L.M. Stark as being officers of the company. Donald McRae, when interviewed in 2001, could not recall any dealings with the above company, yet J.S.L. was obviously involved.

The Rainy (Drizzle) Lake dump used two gin-poles to unload the sleighs onto skidlogs. The skidlogs were positioned so that the ends of the logs projected out from them at both ends. This made it possible to pass cable from the skyline loader around both ends of the logs to be piled on the dump. One small tractor positioned the sleighs while one team removed the empty sleighs (Photo 55).

The skyline jammer, a new machine, consisted of two A-frames each made from two spruce logs about 65 feet long. These frames were situated one at each end of the pile and about 200 feet apart. Two wire cables ran on pulleys between the frames. Attached to the lower cable by means of two crotch lines was a steel spreader bar about six feet long. Directly above the spreader bar and attached to it was a pulley through which the upper cable ran. Tightening of this upper cable lifted the load. From each end of the spreader bar hung two lengths of wire cable. These cables were put around the logs, and the ends of each pair were fastened together by means of a hooking system, which held firmly as long as there was tension applied to it. The cables went slack, however, when the logs were deposited on the pile, and so the hooking system automatically released the cable ends. Power was provided by an eight-cylinder Dodge engine set up to one side of the A-frames. The crew consisted of one man at the engine, two men rolling the logs off the sleighs and two men slinging the cable around the logs. When the hookup was completed, the signal was given and the engine man lifted the logs up and deposited them on the pile (Photo 55).

Bob McRae recounted the advantages of the power jammer: "The power jammer was much faster and piled higher than using gin-

Chapter 6: The Hay Lake Operations: 1942–1952

Photo 55 Skyline Power Jammer—Lifting Mechanism
Two wire cables joined the tall end-supports. The carrying mechanism was suspended from the upper cable by a pulley with wire cables extending to the ends of the carrying-arm. The carrying-arm was moved forward or backward by the lower cable, which was controlled by an operator on the ground.
Source: S. Brackenbury, W. Cleaverley and F.T. Collict, "McRae Lumber Company, Hay Lake, Sabine Township," *U of T Logging Report 125*, 1950.

poles. They were also safer because a man didn't have to take the picks out."[145]

In the summer, the logs were moved by water on cribs towed by a launch to the mill. The procedure, using cribs to move the hardwood logs, was the same as that worked out at the time of the Airy mill. In winter, the Hay Lake mill dump received logs directly from one camp located at the end of Hay Lake. The logs brought to the mill dump during the winter were primarily intended to keep the mill operating between April and the time when logs from the other dump could be towed on the lake. The logs were towed on the ice by a truck to the mill site. To pull the loaded sleighs up a short but steep grade from the lake to the mill site, an extra truck was used. It hitched onto the truck hauling the sleigh, and with both trucks pulling the sleighs were easily brought up the hill. Another power jammer, more efficient than the 200-foot Rainy (Drizzle) Lake jammer, piled the logs beside the mill. The A-frames here were only 150 feet apart

With economically reachable timber coming to an end in the Hay Lake mill cutting area, McRae looked farther south and built the Mink Lake mill to process the timber in that corner of the limit in Bruton Township.

Chapter 7
The Mink Lake Operations: 1952–1957

The Mill

Between Hay Lake and Mink Lake there is a height of land. The hills between these lakes were considered to be too high to haul logs to the Hay Lake mill, even with trucks. Conveniently, there was road access to Mink Lake from Highway 127 at Lake St. Peter, so the Hay Lake mill was dismantled and moved to Mink Lake.

To build this mill, McRae turned to one of his own men, Alex Cenzura. The mill worked well. Alex Cenzura remembered:

> I built that mill at Mink Lake. They thought they would stay three years at the most, but they made a [mistake]. He was there for four years. There was more timber there than what he thought. I was there a couple of years, then they transferred the mill to Rock [Whitefish Lake] and I was there building that. They were at Rock Lake 22 years.[146]

Photo 56 The Mink Lake Mill
In this photo, the jack ladder is to the left of the building and the raised, covered loading area to the right. Two waste bins were used: one is at the far right hand end of the building, while the other exits the mill building at the centre. There was no waste burner at this mill.
Source: McRae Lumber Company

Chapter 7: The Mink Lake Operations: 1952–1957

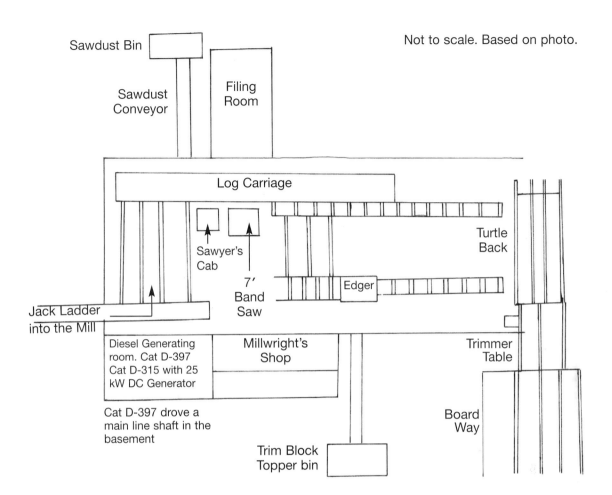

Figure 28 **Mink Lake Mill Layout**

The mill was built with the equipment that was moved down from the Hay Lake site. Like the Hay Lake mill, it was single-sided with a double-edged band saw and edger. Power at the mill was provided by a Cat D-397 that drove a main line shaft and operated the band saw, trimmer, edger, slab saw, planer, conveyors, machine shop and other smaller requirements. A D-315 operated a 25 kW DC generator that supplied electricity for the camp and mill.

Source: Bob McRae

Sawing

Sawing, obviously, is the critical function of a mill, and sharpening and maintaining the saws requires highly skilled technicians. In Lorne Boldt and Ted Kuiack, McRae had two fine workers.

Ted Kuiack was a worker recognized and promoted because of his exceptional talent.

> They made me a fitter for the saws, and I went into the filing room too. There I had a chance to weld saws. Before that, when any saw broke, the filer used to rig the

saw, put it together. But when we moved to Lake St. Peter, it was different. They was welding them. It was really a difficult job. Tension had to be just right. I went to the garage and was working there every three, four weeks, making things, working on electric parts and everything. I was into completely everything in a mechanical way. I had lots of tools of my own, did some welding. And I learned to weld them saws. After that, they said anytime the saw was flat or anything, they say "Come on," and I used to go and weld the saws. I did it for a long, long, time.[147]

Government and Forestry

Perceptions of the best way to harvest the forest changed during the years spent at Hay Lake. While the common attitude of mill owners in earlier years had been that it was best to cut out limits and then move on to fresh forests, the McRaes were now starting to think of using the forest in a different way. By doing so, they would be able to keep the base of the company in the Whitney area. Felix Tomaszewski and later Don George played major roles in the McRae switch to the modern forestry-management methods now mandated by the provincial government.

The *Forest Management Act*, passed in 1947, mandated that companies have a cutting plan approved by the Department of Lands and Forests. This required that a professional forester prepare the plan. J.S.L. had been putting this off, but he hired Felix Tomaszewski in 1952. Felix had been working for Omanique out of Barry's Bay, but he became available when that company went out of business.

Donald McRae gave the background of how the company got into sustained yield cutting:

> Before the war, you didn't have to have a cutting plan. You just put in an application. You'd hire someone to look at the township for you.
>
> Felix Tomaszewski—he was the forester. He set up the cut pretty well on the limits. He came to us about 1952. He had been a prisoner of war. He had been a forester on an estate [in Poland], maybe we would say a regional forester, before the war in Poland. He was taken by the Russians [in 1939]. He was in a Russian camp cutting pulp. There was a deal made for the Polish, and he was with some of the Poles who came out through Persia [Iran]. He was with his family, and the families went to Africa, to Kenya. They had a Polish family camp there for the families. Then he was up through the Middle East and Italy with the Polish Brigade of the British Eighth Army. They were the troops that captured the Monte Cassino monastery from the Germans in a vicious battle.
>
> After the war was over, they came to Canada instead of going back to Poland. He came to us a couple of years later when Omanique folded. He was with us when he died—took a heart attack up at Rock Lake. He had a formal training in forestry in Poland. We set up a 40-year cycle on the limits we had. And that's what we worked on for about 10 or 15 years, until the Park [AFA, 1975] took over.[148]

Felix became more than a mere employee to McRae. The McRaes knew and appreciated the Polish labour that worked for the company. Donald McRae also knew the Poles from his prison camp experience. They were highly trained professional men. Felix came from the same background. His insights into forestry were passed along to J.S.L. and Donald, and they accepted them as ideas from a person worth listening to.

McRae also applied a new method of getting the logs out of the bush while at Mink Lake.

Gravel Roads and Bush Camps

The Mink Lake operation saw major changes in the method of the log haul. Haul roads had traditionally been cut in the fall and used in the New Year after snow had come and the road was iced. At Mink Lake, the decision was made to put in gravelled, year-round roads. Several factors contributed to this change. Mechanical equipment was much improved and bulldozers were now being used to cut out the roads, rather than gangs of men with saws. Trucks with greater carrying capacity made the moving of gravel for the roads as well as the moving of logs less costly and more efficient. Gary Cannon described how the change came about.

> We started using summer roads, gravelling them, in 1952, when we moved up to Mink Lake. They started doing it to get the horses in, the hay loads and stuff like that. Before that we had to have horses and cadge or pull little wagons into the bush. Around Hay Lake, they'd take it up by boat. When we went to Mink Lake, the company decided to gravel one main line into the camps to get the stuff in, and the gangs in and out on the weekends. They found out that when the frost came in December, it froze the roads real hard. Six or eight inches of gravel on the road and it'd freeze just like cement. So they started hauling earlier, about the middle of December. Well, then they found it was workin' out pretty good, so we'd gravel some branch roads where the timber was heavy. If we put a little gravel in there we could draw that in summer. That started at Mink Lake.[149]

Mechanization

While isolation for months in a bush camp had been accepted by previous generations, younger workers now wanted to return home for the weekend. Gravel roads made this possible.

The mechanization of operations continued at Mink Lake. In preparation for the move to Mink Lake, two five-ton GM trucks were purchased. The buying of a Petibone Carry-Lift in the same year marked a significant step in the mechanization of the lumberyard; fewer men were needed now to sort and pile lumber.

By 1954, McRae was committed to the building of year-round gravel roads, as evidenced by its purchase of dump trucks. In 1955, further mechanization of the operation showed in the purchase of two Traxcavator forklifts for piling logs. They were used for loading and unloading log trucks in the win-

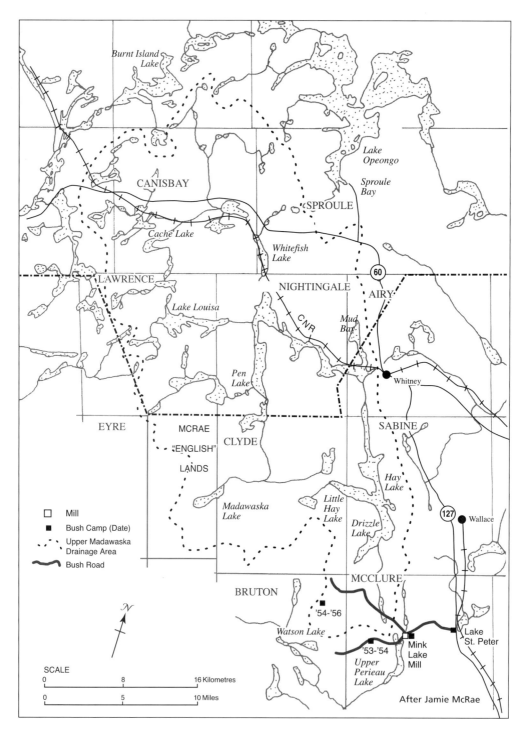

Map 13 **Mink Lake Mill: Bush Camps and Roads**

ter, while in summer, with a bucket on them, they were employed to load gravel into dump trucks during road-making.

In a reminder of the past, in March of 1952, J.S.L. McRae ordered a 34-foot pointer boat equipped with an engine from the grandson of the original builder, John A. Cockburn, of Pembroke. J.R. Booth apparently suggested the design of this famous lumber boat to the English immigrant boat builder in 1854.[150]

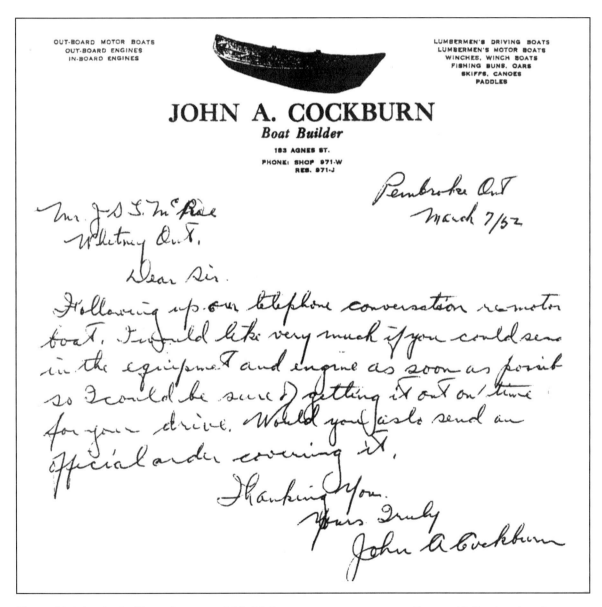

Figure 29 **John A. Cockburn Letter to J.S.L. McRae**   Source: McRae Lumber Company

```
McRAE LUMBER COMPANY LTD.

                                    March 10th, 1952

John A. Cockburn,
163 Agnes Street,
Pembroke, Ontario.

Dear Sir:

Following our telephone conversation to-day you
can go ahead and build the boat 34 ft. long and
2 ft. high, with a 30 ft. bottom and 4 ft. taper,
the bottom to be 5 ft. wide. Put the engine back
to 9' 10" and put in a towing post.

                            Yours very truly,

                            McRAE LUMBER CO. LTD.

JM:IM
```

Figure 30 **McRae Lumber Company Ltd. Letter to John A. Cockburn**
Source: McRae Lumber Company

Photo 57 **GM Tandem Truck**
The two pairs of rear wheels gave this truck its tandem designation. When the truck was not loaded, one pair of wheels did not touch the ground—they were called pup wheels.
Source: McRae Lumber Company

In 1952, two more GM trucks were purchased—four-ton models. By 1954, the switch had been made to GM tandem vehicles

(Photo 57). Gary Cannon remembered that "McRae bought the first tandems in 1954. And that's what we've used until six years ago [1980]. The size increased, from 100 HP to 400 HP."[151]

In the same year, a Lowther tree planter was purchased, followed by another the following year. These machines were used by Tomaszewski to plant red and white pine on the farms in Airy and Sabine townships that J.S.L. had bought during the Depression and World War II.

The maple and birch timber at Mink Lake was big. Cutting it was hard work with a crosscut saw. The very big trees were left until they got the two-man power saw. The first two-man power saw, a Super Irwin chainsaw, came to McRae in the fall of 1953 at Perieau Lake. At the end of August in 1955, an I.E.L. chainsaw was purchased, but it was gone by the end of October as a resale item.

The timber-cutting men were now organized into three-man teams, and the first gang to get a power saw was that of Fred Kmith, Leslie Johnson and Leo Gorgerat. They went to Bancroft with the superintendent, Sandy McGregor. McRae paid for the saw, the cost of which was deducted from the men's wages, with the workers then owning the saw. This procedure became common as the industry felt the workers would give the saws better care if they owned them. Obviously the companies then also had no cost. Gary Cannon remembered:

> We had a three-man gang—the two guys on the power-saw and the teamster. Now say the power saw men had 10 or 15 logs cut ahead of the teamster, they would help the teamster, cutting trails for him,

Photo 58 **Two-Man Power Saw**

The Department of Lands and Forests men in the photograph are, from left to right, Stewart Eady, George Holmberg, Basil Boyle, George Heintzman and John Lacombe.

Source: APM 0960

> help to roll the logs on the skidways and that. Before that, when they were working by the day, there'd be maybe a five-man gang, and they'd have to put up 45 logs a day. There'd be two men on the crosscut, a trail cutter, a teamster and a roller. The three-man power-saw pieceworker gang cut up to 100 logs a day, and then the company would have to add another teamster. McRae would send in a fly team [an additional or extra team] if the skidding distance was long or uphill. McRae wouldn't charge them for the fly team. He'd figure if you could skid two teams of horses a hundred logs a day, why that's pretty good. And they were big logs then, averaging pretty close to 100 feet. I drew a log to Pembroke Veneer with over 500 feet.[152]

The men who were on piecework liked the saws despite their weight. Felling trees and making logs was much easier. The two-man saw lasted two to three years. The lighter one-man saw arrived two years later and quickly replaced the heavy, two-man type. The arrival of chainsaws did exact a cost on the men, as the noise affected their hearing and the vibrations sometimes caused numbness in their hands that became permanent. The coming of chainsaws also meant fewer jobs since fewer cutters were needed.

Some of the birch and maple trees had dieback in them. The tops were dead and the hearts were rotten. They didn't want them if the wood was dark. One day they cut 27 trees that were graded as culls. J.S.L. said, "Don't cut any more of that 'black birch.' You're working for nothing and I'm not making any money!"[153]

Skidding the logs out of the bush still required a teamster and a horse, and this remained a bottleneck in production. Two Blue Ox skidders were purchased in 1956, but were first put into operation at Rock Lake. The problem remained until Timberjack skidders were introduced a few years later and the coordination between cutting and skidding was smoothed out.

Photo 59 Veneer Birch Log
This fine log was about 30 inches (76 cm) in diameter. No flaws were showing and the heart was relatively small at a diameter of three inches (8 cm).
Source: McRae Lumber Company

Photo 60 Gary Cannon
Gary Cannon started with McRae as a very young man. He rose to become a much respected bush superintendent for McRae Lumber Company in 1969.
Source: Cannon Family

But power saws and big trees didn't necessarily translate into profits for the company or bigger paychecks for the men. As Gary Cannon found out, some of the trees were not sound.

Chapter 7: The Mink Lake Operations: 1952–1957

The McRae Lumber Company and the Union

The Lumber and Sawmill Workers' Union was briefly introduced during 1955.

```
McRae Lumber Company,
Whitney, Ontario.

Dear Sir:
This to advise you that the following officers were
duly elected on July 9th, 1955, to the Executive
Board of Local 2537, Lumber and Sawmill Workers'
Union.

President, and Business Agent  .... David MacKenzie
Vice-President ............. Archie Weirmier
Recording Secretary ....... Ken Molyneaux
Financial Sec. and Treasurer . Neil Colquhoun
Warden and Conductor ...... Stan McGill
Trustee     ............... Rene D'Amour - 2 yrs.
   "        ............... Henry Kansas - 2 yrs.
   "        ............... Lucien Roussy - 2 yrs.

                          I remain, Yours very truly,

                          David MacKenzie,
                          President & Business Agent.
                          Local 2537.
```

Figure 31 **Notification of Officers Elected to Executive Board of Local 2537, Lumber and Sawmill Workers' Union**
Source: McRae Lumber Company

During a taped interview in 1986 with Duncan MacGregor and Frank Kuiack, the topic of the union came up. While the union did improve wages slightly, it led to discontent among the men. Some paid their dues; others didn't. There was a lot of bickering.

In the end, the men went to a decertification hearing in Toronto. Duncan MacGregor and Frank Kuiack described the reactions of the men to the big city.

Duncan: They all went to Toronto, you know. There were some of them that had never been to a city before. They seen the streetcar comin', and they didn't know what the hell that was.

Frank: That's true! Eddy Cenzura had never seen one of those dial phones, and Cenzura said "Christ, do ya think we could dial home?"[154]

Duncan MacGregor commented further saying:

> They was all staying at this nice hotel, you know. Old Grant—he was going to take an axe with him to blaze his way back home.

> They voted 90 per cent to get the goddamn union out. They put it out, and then we all had a celebration at the hotel, you know. I can see Teddy Kuiack yet, ordering triples, and I was doin' the same. McRae was payin' for the whole thing, you see.[155]

The McRae Lumber Company has been non-union since 1956.

Cooks and Cooking

When J.S.L. started in the lumber business in 1911, three traditions associated with eating at lumber camps were firmly in place: cooks had to produce quality "salty" pork and good baked beans in quantity, and meals were to be eaten in silence. The McRae operations followed the first two traditions and modified the last one.

Salty pork was an early staple in the bush. Bushmen called it "Chicago Chicken," as much of it was imported from the abattoirs of that city. It came "in rough boxes like they used to put coffins in. The men liked that fat pork in winter because they had to have it to keep warm."[156]

Cooking salty pork was a long process and an art. Boiling the pork for a long time to remove most of the preservative salt was a necessity. The pork was then broiled or fried. Sometimes brown sugar was added and the pork baked in the oven. Most of the grease and salt were then out of the pork, but an attractive salty taste remained.

Bushmen liked salty pork, and "The Boss," J.S.L. McRae, was no exception. Even in later years, when he visited the camps he would seek it out as a treat. Frank Kuiack said, "The old lad'd come down to the camps and sit down and play checkers with a bunch of the men there. That McRae [J.S.L.] liked that salty pork we had. If there was beef he wouldn't be interested, but, oh, he liked that salty pork."[157]

Baked beans came to the Canadian woods in the latter half of the 1800s. They apparently originated in New England and migrated into

Chapter 7: The Mink Lake Operations: 1952–1957

Canada via the Maritimes. Called "fèves au lard" by the Quebecois and "logging berries" or "echo plums" by the English, beans became a staple in the bush. Dried beans could be taken into isolated cutting camps relatively easily and they were a nutritious food. The trick in cooking baked beans was to layer the boiled beans and fat pork and bake them in large kettles that had tight-fitting lids. The fat liquefied in the heat and then merged into the beans.

"The bean pot did its part to develop this country," said Robert J. Taylor of Arnprior, Ontario. "You can't beat an old cast-iron bean pot. I cooked tons of beans but for a special treat I'd take a partridge and put him in the bean pot, cover him with beans, and bake it all. The partridge would come out of there so tender you could just cut it with your fork."[158]

Bush workers required a very high calorie diet: 4,000 calories or more per day was not uncommon. Yet men frequently entered the bush in the fall weighing 180 pounds and were down to 160 pounds by the Christmas break. Salty pork and baked beans were important dietary items.

In most bush camps, the camp cooks enforced the monastic rule of silence during mealtime. This could have been done for two reasons. Firstly, the cooks wanted the men to finish as quickly as possible so they could clean up and then get on with their preparation of the next meal. Talking interfered with eating. Secondly, many bush camps were made up of men who were strangers to one another, as well as being frequently from differing and perhaps disagreeing nationalities. No talking meant no arguments and no fighting according to Gary Cannon. "At meals you couldn't talk. As soon as you said anything, the cookee'd come over. Pass this and pass that, that's all you could say. Cripes, if you talk, you'd never get done."[159]

But this general rule was bent in McRae camps, where most of the men came from the local neighbourhood. They knew each other and chances were that they had been hired by foremen who knew them personally. Visitors noticed a friendlier atmosphere in McRae camps than in the pulp camps further north. Forestry students, in 1949, observed:

> It was very noticeable that the attitude of the cooks was vastly different from that found in pulp camps further north. All the cooks were quite young and had no objection to mid-morning or afternoon snacks and were not insistent on deathly silence at meal times. The men relaxed at the tables while eating and did not bolt their food. There was no rule about talking at the table and any late-comer was cheerfully fed—two cooks and three cookees [assistant cooks] fed 90 men.[160]

Since the work was similar in most lumber bush camps—and the pay was also uniformly low—an important distinguishing feature between lumber companies was the food provided to the men. McRae Lumber had the well-earned reputation of supplying the best food in the upper Madawaska area to its men. This reputation helped the company attract and hold good workers. J.S.L. McRae hired proven cooks, whom he paid well. Many of these men stayed with the company for years. According to Donald McRae, "Each camp had a head cook, a second cook, and cookees. We usually hired the cook. Tom Cannon cooked for Taylor for I don't know how long. In those years, the same cook came every year. The cook hired his own second cook and the cookees."[161]

Photo 61 **Tom Cannon**
Source: APM 2926

Gary Cannon remembered that many Cannon family members of an older generation worked as cooks for McRae: "In latter years, there was Joe, the oldest, and then Tom. But they were all cooks. The four of them worked in the kitchen at the same time for McRae: Joe, Tom, Pat and Bert."[162]

In general, employment in woods work was periodic. McRae broke the typical employment pattern by keeping key men on the payroll year-round for the cut, the sleigh haul and the drive, as well as for summer work in the mill. With jobs difficult to find, particularly in the Depression years, these men repaid McRae with good work and loyalty. Gary Cannon continued, saying,

McRae would take Tom out of the kitchen after he finished at camp, and he'd put Tom in the sawmill. Tom sat on the carriage a lot, and also cleaned the mills. My dad worked in the cookery, drove truck, and he was McRae's edger man. The mill would run from March 'til September, and then the mill men would go to the bush, except for a few key men he'd keep at the mill for repair. Bert was always a cook.[163]

Paul Coulas, Frank Kelly, Ned Bowers and Frank Kuiack were other long-serving McRae cooks. According to Donald McRae,

The first female cooks in the bush camps came in the fifties or sixties. Up at the mill camp at Lake of Two Rivers, we had a man and his wife cook for a couple of years. They had their own sleeping quarters, quartered off. You had maybe 50 men; it was a commuter camp. They had a bull cook, and another cookee who worked with them. His wife waited on the tables; she worked with him, doing some of the cooking too. His name was Charlie Close … He was originally from Calabogie.[164]

Eddie Levean was a bush cook. A bush cook took prepared food out to where the men were cutting. Over an open fire, he would warm up the beans and other food and finish off with a kettle of very strong bush tea.

When the sleigh haul started, I was a bush cook. I'd have a fire goin', and a place to sit—logs to sit on in a circle. And I'd have a big pot o' beans, and all kinds of corn syrup.

I'd leave the camp, maybe about seven o'clock or half-past seven. I'd hang my stuff out there to heat up. I'd hang a pot of beans on a thing there, like two crutches—just so high off the fire to keep them nice and hot. They'd have, too, baloney, and salty pork and the little pork sausages. I'd take bread. And I'd boil tea for them, make it for them—loose tea it was.

Lunch would start about eleven o'clock. It was about half way to the dump. I'd serve to 15 or 20 men. It was quite a gang.

I'd throw away what was left of the beans and tea, and then I'd take the dishes back to camp, and clean them up after supper was done. I'd drop a bar of P&G soap in my bean pot and put all my dishes in there. It'd save me from a big job after supper.

I'd make sure the spot was near a creek for to get the water. And nobody would kick when you had that nice bonfire and a hot lunch. Even on the coldest day.[165]

Cooks put in long, hard days. Duncan MacGregor recognized this:

> The cooks were all good. If he was no good, they'd get rid of him, because you had to please the men. The cooks got paid more than most of the men, but they shoulda got paid more than most of the men—they should've got more still, to tell ya the truth. Feedin' a hundred men, makin' all the bread. Fellas'd eat eight eggs in the mornin'![166]

Donald McRae added,
> The beef came in quarters, and the second cook butchered them up. Either one of them could, but it was usually the second. The first cook did the baking—bread, pies, beans. The food in the camps was always good; and lots of it.[167]

A McRae bush camp cookery made a very favourable impression on forestry students from the University of Toronto. They were studying McRae operations out from the Airy mill in 1931.

> The cookery also presented a contrast to some pulp camps. The inside walls were whitewashed and the building had plenty of window space. Cooking is done on two stoves in the same room as the eating tables. The cook's quarters were in a corner of the building. All the water is carried by hand from the well, and the winter's supply of most food was kept in the meat house and the root cellar. This eliminated the need for any weekly supply system. Each camp kept its own supply of live pigs, but nevertheless more beef was served than pork.[168]

Ted Kuiack knew many of the cooks.

> The cook the first year was a fellow from Barry's Bay, Peter Close. He was there for two years, I think. Then there was Anthony Golcar. He was from Barry's Bay too. Then there was Ned Calman from Whitney, then Ned Lynch from Whitney and Tom Cannon. He was there for quite a number of years. Then there was Charlie Close. He was there for some time. He was from Renfrew. But to tell the truth, I think the best cook we had was Dominic Etmanski from Barry's Bay. He was a clean man. Dominic was the

cook out at Hay Lake. He was a long time with McRae. When he left McRae he went to the high school in Barry's Bay. Tom Cannon was good too.[169]

The Mink Lake mill operated for only four years, but the time spent there was significant for quite a few reasons. A major one was that a new generation of able workers established themselves, replacing older ones that worked through the war and early post-war years at Hay Lake. Able, younger men were identified and took over responsible positions at Mink Lake and proved themselves. The Hay Lake mill was established under wartime constraints and its start-up was problem filled. Difficulties in getting it working well were solved by a young Alex Cenzura. He built the Mink Lake mill, which worked well, and then was put to building the mill at Rock Lake. Lorne Boldt and Ted Kuiack, who were at Mink Lake, became mainstays at the Rock Lake mill, where they looked after the saws and steam. Others such as Fred Kmith, Leo Gorgerat, Fred Parks, Andrew Siydock and Gary Cannon took to the new machinery. The gravel roads were a product of the trucks and bulldozers. The use of such mechanized vehicles as well as the use of the chainsaw meant that fewer men were needed in the bush; at the mill, higher production was also handled by fewer men, with the use of new equipment like the Petibone Carry-lift.

While the increase in mechanization was significant, also important were the new ways of managing the forest that were being developed. The role that Felix Tomaszewski, the forester, played in introducing the concept of sustained yield harvesting can't be documented, but his importance to the company can be understood by virtue of his success at bringing the idea generated by the researchers at Swan Lake to the company.

The Hay Lake mill was the last mill where J.S.L. had daily control, although he remained an obvious presence until his death in 1969. Donald McRae became superintendent at Hay Lake when he returned from overseas. In the three years he was a prisoner of war, he observed and absorbed the management techniques of the men in Stalag III that he respected and admired: Wing Commander Day, the British commander; Bushnell, the leader of the escape team; Harsh, the security head, and the others. Mink Lake was Donald's mill, his operation, and he practised what he had learned. He worked hard, was very fair with his men, recognized and rewarded talent when he saw it, and when he gave his word he stuck by it.

Most importantly, the Mink operation was a happy one. Donald set the tone and this shows in the childhood memories of both Bob and John. Both grew to love the forest and develop a passionate interest in the workings and responsibilities of the business.

CHAPTER 8
The Rock Lake Mill Operations: 1957–1979

With the cut complete in the corner of southern Bruton Township, the Mink Lake mill was dismantled, moved and then rebuilt as what became known as the Rock Lake mill. In terms of processing logs into lumber, little had changed since the days of the Airy mill.

The new mill was located on Whitefish Lake, but, despite this, it was commonly referred to as the Rock Lake mill. This likely came about because the 5.5 km road leading to it from Highway 60 was known as the Rock Lake Road. Near Rock Lake, the road turned and followed the railroad bed of Booth's OA&PS Railway, crossing the Madawaska River to reach the mill site at the southwest corner of Whitefish Lake. The Rock Lake mill was some 32 km from Whitney. Today, a bicycle path built on the OA&PS Railway bed passes by the abandoned mill site, which was rehabilitated and planted in red pine.

The Rock Lake mill was a two-storey structure built on a concrete foundation. The boiler, diesel generator and main driveshaft were on the ground floor.

Photo 62 The Rock Lake Mill on Whitefish Lake **Source:** McRae Lumber Company

The mill at Rock Lake was built largely with used equipment. The following items appear in the McRae Lumber Company accounts for March 1957 as being transferred from the Lake St. Peter (Mink Lake) account to the Rock Lake account.

1	used carriage
20	kiln trucks
400	used rail (purchased)
1	used 8-foot band mill
1	used circular saw sharpener
1	used nigger cylinder
1	new nigger bar
1	used electric motor
1	used cement mixer
57	new kiln trucks
67	new 6" kiln truck wheels
	steel plates and Dutch oven frames
59	new sprockets
1	used CGE generator
1	new Caterpillar D-342 diesel electric engine & Kato generator
1	100 kW, 550 volt, 3 phase, V-belt Caterpillar D-315 generator
1	Rees Model 45HS sawmill KD type burner
1	used horizontal resaw
1	Rees 13FD system
1	1 1/2 HP, 60-cyle, 550 volt furnace motor link belt chain
5	Moore unit heaters with motors

Source: McRae Lumber Company

At first, two wood-fired boilers (Dutch oven types) powered a high-pressure system that drove a steam engine. The steam engine drove the line shaft in the mill, and the surplus steam heated the dry sheds and the hot pond.

The following story reveals much of why McRae Lumber has been a successful business and a special place for many of its workers. Brent Connelly was then a recent graduate of the University of New Brunswick forestry program on his first job. Ted Kuiack knew steam, but didn't have the paper requirements that the government was insisting upon. Kuiack knew he was on the spot. "The boiler engine, all the steam works, water works, all that, that was my affair. A fellow from the Department come down and he knew me quite well and he said. 'You'll pass, I know you will. You put the whole thing together, why wouldn't you be able to get your third-class papers?'"[170]

However, things were not quite so simple. Brent Connelly recounted how this problem worked out:

> You may have heard of Teddy Kuiack … Teddy was McRae's stationary engineer, having run the boiler room for many years. He was exceptionally good at his job and knew steam inside out. Unfortunately, he and Mr. McRae were being told by the Ontario Department of Labour to have Teddy's licence upgraded to the next level or he would have to be replaced. He had tried a couple of times to write the examination but was unsuccessful, as he had little education and, understandably, was intimidated by the process.
>
> It was Duncan who suggested to Mr. McRae that a request be made for me to help Teddy as his scribe. The officials agreed and for the next few weeks in our spare time Teddy taught me all he could about steam and refrigeration so at least I could be familiar with some of the terms…

Chapter 8: The Rock Lake Mill Operations: 1957–1979

The test day finally came, and Teddy and I took off to Toronto in Don McRae's new Buick. The exam took us four hours and the information flowed freely out of Teddy as I scrambled to write it down correctly. We both knew that he had done well that day.[171]

Always on the lookout for means of economizing and improving efficiency, McRae changed the method of power generation. Alex Cenzura said,

> At Rock Lake, we had to change it because that steam engine—it wasn't the goodest old engine. And there was something about the night watchman because you had to fire the boilers and this and that. And McRae thought that with diesel it would be a lot cheaper, they gonna get rid of one man in the engine room. So we took the steam engine out and put in the diesel—no trouble.[172]

The sawmill was then powered by a Caterpillar D-398 diesel engine, which also drove a 250 kW generator that served as the main power supply.

A simple but effective mechanism that warned of mechanical trouble was built into the mill. A taut wire ran the length of the mill's driveshaft. It was anchored at either end to the shaft's supports. This wire vibrated from inputs from the shaft, which in turn was influenced by the vibrations induced by the take-off belts that drove the machinery on the second floor. A rag was tied to this vibrating wire and it constituted a visual marker to the motion in the wire. The stationary engineer could immediately see if there was a problem

Photo 63 **Alex Cenzura**
Alex Cenzura was a brilliant millwright. He had no formal education but his native talent was recognized by J.S.L. McRae.
Source: Caterpillar

on the floor above and either make adjustments or shut off the power. Bob McRae said,

> The most unique feature of the mill was the basement. All the equipment was driven from one main line shaft and a series of jackshafts. Leather belts drove all the equipment in the mill, even the bands. August Ronholm worked in the basement looking after all the belts, pulleys, etc. It was something out of a different era.[174]

The mill, with all its power take-off belts

and machines, was a dangerous place. Fortunately there were no fatalities, just some close calls—some of which were even humorous. Fred Parks remembered: "One day Martin Yantha got his pants caught in a machine and they were pulled off him."[174]

On the second floor there was a double-cut band mill with an eight-foot saw. The carriage, which carried the logs into the saw, was driven by a steam piston affair (shotgun feed). At the Rock Lake mill, the sawyer sat in an open cab right beside the saw. The setters and dogger men, following instructions from the sawyer, turned the log into the position that would yield the highest value of planks. Later the shotgun feed was replaced by an AC/DC drive.

When the mill opened at Rock Lake in 1958, the logs were put in the hot pond and sawed with bark on. Later the logs were put in the hot pond and peeled. Excess waste from the debarker and sawmill was conveyed to a metal-enclosed burner that stood approximately 65 feet from the mill. The burner was conical in shape, with an iron-screen mesh-top burner.[175]

Bob McRae observed, "The tepee burner was a constant fire problem, and the watchmen were

Photo 64 Buddy Boldt at the Band Saw Carriage, Rock Lake Mill
The carriage in the Rock Lake mill was the first one not ridden by the log setter. Buddy Boldt, nephew of Lorne, is seen here in the cab at the controls of the carriage.
Source: Bernd Kreuger

always putting out small fires. It is remarkable that the mill never burned down. Dad and Dunc, they looked after the mill and the yard."[176]

Sawing

Good and efficient sawing is always critical in a mill operation, but this is particularly so in a hardwood mill. The key personnel were the sawyers and saw filers. Donald McRae commented,

> In the hardwood business, for one thing you've got to do a better job of sawing—though the pine's the same. You've got to do a damn good job on the white pine too. But the spruce and hemlock and stuff like that, most of it goes through the planer, and it is sized at the planer. If it's not cut evenly, it's not spoiled. Your hardwood can't be three-quarters of an inch at one end and a quarter at the other, or anything like that.[177]

Lorne Boldt, a long-time McRae employee, ran the filing room, where he sharpened and adjusted the band saws tension. He tells how he found a way to eliminate the need to put logs in the hot pond before sawing in the winter.

> When they convert from pine to hardwood, you change the tension on the saw, the width of the teeth—a narrower tooth for hardwood. You just file it down; you didn't have to buy a different saw. Just change the shape of the tooth and put different tension on it. I've seen filers couldn't cut hardwood at all. No way. Soft wood, it'd go like hell. But hardwood, no way. And they couldn't cut frozen hardwood.

Photo 66 **Lorne Boldt Filing a Resaw Blade**
The McRae Lumber Company has a high percentage of long-time employees.
Source: Bernd Kreuger

For frozen wood, you had to have a very fine swedge on it. I was the first one to cut frozen hardwood for McRae.

Donald said, "We'll cut it all like that." It wasn't any harder for me; you just did it a little different. Now I guess all of the mills is cuttin' frozen pretty well. It's got to be froze solid right through to cut right.

If it's only half froze, you've got a soft spot, your saw will stick, but if it's froze solid it'll go right through. You put a little different gullet for frozen hardwood. The old fella that I learnt from couldn't cut frozen logs at all, at all. And he was a damn good filer! But he didn't believe in a narrow swedge. He had to have a great big one. But it's a lot better with a narrow swedge; you waste less wood with your kerf.[178]

Lorne Boldt had many skills.

If you get a year out of a band saw, you're doin' damn good. Sometimes they'd hit a spike and rip the teeth all out on you. I'd weld the teeth in, and you'd stiffen it by putting heat to it, and oil. I made resaws out of the old bands. Cut the teeth off of one side, cut a chunk out and weld her together. Resaw was 10 inches and the band saw was 14. Here they're using 10 inch because the logs are smaller now. The bush is getting run out. They've got to take what the forestry [MNR] lets them take.[179]

Bob McRae commented on the sawing procedures that finished the boards.

The mill at Rock Lake had a seven-foot flatbed resaw that processed slabs and doubles from the carriage. The boards were edged on a manual setting edger. The trimmers were conventional Canadian-style and the boardway was open with just the chain and walkways covered. It would have been cold in the winter. The production was tallied by Alcid Martin in a little hut beside the boardway. Most of the better lumber was put up on sticks in the yard or drying sheds and then shipped air-dried.[180]

To put the lumber "up on sticks" or pickets is to separate each row or level of lumber in a pile. This allows sufficient airflow under and over the boards to dry them. Most good lumber now produced at the mill is shipped green within a day or two, but if any has to sit in the yard for as little as a weekend it is put up on sticks. This prevents the wood from spoiling. Any other wood, like hardwood frame stock, is put on sticks, as it may sit in the yard for a long time (Photo 67).[181]

Photo 66 **Robert Recoskie Edging, Rock Lake Mill**
Source: Bernd Kreuger

The lumber drying sheds were well removed from the mill, as were the other camp buildings. The lumberyard was 200 to 300 feet from the sawmill on land with a good access road. According to Bob McRae,

> In the yard, two Petibone lift trucks handled all the lumber from the mill, supplied the shipping gang with lumber, stored the lumber in sheds, etc. In the log yard, one Cat 966 with log forks dumped all the trucks, fed the hot pond and later took logs from the hot pond and put them on the peeler deck.[182]

The office, camp buildings, garage and blacksmith shop were heated independently by wood-fired heaters, with some oil-fired hot air furnaces. The machine shop, blacksmith and vehicle maintenance operations were housed in detached buildings.

A Change in the Quality of Wood Coming from the Forest and Consequent Changes in Mill Operations

The moving of mills in the past, from Airy to Lake of Two Rivers to Hay Lake to Mink Lake, had been undertaken when the cut at one site had been completed and a fresh area of timber was needed. This was not quite the case with the building of the Rock Lake mill. With this move, the McRae Lumber Company was near the centre of its limits, which was an advantage, but much of this land had previously been cut for timber. The more easily accessible prime timber had there-

Photo 67 **Gary Cannon, Duncan MacGregor and J.S.L. McRae at Rock Lake**

These three men, along with Donald McRae, were cornerstones of McRae Lumber for many years. Gary Cannon became bush boss in 1969.

In the pile of boards, note the "sticks" separating the lumber. The sticks created spaces for the air to circulate and dry the boards.

Source: McRae Lumber Company

fore been cut and the growing period had not been sufficiently long to produce numerous, top-quality trees. Mechanization did make it possible to reach some previously by passed areas and the private English Lands in Eyre Township, where a fresh area yielded high quality timber. Yet it was obvious that the quality of the tree stems coming out of the forest was decreasing, as a high percentage of what would formerly have been left standing as cull was now being cut and charged to the company as part of stumpage fees.

The Influence of Felix Tomaszewski

Felix Tomaszewski, the Polish-educated forester McRae hired at the Hay Lake mill, continued to produce the cutting plans required by the government, but in addition he established a close but informal relationship with the researchers at the Swan Lake research station. There he talked to Mac McLean and Harvey Anderson, two key researchers. The forestry concepts and research results he learned from them were passed along to J.S.L. and Donald McRae. Very important among these was the concept of perpetual harvesting, something sharply different from the traditional practice of high-grading an area and then moving the mill to another fresh area when the trees ran out. McRae Lumber didn't have a fresh area to move to. The McRaes also wanted to continue milling in the Whitney area. The new ideas brought by Tomaszewski began the process of conditioning and educating the McRaes, thereby making them interested in and receptive to the continuing research results coming from Swan Lake.

Professionally, Tomaszewski was very progressive. He began a plantation experiment with yellow birch at the Rock Lake mill site, where he ran into the same problem as was encountered at Swan Lake: herbicide, of the four-legged variety—white-tailed deer that jumped the eight-foot fence and ate up his thriving saplings! He was more successful in reforesting in pine some 800 acres of farmland previously bought by J.S.L.

Tomaszewski was also part of the management team that led McRae Lumber Company into the production and marketing of wood chips.

Photo 68 **Donald McRae and Felix Tomaszewski at Rock Lake**

Tomaszewski was not only the respected professional forester at McRaes, he also became a family friend, and was the favourite bridge partner of Duncan MacGregor. His sudden death in 1962 was a deeply felt loss.

Source: McRae Lumber Company

Chips: A New Product— and Also a Gamble

With progressive leadership, the company moved into the production and marketing of wood chips to paper companies. This was done to utilize the fibre present in the poor quality, formerly ignored cull wood coming out of the bush. In the spring of 1960, a small Forano chipper was added to the mill. This machine had a wheel with blades on it that chewed up slabs and much of the formerly waste wood that had been sent to the burner.

McRae sought out and developed this market. Bob McRae commented on the problems faced by his father.

> In the early sixties Dad put in a peeler and chipper. The logs were then peeled before going into the mill, and all the waste was chipped. Initially, the chips went into Domtar at Cornwall. When the pulp mill opened at Portage du Fort in Quebec, the chips went there because it was closer. When the chips went to Cornwall, the chips went by a high-sided tandem truck to the rail siding at Whitney. The company built two self-dumping bins out of wood beside the mill at Rock Lake. The chips came out of the screen and were raised into the bins by an elevator conveyor. The truck drivers then loaded their own loads and took the chips to Whitney. At the Whitney rail siding, the company built a ramp. The truck drove up on the ramp and dumped the chips into the chip rail car. The two chip drivers were Eddie Cenzura and Carl Dubreuil. Eddie's son is now the head detective with the Metro Toronto police. All I can remember about shipping chips by rail was that the CNR had a terrible time getting the number of cars right. The siding was either full of empty cars or Dad was phoning to get CNR to bring cars.[183]

A story concerning the installation of the chipper reveals quite a bit about J.S.L.'s personality and his relations with his son, Donald. Alex Cenzura, the millwright, was a witness to it.

> I remember at Rock Lake when we put that chipper in there and this and that. Donald wants it this way and the Old Man wants it the other way. Donald tried to explain to the Old Man, because the Old Man didn't know much about the mill. He knows how to sell lumber, the prices and this and that, but how it was to be built was something he didn't know. And Donald thought he knew a little bit more. I was in the office one time, Donald wants it this way and the Old Man wants a different way.
>
> "Goddamn," the Old Man says, "I'm behind the shotgun. I'm not planning on going yet. You keep your goddamn mouth shut. The way I want—that's the way it's going to be."
>
> Poor Donald, there was nothing he could say. "All right, Dad, have it your way. You'll see, you'll see."
>
> A lot of times Donald was right about the mill. Old Man McRae didn't know much about the mill. Still, he owned the lumber mills and in his lifetime he runs a good business. It didn't go broke. So he had a good head for that.[184]

Photo 69 Bark Conveyor, Bin and Hot Pond at Rock Lake Mill

The frozen logs were thawed in the hot pond and then debarked before being sawn. The conveyor brought the bark back from the mill and deposited it in the bin at the end. From there the bark was loaded into trucks and taken to a buyer.

Source: Mark Webber

Lumber is the major earner of income for the McRae Lumber Company, but the sale of chips more than covers the expense of stumpage for less-than-second-grade wood and the cost of converting it to a saleable product. Donald McRae gambled that it would work out this way.

The Selling of the English Lands

During the time of the Hay Lake mill, J.S.L. had bought land in Eyre Township. An English company had gone bankrupt and the land had reverted to the township. In 1961, McRae Lumber was approached by the Property Branch of the Department of Public Works to sell "The English Lands": Concessions 4 to 14, containing lots 21 to 34, of Eyre Township. These lands drained into Algonquin Park through the Galipo Lakes, as well as from Madawaska Lake, the headwater lake of the South Madawaska River, and the government wanted them under Park control.

J.S.L. McRae, through Renfrew MPP and lawyer James A. Moloney, negotiated the sale of these lands while the company retained the rights to the timber.

In 1993, a long time after the sale, this land, known as the McRae Addition, was added to Algonquin Park.

Bush Operations: A Time of Great Change

McRae had started gravelling roads at Mink Lake. In time, gravelled roads made possible the regular and year-round delivery of trucked wood to the mills. In 1986, Gary Cannon said, "Since 1950, McRae has built over 500 miles of gravel roads all through the limits." The roads were gravelled with equipment consisting of three to four old dump trucks, a D-6 bulldozer to spread the gravel and a Caterpillar 950 loader with a rock bucket to load the gravel trucks.[185]

The McRaes later extended a road from their Whitney mills, built in the seventies to join the Hydro Road and the network of roads previously developed from Rock Lake. These roads are managed now by the AFA. The McRaes also use some of these roads to access their leases and private lands in Eyre, Sabine and McClure townships (Map 2).

Today there is a comprehensive set of criteria applied to road construction by the MNR/AFA that regulates both the amount of use the roads get and their impact on the environment, particularly where they cross streams. These roads are not open to the general public.

Roads in the forest are classified as primary, secondary and tertiary, with the last two being refurbished on a current need basis. Skidding trails developed for previous cuts of the forest are used where possible during current operations.[186]

Photo 70 Caterpillar Bulldozer Breaking Out a Road
The bulldozer and other types of machinery made the building and use of roads much easier.
Source: Caterpillar

The Search for a Machine to Replace the Horse

In the woods industry, the replacement of the horse by machines was a difficult process. It took time and considerable experimentation, involving both industry and the government. Two developments spurred the search for a mechanical skidder. First, the supply of horses fell precipitously as they disappeared from agriculture and the transportation industries. After 1944, the number of animals declined

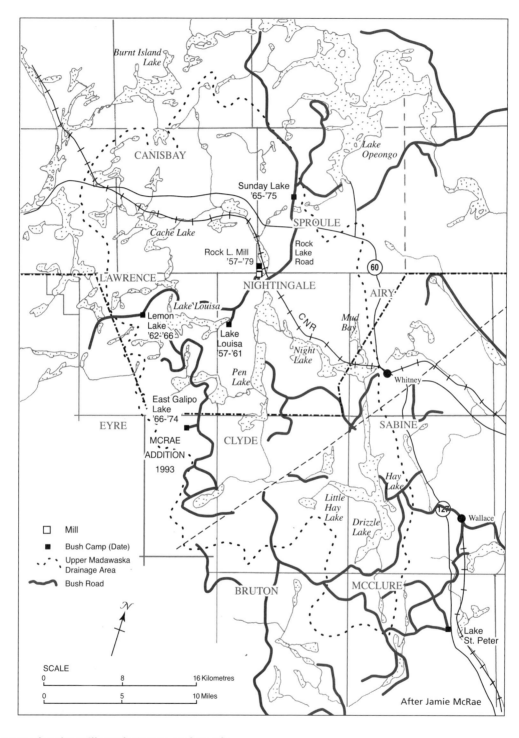

Map 14 **Rock Lake Mill: Bush Camps and Roads**

across the country at a rate of some 100,000 per year. With this decline, woods operators found it more difficult to get qualified teamsters, supplies of harnesses, collars and other essential paraphernalia.[187]

Second, the efficiency of the chainsaw created a bottleneck in productivity in the woods. Woods superintendents took a long time to adjust to the efficiency of chainsaws in cutting trees, and traditional skidding with horses could not keep pace with felling by chainsaws. The early machines did not prove to be satisfactory.

The problem became one of getting the right machine. In 1956, McRae ordered two Blue Ox skidders for operations out of the new Rock Lake mill.

This machine

> was an open-cab, four-wheeled-drive vehicle powered by a 97-horsepower engine. The tires had been specifically designed to withstand punctures, and to minimize damage to the forest floor. The machine had eighteen inches of ground clearance as well as protective plates on its underside. Like the crawler-tractors, the Blue Ox had a fairlead, and a winch mounted at the rear.[188]

A fairlead was a V-shaped steel arm that raised the cable coming from the winch to some six feet above the ground. When the logs were winched in to the Blue Ox, the leading end of the logs would be lifted a foot or so off the ground. This avoided snags and made for easier towing.

Fred Parks didn't think much of the Blue Ox. "The Blue Ox was 'a lemon'—nothing but a two-wheeled wagon. It needed more power. You needed to do a lot of winching."[189]

Gary Cannon, who worked with the Blue Ox as a young bush worker, commented:

> I drove a Blue Ox. It was a real son-of-a-gun. If we were working in 1958 and 1959 the way we are now, we'd been okay. But they'd get three guys cutting to two Blue Oxens. Me and a fella by name of Danny Yakabuski drove them in '58 and '59. They'd send the cutters in three weeks before us, and the cutters'd go ahead and slash. When you'd come in to skid these logs out, they were covered with tops and everything was criss-crossed really bad— you couldn't do much.
>
> The company wanted a hundred logs a day from each machine. So what you done was a man would choke for you, hooking the cable on with, say, five or six logs, and away you'd go to the skidway with it.
>
> Now it wasn't bad in the dry weather, but we didn't have any chains on them. It was just a four-wheel tractor with a steering wheel in it, no blade in front or anything. It just drove like a half-ton, eh.
>
> We had terrible problems, but if we had of gone to the bush the same day the cutters had gone to the bush, and we'd have all worked together, why our production would have been up a third now. Now we get 123 trees a day per skidder, but if we put the cutter in three weeks ahead, you wouldn't get 70 trees a day because your trails would be blocked. There are only certain places you can skid in the bush. If you're on a hill or in a gully, there are only certain trails you can

take. We had so much trouble they had to put the horses in with us. The horses'd take the close skidding. In fact, Albert Perry was the teamster that year with me. Those Blue Oxes would get stuck all the time, and you'd break an axle, break the hubs. And they steered terrible.[190]

In 1959, McRae tried John Deere crawler-tractors for skidding. Gary Cannon was not impressed with them either.

> When you worked with tractors rather than horses, it was easier for the operators. If it took you any more than three minutes to bring out three logs, you were skidding too far—if you had steady going.
>
> Now if you were in rough terrain, you might only be skidding five or six hundred feet. You'd have about 65 feet of cable on your tractor, and you were in soft stuff where the tractor couldn't pull it, or you took too big a load, which was a waste of time. You'd just go 65 feet, stop, and winch it in; 65 feet, stop and winch it in. So what you were better to do was take two logs, go and come back. A lot of guys wanted to take out three or four and do a lot of winching, but it was hard on the winch, hard on the final drives, hard on everything.
>
> Our count with the tractor was 75 to 85 logs a day. The log maker could usually cut that much, but the skidding part was sometimes the toughest. What they done then was change the roads a bit. The horses could never skid uphill.[191]

McRae then tried the International crawler-tractor for skidding in 1962. The crawler-tractor had been in use for some time and was found to be satisfactory for hauling wood trains out of the bush over poor roads, but it could not operate effectively where rock was common on the surface. Built into the tractor were hydraulic lifting arms and cables on winching machines that led through an arched fairlead from a height. Both were useful additions, but the total machine did not skid logs from the bush efficiently.

Gary Cannon didn't think much of the crawler-tractors. "I remember mine was a 440, two-cylinder International. We run them that fall, that winter. And oh god, they had a lot of trouble with the drives in them. They were just unbelievable, the cost. They were little wee caterpillar tractors with a little winch on behind."[192]

Photo 71 **International Crawler-Tractor**
Source: McRae Lumber Company

Like other operators, McRae found neither the first wheeled skidder nor the tracked vehicles proved to be the answer for skidding logs in the bush. In *Bush Workers and Bosses: Logging in*

Northern Ontario, Ian Radforth wrote:

> In comparison with the crawler-tractors, the skidder could travel faster and they had a superior availability record, as well as low maintenance costs. However, crawler-tractors could negotiate steeper slopes, work in a wider range of ground conditions, and carry heavier loads. The chief limitation of the early skidders was its wide turning arc, a great handicap in the boulder-strewn, Shield country. The skidder was also cheaper to maintain than a crawler-tractor, but it had a far higher initial cost. Woods managers considered that neither machine represented a satisfactory piece of skidding equipment for the Canadian Shield."[193]

The solution was finally found when the Woods Section of the Department of Lands and Forests and the industry worked together to develop a prototype articulated skidder, a vehicle that was hinged at the centre and steered by powerful hydraulic arms. The machine had "a small turning arc, and by wiggling back and forth, it could duckwalk across muskegs and through deep mud."[194]

The Timberjack was created by Bob Simmons, Wes McGill and Vern King of Woodstock, Ontario in 1960. Its small turning radius gave it good manoeuvrability, and a winch, operating through an arched fairlead, gave great pulling power. This machine was an instant success and sold worldwide.[195]

McRae bought its first Timberjack in 1964. It took a while for management to learn how to employ the new machinery efficiently. Gary Cannon commented,

Photo 72 Timberjack Skidder
Note the choking cables gathered up under the fairlead. A blade plow was attached to the front of this machine.
Source: D. Lloyd

> I didn't think anything could skid uphill. But the foremen then said, "Okay, the tractors are a little different than the horses." So they made the roads a little further apart, a little uphill. But as the roads got further apart, you were losing production; you might average 75 logs a day. The foremen were older chaps and they didn't understand equipment. They understood horses.[196]

Timberjacks were not without their problems either according to Cannon.

> When the Timberjacks came in, they thought they'd build great big landings, maybe two acres big, and maybe a half-mile apart. And have this piece of bush come to this landing and this piece of bush come to that landing. But they weren't thinking of the hills between the landings, if you had to skid up from behind. So the first year we had Timberjacks, and in '64, in the first two

months, we had to change every winch on them. We burnt the winches out. We had 75 feet of cable on them, and we'd go 75 feet and winch. But the gears were working too hard. They were skidding over half a mile. I was one of the guys running the skidders at that time.[197]

While trained mechanics were necessary for large repair jobs, operators were expected to perform minor repairs such as clearing a frozen fuel line. One skidder operator had this experience:

> He returned to the skidder, unhinged the protective motor shield so that he could get at the fuel line, and unbolted it from the pump, the icy wrench leaving a white cold-burn across his hands and fingers. Wriggling himself into the space between the huge tire and the frame, he grasped the fuel line and placed his lips over the end. He sucked hard, drawing a mixture of fuel, ice and debris into his mouth. Quickly he turned and expelled the foul fluid, spitting several times in an attempt to clear the taste that, regardless of his efforts, would linger most of the day. He had to repeat this procedure a number of times until he obtained a mouthful of clear fuel. Hoping the line would remain ice-free now, he reattached it to the pump. "God, what a way to make a living!" he declared to no one in particular.[198]

Timberjacks required the coordination of fellers, a skidder-operator and sometimes a chokerman to operate efficiently. The use of skidders caused health problems for the operators as a result of high noise levels, vibrations and jolting. This led to bush operations being conducted by brokers or jobbers, separate from the company because "the highest production rates and lowest levels of machinery downtime were achieved … when operators owned their skidders and were hired on a contract basis."[199]

This also enabled the company to avoid the high capital cost of the machinery, and, just as important, avoid the problems and costs of workers' compensation in an accident-prone phase of logging. Compared to horses, the Timberjack was tireless. It could be used in year-round operations and could handle tree-length stems. All of these features led to the regularizing of the flow of wood to the mill.

Further Mechanization in the Woods: The Replacement of Gin-Poles and Jammers by Loaders

About 1960, McRae bought two Hopto loaders. These were the first log loaders and they replaced the gin-poles and skyline power loaders. As one of the first machines for this purpose, it had problems. It had a separate diesel engine coupled to the hydraulic pump without a clutch. It was hard to start in winter, and the lack of a clutch made it hard on the hydraulic pump and the hoses that operated the lifting arms. Bob McRae commented, "Even today, with hydraulic fluid having a bottom range of -35° C and synthetic oils, starting engines in the winter continues to be a problem. In 1960, the impact of severe winter temperatures must have created a nightmare."[200]

In 1966, the company sold the Hopto loaders and upgraded to Prentice loaders mounted on trucks. This machine ran the hydraulic pump from a PTO (Power Take Off) gear on the truck engine, so the pump could be gradually engaged by a clutch. In winter, this meant that the truck engine could be started first and then the hydraulic pump could be gradually engaged and warmed up.

Photo 73 **Prentice Loader**
The Prentice loader was a very effective machine. Fred Kmith made the cab for this loader.
Source: Paul Cannon

The Cutting System of the 1960s: Hot Skidding

The cutting system used in the early 1960s was designed to produce good logs for the mill, but in reality was very wasteful of wood fibre. Any tree that wouldn't produce a sawlog was left standing, and, after slashing, there was further waste as each landing had a pile of butt ends, tops, etc. The skidders were small (Timberjack 205s) and none of them were equipped with chains. The cutterman topped his own trees and usually topped the tree length where any hope for a sawlog ended. On the landing, the skidders pulled the hitch of tree length stems onto a crosser log in front of the Prentice loader. The slasher man on the landing cut the tree length into logs using an eight-foot measuring stick and a power saw.

Photo 74 **First Slasher: Tower Hill near Madawaska Lake**

Bob McRae commented on this slashing system:
> With the Prentice loader, one operator could do the work of three or four gin-pole teams of four men each. These machines, with powerful hydraulic arms, held the stems that were then slashed into logs. The logs were then loaded into trucks that went directly to the mill in a system termed "hot skidding."[201]

Source: Robert Cannon

This system basically stayed in place until the advent of tree-length stem delivery to the Whitney mills in 1980.

The cut logs were then loaded onto tandem trucks. Figure 32 shows the typical chaining of a load of logs.

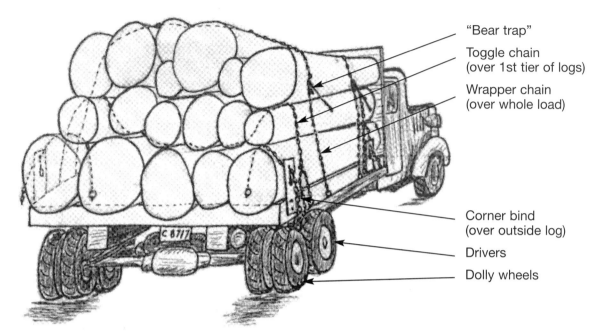

Figure 32 **Method of Binding Logs on Trucks**
Tandem trucks had two pairs of rear wheels
Source: L. Cahill, D.W. MacGregor and S.W. Lukinuk, "Shier, Clinton Township, Haliburton County," *U of T Logging Report 121*, 1949.

Sandy McGregor: Bush Superitendent

Sandy McGregor was the bush superintendent who had been with J.S.L. McRae since the Airy days. A bush superintendent had a very responsible position. It was his job to inspect, or walk, the bush and assess the timber found on it. What were the sizes of the trees and what species were present? Was there a good location for the haul road and could a reasonable tote road route that avoided steep hills be found? A good source of potable water, removed from drainage from the horse stables and latrines, was also a necessity. The answers to these questions went into the very important decision of whether to bid on a timber licence and at what price. In the bush, camp foremen reported to the bush superintendent. Sandy was a very good man when assessing an operation based on horse power but he did not adjust to the changes needed when mechanization of the operation took place. Skidding trails and the placement of roads caused the biggest problem.

Times had changed by the early 1960s and Sandy was retired much against his wishes. Brent Connelly, the replacement forester for Felix Tomaszewski, drove him back to Huntsville after his last day on the job; it was a long, uncomfortable drive. Sandy didn't have much luck in retirement. While in the bush to the west of the Park assessing a stand of trees with another man, he was attacked and badly bitten by a wolf. The animal was rabid and

Sandy had to undergo the very painful, deep needle treatment for rabies. He survived that ordeal but soon after his health declined and he died.

Brent Connelly then moved on to another company and the McRae forestry work was taken up on a private contract by Bill Hall and his company, Opeongo Forestry Services of Renfrew. Hall prepared the necessary fire plans and professional forestry work required by the government. Paul Kuiack supervised the bush operations.[202]

The Last of the Horses

The last horses worked in the bush at Lemon Lake, north of Lake Louisa, in 1962. Gary Cannon said, "I think Frank [Shalla] was the oldest teamster there. As long as there were horses in the bush, McRae left him on them. His brother Maxie was also a teamster with Felix Luckasavitch and Felix Voldock at the end."[203]

The partnership between a teamster and his horses was frequently one of great affection. This shows in the following story told by Felix Voldock:

> I was coming down a straw-covered hill with my team and a full load. The hill had three curves on it. I made the first two okay, but I was picking up speed, and it felt as if the horses' feet were only touching the ground every other step. All the men were yelling at me to "Jump, jump, jump off, Felix. You're not going to make it." But I stayed right up there on top of the load of logs, trying to make sure the horses got down safely. We did, and I got off and had a good little talk with those horses, telling them what a good job they had done.[204]

The passing of the horse marked the end of an era. A horse had to be given rest and this also gave the men a chance to pause. A machine was tireless and could move far more logs, so fewer men were needed. There was no sentiment exchanged between man and machine. Ian Radforth wrote, "Old-timers claim that the disappearance of the horses changed the camps more than anything else in lumbering history."[205]

Bush Camps and Food

In 1971, McRae's last two bush camps consisted of two sets of double trailers. Each camp accommodated 22 men, at two a room, with a shower. There was a separate trailer for the office, with accommodations for the office staff and scalers. Gary Cannon commented:

> It was good going to the camp when we were young lads. There were lots of beans. The food was pretty heavy, but it was good. The last camp we had was cafeteria style. As you walked in, you picked up your tray, and they had big warming pans there, and you picked up what you wanted. There was fresh bread, donuts and cookies every day. Thursday was steak night. One night it would be pork chops, another chicken or turkey. Following tradition, there was salty pork available—and all for board of $2 a day.[206]

But these camps were the end of an era. The last one closed in 1974, and the men then commuted to work every day. The advent of all-weather gravel roads led to the closing of bush camps.

The bush camps closed without a whimper. The nostalgia came later when the old timers met. In Ottawa Valley lumbering history, there had been over a hundred years of going to camp. Walking in in August and coming out in April. If your camp was close enough, you walked out for Christmas. Looking back over his years in the bush, Felix Voldock said,

> The days in the bush weren't always good, and you were usually there because you needed the money, but the men you worked with were men that would always be friends. All the cold days and nights were worth the celebration you had together in the spring when you came out.[207]

Donald McRae as Manager of the McRae Lumber Company

In June of 1969, J.S.L. McRae died. Donald McRae and his sister Marjorie inherited the McRae Lumber Company. Donald took over management of the company with the support of Marjorie.

Donald assumed the position of leadership of the company with a solid background of experience. Before the war, he had begun full-time work as a clerk in the bush camps out of the Lake of Two Rivers mill. This involved working with a foreman in directing the cutting and managing the camp. Management of the mill followed. His leadership qualities showed, as he was commissioned as gunnery officer in an RAF bomber squadron. After 33 months as a POW in Stalag III exposed him to the leadership techniques and manner of the British commanding officer, "Wings" Day. Donald held him in high regard. In the camp Donald came in contact with Polish fliers who had joined the RAF. The men were fine warriors and well respected in the camp. Two of these highly educated men designed the hidden entrances to the escape tunnels. This respect later facilitated the acceptance of the Polish forester Tomaszewski into the McRae Lumber Company family. Tomaszewski's progressive ideas on forestry methods together with those Tomaszewski gathered from contact with the scientists at the Swan Lake Research station were passed on to J.S.L. and Donald and began their education into new progressive forestry methods.

An early move by Donald was the promotion of Gary Cannon to woods superintendent. Gary's comments are revealing of Donald's management skills—the ability to choose able men, to support them and then let them do their job.

> Dunc and I got along pretty good. Dunc'd say, "Jeez Gary, I think when Mr. McRae goes you and I are going to go too. Donald's going to fire you and I as sure as Christ."
>
> So Mr. McRae died and Donald come over here that weekend after the funeral and said, "I'm gonna change the operations around a little. Maybe you'd go into the bush and help me look after the bush."

I was glad that he did that but I was a little scared, eh. I knew the operations, the way they should be, but we were having trouble up there—too many men for the amount of work that was being done. So he said, "What I'd like you to do is see if you can cut down on the manpower a bit because we're not getting the production we should." He said, "You see what is going on and we'll talk it over, and the guys we should get rid of we'll get rid of."

It took us about a year to get that cut back, but he stood behind me the whole time. If I needed somethin' or I was getting down, he'd help me, explain what I should do. And then, after a few years, you get used to it, and he'd say, "Okay, go ahead and do it the way you think." He really backed me up, Donald.[208]

This appointment was a popular and very effective one, for Gary had a great feel for the use of machinery in the bush, but most importantly, he brought a warm and friendly personality that all were attracted to.

By the 1970s, building roads in order to keep ahead of cutting was becoming difficult and expensive; as they were getting fewer good trees per acre, the roads had to be extended faster. This was the result of two factors: the forest had not had enough time to grow good trees on previously high-graded areas, and selective tree marking, started in 1968. The partial cutting methods mandated by the government in the Master Plan of 1974 resulted in a high percentage of the stems delivered to the mill being trees that would have previously been graded as culls. Lower grade stems yielded lower value products.

Gary Cannon noted the impact of partial cutting on the mill:

> We didn't have a big chipper. I was trying to build roads to get them ahead, but I couldn't seem to get them ahead because we were using too many acres. They were marking the trees we should cut, but they weren't marking enough of them. They were doing what they thought they should do. I was thinking they were doing the wrong thing because I wasn't getting the volume off an acre. If you are taking out a thousand pieces per acre, you might as well forget about it because you wouldn't be paying for the roads. And that's what we were getting. For about two years we had it pretty tough, and then things started to get a little better. When I used to talk to Donald, he'd be pretty upset.[209]

This radical change in the quality of wood coming out of the forest to the mill had a great impact on the company. Expenses were much higher and the production of lumber was lower. Despite the difficulties, McRae saw the problems through and, in time, found solutions to them.

The all-weather gravel roads also produced a domino effect on other aspects of operation in the bush, for they contributed to the retiring of the horse, the beginning of commuting to work, and, thus, the end of the bush camps and all the marvelous food that came from the cookeries.

Donald McRae's Response to The Master Plan of 1974 and the Algonquin Forestry Authority in 1975

Donald McRae then led the company through the very upsetting times that led to the declaration of the Master Plan for Algonquin Provincial Park of 1974, and the establishment of the Algonquin Forestry Authority in 1975. It wasn't easy.

In 1974, the controversy over whether Algonquin Park would continue to be timbered was brought to a head. The Master Plan of that year decided that it would—but under new rules. A major decision of the government was to cancel all timber licences in the Park and to create a government agency called the Algonquin Forestry Authority. This organization was charged with delivering to the mills, that previously had their own timber licences in the Park, the wood they needed to stay in business. The Master Plan also directed that all mills be removed from Algonquin. Donald McRae had anticipated this situation, and he built the low-grade scragg mill in Whitney on company-owned land in 1973. Then he planned the high-grade band mill which was built on the same site in 1979. The Rock Lake mill closed in 1980 and all operations were located at Whitney. The Whitney Boy Scouts later planted the Rock Lake mill site in red pine. Today, a bicycle trail leads from Rock Lake to Lake of Two Rivers along the old J.R. Booth Ottawa, Arnprior and Parry Sound Railway right-of-way. At the old mill site, there is a plaque containing several pictures and information about the mill that once stood there.

The Apprenticeships of Bob and John McRae

Bob and John could not have escaped exposure to sawmilling while growing up in Whitney. While their grandfather, J.S.L., was still very much the Boss, their father, Donald, managed the mill and their great-uncle Duncan MacGregor looked after the yard. It would have been a life of constant exposure to bush operations and milling.

On vacations from Lakefield College School, both grandsons took summer jobs with the company. The jobs were varied: taking pine logs from the lake to the mill, scaling logs cut by jobbers in the bush, grading and piling lumber under the expert direction of Duncan MacGregor. In spending their summers this way, they absorbed what was involved in running a sawmilling operation. They developed an appreciation for the hard physical labour required of the working men, and, as well, an appreciation of the objectives of forest management.

Ian Radforth recorded the memories of Bob and John in the summer of 1986.

Photo 76 **Bob and John McRae**
Source: Ottawa Citizen and McRae Lumber Company

Bob McRae

My earliest memories of the lumber business consisted of going into the bush camps at Mink Lake with my father when I was seven to nine years old. When I was 15 years old, I worked for a month or so at the mill at Rock Lake. My first job was helping Joe Lavally[210] and Gary Cannon take pine logs out of the lake. The company used to store the pine logs in the water for two to three years and then cut all the logs at once. I got much better at poling the pine logs to the Prentice loader that put them into a tandem truck to take to the mill. I fell in a few times. Later in the summers, I would work in the yard helping Dunc McGregor ship lumber, and I scaled logs in the bush. As I remember back, the lumber business was very different from today.

In the bush, the business had a company-run operation. The jobbers were paid on their scale for the day, shared between cutter and skidder operator. I had to count the logs accurately and scale them. Paddy Roche gave me a short scaling course before I went into the bush. I would imagine now that it took a while to get it accurate. Paddy made up the tallies. I scaled in the bush for five or six summers.

I often helped the slasherman measure the logs, and at the same time counted and scaled the logs. It helped being young and athletic. The chief menace in June was the mosquitoes but, after that, July and August were great. Later on, I travelled in the bush with Bill Hall and Joe Lavally doing the forestry for the forestry plan. In the pre-AFA days, the company had to make up a forestry plan for the Department of Lands and Forests, outlining the cutting areas, proposed roads, etc. What I mainly remember is taking a small canoe with motor and accessing the areas. Then we would do sample cruises of so many chains, shouting out the tree's diameter at chest height, [e.g.] "maple, 16 inches." Bill Hall was a great fellow and a forester by trade who ran his own forestry consulting business.

I worked for the company in 1970, then went travelling before returning to Queen's University. After Queen's, I worked in Ottawa for two years and didn't come back to work at the company again till 1975.[211]

John McRae

John McRae remembered when he first went into the bush with his father:

The woodworkers were still working Saturdays. I'd go into the camps with my father. It was great excitement to go with him and have breakfast in the camp. They worked half a day on Saturdays. I remember they kept pigs at the mill; that's how they kept their meat, how they had fresh pork. I recall the tales about the bears at the cookeries, those sorts of things. The first mill site I remember was at Mink Lake. There was a county road, a reasonable road, going into it. I don't remember when they built the mill at Mink Lake, but I remember going there as a kid.

I went to public school in Whitney, three classes with 25 to 30 in each, and a mixture of grades. I'm the third child: Janet's six years older, Bob [is] four, and Cathy [is] a year-and-a-half younger. I went through to Grade 7 here in Whitney, and then I went to Lakefield College School, just north of Peterborough; and then to University of Western Ontario for a Commerce degree.

I think I was 15 when I first started to work for the company as a summer job. The first job I had was basically cleaning up around conveyors. At 15, I was doing that or piling lumber. Then one morning they needed someone to go to the woods to tally sawlogs at the landing. So my grandfather took it upon himself to volunteer me to actually start scaling, measuring material that was dragged into the landings. He sent someone in for about two days with me, and then after that I was expected to sink or swim. I guess I swam rather reluctantly! It was interesting, but I was young, and initially I was not very confident that I was doing the right thing. But it came. When you start, even to recognize what various species are, even without the bark on, is difficult. But after a while, it becomes rather commonplace. I enjoyed it that summer. It was interesting to see how things were done—the whole process. I commuted everyday from Whitney. I used to go to the mill site with my father, and we'd go in by half-ton. The greatest difficulty working in the woods were the flies in springtime. Other than that it was fine. It was interesting—just learning.

I learned how to handle the equipment, a power saw, and so forth. I gained a pretty good understanding of the whole process: the objectives, how the forest was managed, and an appreciation of the difficulty of logging operations in general. It's hard physical labour, and you're at the mercy of the terrain and the weather to a degree. You're not isolated from these things; you have to work with them. And often you have to work against them. For instance, if it happens to be a wet summer, it can get awfully muddy. That means it's difficult to skid, to pull the trees out. Machinery gets stuck; it's difficult to build roads. It's almost impossible to maintain the cutting operation when it's pouring rain. It's too dangerous. It's slippery when things get wet, and visibility's not good. Probably the most valuable thing it did was give me an appreciation for the fact that it is a hard task. And it gave me an appreciation of forest management and what it attempts to accomplish.

Most summers at high school and university I worked in the woods. Two summers I stayed in London and painted houses. And one summer I was the shift boss of the night crew at Rock Lake. In previous summers, at least while I was at university, I worked around the sawmill in the spring, in May, before the woods operations got going in June. At other times I looked after crews planting trees and things like that.

I hadn't really always thought about coming into the business. I know the thought was in my father's mind. Oh, I'd thought about it, but I really hadn't come to terms with it. I didn't see summer jobs as an apprenticeship. It was a good way to spend the summer though, in that you were outside most of the time.

When I started full time after university, I

came to the sawmill we had here. It was built in 1973, and I started to look after it, to get the feel for running a full operation. I was in charge of that operation, and I started to dabble in sales. At that time, my father was very active in the business, and the office was at that house where I live. I'd spend some time at that office and some time at the office here. I got a good understanding of what's involved in running a sawmilling operation, an appreciation of the problems. It was pretty time-consuming. We were running double shifts then, day and night. Of course, I didn't have many demands on my time besides work, and so I really didn't mind.[212]

In both of the above accounts what comes through is a willing acceptance of work. This even shows in a humorous way when J.S.L., John's grandfather, in a manner typical of him, "volunteered" John to go into the bush to scale logs for the first time. But perhaps their fate was sealed much earlier, when as children they happily went into the bush camps with their dad. John said, "The cookery stands out in my memory ... the concept of a cookery—the huge pans or tins they used to have filled up with cookies. They used to have four or five kinds, and you could just go up and get some. That was a big thrill."[213]

Bob also recalled: "My earliest memories of the lumber business consisted of going into the bush camps at Mink Lake with my father when I was seven to nine years old. All I remember was the cookery at the mill—they had good cookies!"[214]

Obviously, the cookies got them!

A Smooth Change in Management

The time at Rock Lake mill was the culmination of Donald McRae's very successful career in lumbering. This was a big operation, both at the mill and in the bush, and both aspects ran well. Under Donald, the very difficult times leading to the Master Plan in 1974 were survived and adjustments to the AFA were made.

Donald's contemporaries remember him with great respect.

E. Ray Townsend, retired operations supervisor of the AFA, remembered doing business with Donald. At that time, Ray was with Wood Mosaic, and they needed veneer logs and McRae Lumber needed pine. "Donald was all business—a hard bargainer, but if a verbal agreement was reached, when the paper work arrived it was exactly as agreed upon, for he was a man of his word"[215]

Brent Connelly's first job as a forester was at Rock Lake in the early 1960s. "At the mill, Donald had a lot of decisions to make every day relative to the bush, mill and the office, and he did so as a very competent business man. He had futuristic insight and a willingness to try the new. Donald worked hard and was well respected by the men."[216]

Donald brought his sons Bob and John into the company in a way that was very different than what he had experienced under his father, J.S.L. While Donald made certain they both experienced the different facets of the business, this was done without the frequent outbursts he suffered from his father. At the end of the 1970s, with the new band mill in place, Donald McRae retired. Since then, Donald has had the pleasure in seeing Bob and John run the company very well.

Bill Brown, now a retired general manager of the AFA commented: "I give top marks to Don McRae in the way that he managed the transfer of responsibility of McRae Mills to Bob and John."[217]

Tried and True Remedies for Black Flies

"I have never know it to fail: 3 oz. pine tar, 2 oz. castor oil, 1 oz. pennyroyal oil. Simmer all together over a slow fire, and bottle for use. Rub it thoroughly and liberally at first and after you have established a good glaze, a little replenishing from day to day will be sufficient. And don't fool with soap and towels. A good safe coat of this varnish grows better the longer it is left on. Last summer I carried a cake of soap and a towel in my knapsack for a seven week tour and never used either a single time. When I had established a good glaze on the skin, it was too valuable to be sacrificed for any weak whim connected with soap and water."

A good mixture agaist mosquitoes and black flies can be made as follows: Take 2 1/2 lbs. mutton tallow, melt and strain it. While still hot, add 1/2 lb. black tar. Stir thoroughly. When nearly cool, stir in 3 oz. oil of citronella and 1 1/2 oz. pennyroyal.

Plain kerosene is certain death to all sorts of insects so long as they have not burrowed beneath the skin.

Source: Cynamid of Canada Co. Ltd.

Section III: Interaction with the Province of Ontario

The McRae Lumber Company has always been dependent on timber from Crown land, and Algonquin Provincial Park has provided much of that wood. Therefore, the company has been subject to the changing provincial rules and regulations regarding that harvest. Since World War II, the rules of operation have become increasingly restrictive, as both new science and changing government policy (often driven by public opinion about appropriate Park management) have been incorporated into procedures and regulations.

In 1947, the Ontario government passed the Forest Management Act. *It required that cutting plans of each company be approved by the Department of Lands and Forests. To supervise this, the Department of Lands and Forests was organized by Deputy Minister Frank MacDougall into forest management units. Photographic and field surveys covering the forested lands of the province were undertaken, from which maps of tree types and their condition were made. The objective was to have the provincial forests managed on a sustained yield basis. Achieving this objective has proven to be a difficult, ongoing task.*

A growing, mobile and affluent urban society led to another problem in Algonquin and other parks. Some of the canoeists and recreationists, who began to use the Park in large numbers after the war, were horrified to find timbering in Algonquin Provincial Park, despite the fact that it had been going on since long before the creation of the Park. They viewed parks as "wilderness sanctuaries," and demanded that timbering cease in them. These problems centred on Algonquin Park and became very pressing in the late 1960s and early 1970s.

Brokered by the government, the Algonquin Park Master Plan of 1974 was the legislated compromise between the logging interests and the recreationists. The Park was first divided into zones; logging was prohibited in areas designated as Natural, Recreational and Primitive, as well as at Historic sites (Map 19). These zones included significant areas where individual companies held timber licences. If prevented from timbering there, these firms would be put out of business. Several steps were necessary to maintain the mills dependent on Algonquin timber: all company licences in the Park were cancelled; the AFA, a Crown agency, was created and given the sole licence to timber in the remaining Recreational/Utilization zone. This zone, fortunately, included large areas not under licence and the wood available in the total Recreational/Utilization zone was sufficient to cover the needs of all previous licence holders. The AFA was charged with the responsibility of delivering the same amount of wood to the mills of the former timber licence holders as they had previously cut.

In the last 30 years, the province has enacted further legislation extending its control over all Crown forested lands, with the objective of having industry manage and pay for sustainable forest management on these Crown lands, while still allowing the government to maintain provincial stumpage revenues and ultimate control. Since McRae Lumber Company has timber licences outside the Park on Crown land, the company has had to comply and adjust to these regulations as well.

CHAPTER 9
Reconciling Timber Harvesting with Recreational Use in Algonquin Provincial Park

Algonquin Park evokes passionate feelings in many people. During the over 30 years explored in this chapter, a pitched battle was fought over how the Park was to be used. The "Battle for Algonquin" pitted park protectionists, who held the philosophic position that timbering had no place in a public park—that parks should be sanctuaries, totally free of all signs of man—against the wood mill owners of the area, who had long-standing legal licences to cut the trees under government regulations. The Department of Lands and Forest, which managed the Park for the province for both recreational and timbering purposes, was caught in the middle. So also were the members of the Ontario legislature, where, generally speaking, politicians from urban ridings supported the protectionists while those MPPs who represented rural ridings were sympathetic to the timber interests. When the Master Plan was developed in 1974, it confirmed MacDougall's previous policy of multiple use.

Many fine and able people filled the ranks of both sides: mill owners and intellectuals, naturalists and scientists, politicians and lumberjacks. Most importantly, under this compromise Algonquin Park has continued to spread its magic, and allow a wide range of uses.

The Frank MacDougall Years as Deputy Minister of the Department of Lands and Forests: 1941–1966

While he was not on stage when the final act was presented, Frank MacDougall dominated most of the years when the debate over the uses of Algonquin Park was at its peak. The people who eventually quietened the problems were either his men or were greatly influenced by him.

Following World War II, the government, through the Department of Lands and Forests, realized that the forest resources of the province were in bad shape. Authorities knew that harmful practices such as heavy high-grading had taken place. MacDougall advocated increasing the budget of the Department of Lands and Forests to bring management to the level where it could supervise permanent forest cropping and thus sustain the pulp and paper and lumber industries.[218] Political action followed.

Major-General Howard Kennedy was appointed to head a Royal Commission of Enquiry in 1946. His findings were very critical of "the tremendous waste he found in single purpose operations in forest road construction and inefficient cutting; in the flooding of great tracts of standing timber for power dams and in grinding valuable sawlogs into pulp."[219]

To avoid the abuse of provincial forest lands, Kennedy advocated policies and procedures to be followed.

> Kennedy based his recommendations on the same basic objective that the Department had adopted: that of achieving a sustained yield. This meant managing the forests so that the amount of timber cut each year should be replaced by new growth. It meant applying forest regulations to all equally; encouraging those who aimed at sound silvicultural practice; removing all useable species up to the limit that the market could absorb; and starting new growth on cutover areas equal or better than that removed. Without such a policy, a continual lowering of the forest's quality would be inevitable.[220]

Wanting to maintain their virtual autonomy on their company licences, the forest companies resisted the proposed controls, but Abitibi Power & Paper, then under considerable economic stress and government pressure, capitulated and accepted conditions that required it to submit operational and management plans for its limits for government approval before the cutting season began. Other companies followed. The development of ways and means to supervise and control company forest operations by the government through the Department of Lands and Forests followed.[221]

Major-General Kennedy's investigation led to the *Forest Management Act* of 1947, which required the submission of plans and maps for approval by the Minister, who could cancel a licence or stop operations if a company did not comply with regulations. This act forced companies that were logging Crown land to hire professional foresters to produce the required cutting plans. By the 1950s, the Department of Lands and Forests had established management units, and the field supervision organization had been strengthened. The province took the company management plans, analyzed them, and gradually worked out a management scheme that was acceptable to both the company and government. In theory, the province had achieved complete control over the industry.[222]

Before these controls could work, however, the province had to know the condition of the forest. To achieve this, a provincial forest inventory was carried out under J.A. Brodie of the Timber Management Branch. In 1946, an aerial photographic survey was made covering all accessible forest land from the Kawarthas to 60 miles north of the CNR transcontinental railway line. Bush teams of timber cruisers and tallymen then measured tenth-of-an-acre plots as field checks of aerial photo interpretations. The federal government paid one-half of the cost. This photographic work was completed by the early 1950s.[223] Photo interpretation can produce maps of the forest by species and size. Harvesting and silvicultural plans can then be made using this information. Photographing Crown forest lands every 20 years is now a standard procedure, directly paid for by billing the private sector.

While Crown lands belong to the people of Ontario, logging companies traditionally viewed their leased land as "private" property, and they more than occasionally proceeded to do as they wished with the forest, cutting to maximize immediate profit. Trespass cutting on an adjoining lease, not obeying diameter limit restrictions, and not using cull trees were common, and these were just some of the infractions that were charged to lease-holding companies. Control was to become much tighter with the development and training of field staff under Deputy Minister Frank MacDougall.

There was, however, a severe shortage of technicians, such as the scalers who measured the wood cut, and others who supervised company operations. The Forest Ranger School (later called the Frost Centre) established by MacDougall at Dorset in 1946 addressed this situation in part. The universities of Toronto and New Brunswick produced degreed foresters who were employed by industry and government. Both sources of manpower contributed to the strengthening of the department's field organization to the point that it was capable of supervising forest users. The staffing and role of the Department of Lands and Forests in Algonquin Park as developed after World War II serves as an example of this control.

The Pembroke Administrative District

The Pembroke Administrative District of the Department of Lands and Forests covered most of Algonquin Park in 1967 (Map 15). The district office was located in the town of Pembroke. Appendix V shows the positions and distribution of staff under this office and provides a detailed example of the extent to which the Department of Lands and Forests developed.

Between 1941 and 1945, the forest industries clustered at the mouth of the Petawawa River requested assistance in accessing future supplies of accessible timber, as merchantable wood located close to drivable water was in short supply and roads built individually by these mills into their limits were inefficient from a larger regional perspective. To remedy this situation, the Petawawa Management Unit was established in 1945 by Order-in-Council. In 1951, this unit was the first to receive ministerial approval of a management plan for an area of commercially operable timber when the companies holding leases in the area agreed to work under government regulation. The objective was to have the area yield a continuous supply of wood to the mills of the region.[224] In this major undertaking, the Department of Lands and Forests built access roads, assigned short-term cutting leases of two to three years to individual mills, planned the cut and silviculture, and, in some cases, even harvested the wood. The mills paid for these full-scale services in the price of the wood they received. In some ways, the government presence in the Petawawa Management Unit foreshadowed the much later creation of the Algonquin Forestry Authority.

After the war, trucks replaced river drives for the movement of most logs to mills. The construction of the Achray Road (Barron Canyon) by the Department of Lands and Forests was necessary to open up a major part of eastern Algonquin to modern logging. It

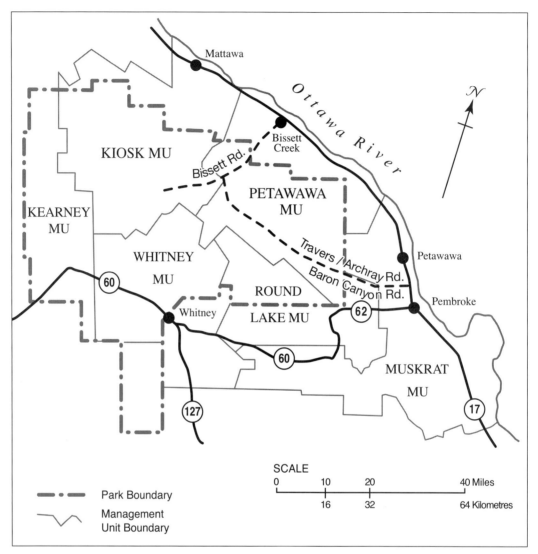

Map 15 Management Units: Pembroke Administrative District, 1967
The administrative districts and management units generally followed timber licence boundaries and did not correspond to the boundaries of Algonquin Provincial Park.
Source: Ontario Department of Lands and Forests

was a major undertaking. Initially, this road exited the Park via the Petawawa Military Reserve at Montgomery Lake to take advantage of an existing bridge across the Petawawa River. In 1958–1959, the Achray road was extended from Ignace Lake southward through the Park to the Alice town line. This extension avoided the Petawawa military base, and permitted the road to be opened up to both logging and recreational users.[225] This greater accessibility was not favoured by some groups who wanted to preserve the wilderness aspect of the Park.

To open up the unit's more inaccessible areas, two other sections of primary road were constructed: from Mackey Station on the Ottawa River, some 13 miles of road were built into the northern edge of Edgar Township in 1964; from Highway 17 near Bissett, a 25-mile road was built to Radiant Lake between 1965 and 1966 (Map 15). A six-mile extension was also built to connect this road up to the Achray-Lake Travers Road (Barron Canyon Road). From all of these primary access roads, secondary and tertiary roads were pushed into the forest to permit harvesting and silvicultural work.

At this time, the Department of Lands and Forests was greatly influenced by the success of large scale clear-cutting and mechanized harvesting in the northern boreal forest. The obvious thing for the Department to do was to transfer these successful methods, which had been tested in the northern forests, to the coniferous stands in the Petawawa Management Unit in Algonquin Park. Unfortunately, the results in the Algonquin forest were not positive.

In the 1950s, Dr. Walter Plonski initiated the clear-cutting of pine in 200-yard-wide strips. Oriented parallel to the prevailing winds, the shaved land was supposed to reseed itself in pine. These clear-cut strips, which could be seen for miles from the air, came to infuriate conservationists. Wilderness canoeists said they desecrated their park, even though most of them couldn't be seen from ground level. More importantly, by the mid 1960s it was clear that the expected regeneration wasn't taking place successfully, except in areas where soils derived from glacial outwash sands occurred—there the white pine seeded very well. Where the clay fraction was higher in the soil, the rapid growth of poplar caused problems, though the poplar did provide shade, which helped reduce the spread of pine weevils. The Petawawa unit came in for a great deal of attention from both conservationists and the Department of Lands and Forests.

White pine weevil became a major problem for regeneration of pine on the cleared strips. The weevil commonly feeds on the lead terminal shoots of young pine, thereby killing them. This forces a lateral branch to become the leading shoot and it frequently succumbs as well. In the end, the growth of the pine is distorted, preventing the development of the single trunks that produce satisfactory sawlogs. The weevil thrives on the young pines found in open sunlight, such as those found in clearcuts, where protective shade is not present.

Another problem for all white pine came from the United States, into which it was likely imported in 1898 with a shipment of white pine grown in a German nursery. The white pine blister rust spread, and by 1937 pine in Algonquin were infected. White pine blister rust has a life cycle that involves wild currant bushes, which are common in the Park. An infected wild currant bush produces spores on the underside of its leaves. These are borne by the wind and infect the tips of the needles on the white pine, whereupon the needles die and turn brown. The rust fungus grows into the twig, then into the branch.

White sap oozing from infection sites on white pine signals the likely presence of this malady. Eventually the infection penetrates the trunk, killing the tree. All the while, windborne spores produced on the pine spread and infect more currant bushes, thus continuing the cycle. Eradication of currant bushes in an area the size of the Park, which would allow the cycle to be broken, is impossible.[226]

The Whitney Unit of the Pembroke Administrative District and the Work of Don George

This unit supervised forest operations on licences both inside and outside the Park (Map 15), an area which included a good portion of the timber licences held by McRae Lumber. The forester heading the Whitney Management Unit had a great deal to do with the McRae Lumber Company.

After graduating in forestry from the University of New Brunswick, Don George worked for Price Brothers in Quebec. There he became bilingual and earned a degree as an *ingénieur forestier* (forest engineer) through Laval University. Quebec law required that any timber licence-holder employ a forester as the firm's designated authority for a variety of policy transactions. It also defined the tasks that could be performed only by an *ingénieur forestier*. Not only was normal forest training seen as necessary, but the social responsibility of the forester was stressed also. This is well illustrated by the Code of Practice for the Quebec forest engineer. Among other items, it spelled out a set of "duties and obligations."

> The forest engineer must focus attention on the prime importance of the forest to the economy of Quebec. He must support every measure aimed at improving the forest heritage and welfare of society. He must also inform the public of the order of any forest policy measure or provision that he considers prejudicial to the common good of society.[227]

Don was impressed by this admonishment and was cognizant of this duty throughout his career. Don also holds Registered Professional Forester qualifications for Ontario.

Before coming to Algonquin, he worked for Georgia Pacific, an American firm that had large, mainly coniferous holdings in southwestern New Brunswick. This progressive firm treated the forest as a long-term, sustained-yield holding and harvested it by a selection system, not the commonly employed clear-cutting method. It also had a firm policy of avoiding damage to the growing stock when harvesting and skidding. Here, Don George learned and practised how to get a partial or selection cut up and running with minimal-damage skidding.

Don George assumed the position of management forester of the Whitney Management Unit in 1968. William Hueston was Park superintendent and the district forester for the Pembroke District at this time. In his position, Don was responsible for overseeing all forest operations in the unit. To become familiar with his new territory and to bring its maps up-to-date, he designed an aerial camera system, mounted it in an MNR de Havilland Beaver, took his own stereo photographs, and then made maps specific to his needs. While Don did not invent aerial photography, this procedure was an example of the energy and progressive thinking that he brought to the unit. The Whitney unit established a reputation as being the most advanced in the Park.[228]

Don started applying the methods that he had learned at Georgia Pacific while working with Andrew Wojcik, a jobber who was cutting and skidding a small licence east of Madawaska in Clancy Township. Don took Park superintendent Hueston to see two operations—one on the former Weston property on Victoria Lake in Murchison Township, which was harvested by traditional methods; and the other, Wojcik's cutting and skidding operation, which employed the partial-cutting selection system and minimum-damage skidding (Appendix IX). When he saw Wojcik's work, Superintendent Hueston said, "Let's do it all that way!"[229]

Don George established a good working relationship with the McRae Lumber Company and convinced McRae to adapt its mill and develop markets to use what the bush was then capable of providing. Even crusty and traditionally inclined J.S.L. was impressed by Don George. Don directed the first large-scale tree marking and selection cut on a McRae timber lease in Sproule Township in 1968. The McRaes took a liking to Don George. Don acknowledged: "I would have accomplished nothing without that."[230]

Because of his accomplishments with the Whitney Management Unit, Don George was promoted to management supervisor in all phases of forestry for Algonquin in 1973. This was not an easy posting.

The Pembroke, Kiosk, Muskrat and Round Lake management units of the Pembroke district were still harvesting in the high-grading way. In one of his first acts, he ended the strip clear-cutting of pine. Don then set out to extend the shelterwood system of tree marking and cutting to most hardwoods and all white pine across the Park, along with minimum-damage skidding procedures. It was very difficult to change old ways. Some caught on faster than others.

One day in the woods, he was with a distinguished visiting forester who was promoting clear-cutting. The visitor asked the workers which method of logging they preferred—clear- or selective-cutting. The loggers replied that they liked the marked selective-cutting much better because with it they were leaving trees to grow that either they or their sons could cut. This made Don's day!

In 1974, the Master Plan legislated that the selection and shelterwood cutting systems be followed. The AFA, created in 1975, had to follow the new regulations. The new methods required the training of tree markers and supervisory staff. All of this took place against a history or tradition of high-grading, the cutting of only the best trees to a diameter limit. The best trees were what the mills expected and were adapted to receiving, as well as being what the bush workers were used to cutting.

In the early years, Don George wrote out more charges against the AFA than were incurred by all the big pulp and paper companies operating in northern Ontario. Joe Bird, the first general manager of the AFA, once spent a very intensive week preparing for a hearing before a legislative committee to explain a series of charges. But Don George insisted that the new procedures be followed. He persisted in seeing that such was done with the backing of the Park superintendents. It simply took time to get the skills in place and all parties on the same page.

Don George retired from the MNR in 1988. He has been ranked, along with Frank MacDougall, as the most influential forester to work in Algonquin Park. I would also add the three research scientists of the Ontario Forest Research Institute of the OMNR at Swan Lake: Don Burton, Mac McLean and Harvey Anderson. These men, whose work is detailed below, carried out research on yellow birch and sugar maple, and their findings have been applied well beyond the borders of the Park.

Don George still maintains his great interest in Algonquin Park forestry. In August 2005, he returned to the Park with his former technician, Dave Harper. They visited Louisa flats where they had carried out a marked partial selection cut in the 1970s. The AFA, just previous to their visit, had made another selection cut. Don was pleased with the condition of the woods.

Research in Algonquin Provincial Park

By the 1960s, Algonquin Park was the location of five significant research stations. A Fisheries Research Laboratory on Lake Opeongo was established in 1936. This facility was renamed the Harkness Laboratory of Fisheries Research in 1960 in memory of its founder and first director. The 31-square mile (8,032-ha) Wildlife Research Area based on Lake Sasajewun was created in 1944, and the 9.28-square mile (22.3-ha) National Research Council Radio Observatory at Lake Travers was built in 1959. An additional 54.7 hectares were added to this complex in 1966 to accommodate a radio telescope. The federal government is still operating the telescope and buildings. The Natural Science and Energy Research Council ended its responsibility for the site in 1995. Other than being located there, this last facility had little to do with Algonquin Park. It

Photo 76 Don George and Dave Harper, Louisa Flats

Note the variety in size of trees and the quantity of small tree growth. Many of these small trees will not reach maturity, but will die off due to competition for light.

Source: Courtesy of Don George

did play a significant role in early research and space travel monitoring in cooperation with NASA. The Swan Lake Forest Research Station began operation in 1950.[231] The Algonquin Visitor Centre (1993) houses the Algonquin Park Museum Archives and a fine library that has been used by many researchers. The establishment of these facilities relate back to one of the original purposes of the Park—to promote research.

The Swan Lake Forest Research Reserve

Studies at Swan Lake at first focused on the problem of why yellow birch was not regenerating. Don Burton was in charge of Swan Lake at first (1950–1956); George Sinclair replaced him in 1956 and Mac McLean ran the station from 1957 to 1987. Harvey Anderson succeeded McLean, running things from 1987 to 1994. Retired now as Scientist Emeritus, Ontario, Harvey Anderson remains active as a consultant. Al Gordon, another notable researcher at Swan Lake, worked on coniferous trees—especially spruce.[232]

Yellow Birch Research

During World War II, yellow birch was heavily cut for use as veneer plywood in Mosquito bombers, and birch timbers were used for constructing naval corvettes. In the 1940s, it became apparent that the yellow birch was in trouble. Dieback, an incompletely understood condition similar to that found first in trees in New Brunswick, was affecting older trees in the Park. It also became apparent that regeneration sufficient to maintain yellow birch as a commercial tree was not taking place. The province did not want to lose this resource, hence the justification for setting up the research station. The original research reserve was established in 1950 with 1,120 hectares (2,964 acres) and expanded in 1960 to 3,480 hectares (8,596 acres) (Map 18).

Photo 77 Yellow Birch at Swan Lake Naturally Regenerated in 1961 Following Prescribed Fire and Group Selection Harvesting (ca. 1992)
Left to right: Don Burton, Alan Gordon and Harvey Anderson
Source: Ontario Forestry Research Institute. Courtesy of Harvey Anderson.

Research found that yellow birch experienced good seed crops every three or four years, but that even then regeneration was poor. Yellow birch seeds are small, having only 1 per cent of the mass of maple seeds, and do not have the stored energy sufficient for their rootlets to penetrate the thick leaf litter produced by sugar maple the previous fall. Research conducted at the station discovered that most of the seeds that did succeed in germinating then died from lack of sufficient light. To regenerate well, yellow birch needs a good supply of seed, an exposed seedbed mixture of humus and silty-sand mineral that is fresh to moist, with 40 per cent light. These conditions enable them to outgrow competing maple seedlings. The yellow birch seedlings require more light because they rely on current photosynthesis for growth rather than the stored energy that maple seedlings derive from their seed.

With the cooperation of Muskoka Wood, which held the timber lease in the northern portion of the original research area, experimental harvesting was carried out in 1953. The canopy was opened up to permit varying amounts of light to reach the ground, and scarification was used on specific plots to remove leaf litter. With scarification, the surface of the ground is stirred up, exposing the mineral soil. This greatly improves the chance of yellow birch seed landing on exposed soil and so increases its chance of growing successfully. Some 16 years of study on sample plots was carried out. Fencing was found to be necessary to exclude the white-tailed deer, then common in the Park.

Photo 78 **A Discussion of Site Classification Methods at Swan Lake**

Mac McLean (*right*) and Harvey Anderson (*second from right*) discuss site classification methods at Swan Lake with Site Specialists Keith Jones and Ted Taylor of Guelph University (ca. 1984).
Source: Ontario Forestry Research Institute. Courtesy of Harvey Anderson.

In 1958, experiments were initiated using prescribed burns to prepare seedbeds for birch regeneration (probably the first such studies in the province). It was found that controlled ground fire to remove the leaf litter of the previous fall, if followed by a good seed year, produced good seeding and development of saplings, if they were given protection from browsing deer. Interestingly, where yellow birch seeded densely, observations suggested that chemicals released by the seedlings inhibited the seeding of sugar maple.

It was found that the establishment of yellow birch as a renewable resource required encouragement to avoid dominance of it by sugar maple and beech. Research demonstrated that careful tree marking, followed by group selection harvesting, could lead to the satisfactory regeneration of yellow birch.[233]

This tree, however, ideally needs tending at the sapling stage to achieve ideal growth. Since this procedure is expensive, labour-intensive work, it is frequently not done.

Sugar Maple Research

In terms of the area it covers, sugar maple is the dominant tree of the Algonquin Highlands, and it forms the largest component of the hardwood cut. Maple wood is commonly used in flooring and furniture. Unfortunately, many of the stems are flawed by the presence of mineral streak and rot. "Brown heart," typical of sugar maple, is also a problem, as it can be a large percentage of a stem's diameter. Heartwood is less desirable and can be sold only as cheaper construction wood. These flaws are generally blamed on the inferior soils found on the Canadian Shield.

Curiously, research has shown that some superior trees can be found in association with defective trees. Soils, it seems, only indirectly affect the wood. Further investigations have shown that a higher incidence of flaws is more frequently associated with slow stem or trunk growth. When trees having flaws are exposed to greater light, increased diameter growth occurs, and this raises the quality of the wood. Further research has not discovered all the answers to these problems. But researchers have discovered that wood quality in sugar maple, a shade-tolerant tree, could be enhanced if the tree received more sunlight. This prompts faster and clearer wood growth. In order to accomplish this, increased thinning of trees, which allows for more sunlight in the forest, was built into tree-marking instructions in sugar maple stands.[234]

Over time, the forestry research scientists found the way to both re-establish yellow birch, and to boost the productivity of sugar maple. The research carried on at Swan Lake has contributed greatly to the health and productivity of Ontario's forests.

The Spreading of Ideas

The Swan Lake research was not a secret, for its findings and facilities were well known to the forest community in Algonquin. Donald McRae and his forester Felix Tomaszewski, for instance, were familiar with the work of Burton. Mac McLean marked a hillside behind the Rock Lake mill, which McRae cut. Subsequent growth of the stand was very successful. Don George, the forester at the Whitney Management Unit, was also a regular, "informal" but active participant in the research with Mac McLean and others.

Don George wrote:

> To maximize growth potential, crop trees had to be tended carefully. Competition for light was necessary to produce long straight stems free of branching, yet not overdone since too much light would have produced "bean pole" stems. Too much light at a site caused by a lack of competition produced heavy lower branching which was not wanted, and the removal of some side competition was necessary to promote diameter development. Trial plots with varying light conditions determined the optimum spacing of the trees that maximized growth and

diameter. These conditions were then taken into the field and applied by tree markers.[235]

Both Mac and I were strong believers in the selection system as a means of providing high value hardwood timber. In fact Mac had also read Eyre and Zillgitt, and he was doing field trials at Swan Lake to assess the impact of their system in the Park.[236] So I suppose it was natural that we would both see it as a solution to the Park crisis.

It was also important that the forest base was well suited to this approach … the times were right. So we went up to Mac's research station and got started. We also brought in Doc Raymond, the bio-mathematician who worked at the Department's research station at Maple. We had to develop our own yield tables and growth curves to guide the selection process. We also developed a new quality class cruising system so that we could judge whether sufficient stock was available on a tract to justify a selection cut.

We did our first cuts right there in 1968. When we returned 20 years later for the second cut, the cross-sections of the trees we had released provided very dramatic evidence of the growth gains achieved by the system. Although it took some time to get the system up and running, it seemed as if the hardwood problem was on its way to being solved.[237]

Later, tree marking became a required procedure on Crown timberlands. Dave Duego, a forester responsible for the Almaguin area of the Ministry of Natural Resource's Bracebridge District, said that tree markers and resource managers now routinely incorporate yellow birch management into their day-to-day work in tolerant hardwoods. He credits the research at Swan Lake by Don Burton and Mac McLean, with Harvey Anderson putting it all together, for making this happen in the field.

The strength of Harvey's work was that he could display it and get endorsements from the scientific community because it was extremely solid work. He could also take that same work and translate it into language that lay people could use; the technical people who are actually doing the work in the field, for which his work was designed.

Harvey had gathered up a great deal of information and experience, and he shared that with us. We took as much of that as we could practically use, and we are putting that to work in the field. That is going on each and every day that our tree markers are out there working. Those of us that are planning forest management are working it into our planning document.

There are some efforts where we are putting it on fairly large areas, 75/100 hectares at a time. We are also creating yellow birch on much smaller areas, anywhere from one-half a hectare to five to six hectares as an integral component of the single tree selection system. Yellow birch is being scattered into the single-tree selection system, which normally grows sugar maple and beech. It is increasing the amount of biological diversity that exists in our forest areas, and certainly increasing the value because the yellow birch lumber and veneer is worth more than the other species that it competes with. It could represent upward to 10 per cent of the area that we do in any given year, whereas in times prior to our actually going after it, it occurred only by accident.[238]

Continuing Research at Swan Lake

Harvey Anderson, Scientist Emeritus, OFRI reported the following events at Swan Lake in 2005.

1. Two single-tree selection cutting trials, established at Swan Lake in 1967, were remeasured and silvicultural cutting prescriptions prepared. This will be the fourth harvest for the "Parkside Gully" trial and the fifth for the "Scott Lake" trial. They are both on a relatively short 12-year cutting cycle because of the superior site, the high quality of the stands, and their demonstrated high productivity rates. These trials are considered to be benchmarks for tolerant hardwood management.

2. Also remeasured was an old-growth tolerant hardwood stand which serves as a control for one of the cutting trials.

3. A new digital imagery/LIDAR (radar) project was initiated that will provide a new digital topographic base map and inventory of the Reserve to aid future forest and road management strategies.

4. Work was begun on demolition of some outdated buildings at the Swan Lake camp and the follow-up site restoration. This will make the facility more efficient to operate.[239]

The Rise of Conflicting Use Problems in Parks

The notion of protecting scenery and attractive vistas in parks goes back in the case of Algonquin to George Bartlett, who was superintendent between 1898 and 1922. One of his regular complaints with the lumber companies concerned their flooding of land early in the spring to provide water to flush their logs down the rivers and out to the Ottawa River or other surrounding waterways. While this was damaging, it was the retention of high levels after the log drives had gone through that was the major cause of shoreline tree die-off, resulting in the drowned, dead landscapes painted by artists such as Tom Thomson.[240] Other protectionist-inspired incidents took place in the early years of the Park. Among them was the Highland Inn tourist protest that resulted in the government buying out the Munn Company and its timber leases in 1910. In 1929, protests by Cache Lake leaseholders led to shoreline reserves around that lake by Minister Finlayson when he granted McRae Lumber a timber licence in Canisbay Township in 1929.[241] Ten years later, Superintendent Frank MacDougall felt it appropriate to place a standard system of shoreline timber reserves across the Park.[242]

In the late 1950s and into the 1960s, a multitude of problems descended upon the Parks Branch of the Department of Lands and Forests. While problems varied depending on the circumstances of a particular park, they included overcrowding, rowdyism and the destruction of natural environments, along with overcrowded canoe routes, excessive litter and conflict between canoeists and motorboat users.[243] Interestingly, the presence or absence of timbering activity was not mentioned as a problem by Park visitors at this time. One particular issue, however, was solved quickly.

A shadow over Algonquin was the possibility of uranium mining interests expanding northward from the Bancroft area and potentially impacting on the Park. Premier Frost solved this issue, as it concerned him personally. North of Bancroft lay his favourite fly-fishing area, which was centred on the York River. He simply had Clyde and Bruton townships immediately reserved from mining in 1956. In 1961, these townships officially became part of Algonquin, where mining is prohibited.[244] Trapping and hunting in Clyde and Bruton, licensed previous to the inclusion of these townships into the Park, were permitted to continue (Appendix VII).

Logging in Algonquin and other provincial parks soon became an acute issue, and it proved to be much more difficult to solve.

The Roots of Protectionism for Parks in the 1950s and 1960s

Logging took place in Algonquin Park and on Crown land throughout the province at this time under the terms of licences granted to companies. But methods of logging had changed over the years. In the past, most cutting activity in the bush had taken place in the autumn and throughout the winter; this was followed by an early spring drive of logs to the mills, which in turn was followed by a relatively quiet summer. Cutting by chainsaw was now taking place nearly year round, and the logs were being removed by noisy, diesel-powered skidders and trucks on high-speed, gravelled roads that spread throughout the Park. Tourists, and the interior campers who were visiting the Park in greatly increasing numbers, expected to have a quiet "wilderness experience." Timbering had become obvious and very noisy. The two major users of the Park clashed head on—the recreational canoeists and park protectionists versus the timber interests.

Fuelling this situation was a growing public awareness of the quality of the environment. The ecological awareness that developed in the 1950s was greatly influenced by Aldo Leopold's *A Sand County Almanac*. Leopold exhorted his readers to develop an "ecological conscience," and "to quit thinking about decent land-use solely as an economic problem. A thing is right when it tends to preserve the integrity, stability, and beauty of a biotic community. It is wrong when it tends otherwise."[245]

Following this lead, the Federation of Ontario Naturalists (FON) advocated that legal provisions must be made to set aside natural area sanctuaries and research areas.

W.B. (Ben) Greenwood, the first head of the Parks Division that was created in 1954, welcomed the FON's "Outline of a Basis for a Parks Policy for Ontario."[246] Premier Frost personally drove the Ontario Parks Integration Board to formulate a policy for dealing with all the parks across the province.[247]

Protection became a major objective of

Ontario park planners and three new classes of parks were created—Nature Reserves, Wilderness Areas and Historic Sites. The *Wilderness Areas Act* came into effect in March 1959 and was the direct result of the efforts of the FON. The protected areas were, however, limited to areas less than one square mile—or 640 acres (259 ha). This small size for wilderness areas was not considered to be a problem at the time. Two exceptions were made: Cape Henrietta-Maria, with some 1.8 million hectares (9,500 sq. mi.) on Hudson Bay was renamed Polar Bear Provincial Park and Pukaskwa on Lake Superior eventually became a federal park.

Yet in the late 1950s, the Department of Lands and Forests had no "general policies, management guidelines and master plans."[248] In 1957, the Timber Branch pressured the Department of Lands and Forests into announcing a deer hunt in the Bonnechere and Petawawa valleys in Algonquin because browsing by the numerous deer had caused severe damage to stands of young yellow birch and pine, and was ruining regeneration efforts. A noisy outcry by naturalists and sportsmen's clubs, saying that a hunt would betray the Park's status as a sanctuary for wildlife, brought an end to that idea.[249] The opening, for the use of fishermen, of logging roads to northern and eastern sections of Algonquin in 1958 brought protests by the Conservation Council of Ontario. The Quetico Foundation joined the protest, fearing that roads would be opened in that park. Lack of a firm policy was becoming increasingly apparent. Later, in 1958, the Quetico Foundation petitioned the government for the development of a general policy on parks, and the reservation of over 1.3 million hectares (5,000 sq. mi) of additional parkland in northern Ontario. The FON followed before the year ended by submitting an "Outline of a Basis for a Parks Policy for Ontario," for the first time citing ecological principles—"the interrelationships among all forms of plant and animal life in time and space."[250]

Omand's Attempt at Planning

Pressure by the Parks Branch to resolve these conflicts became a huge headache for the politicians and the Department of Lands and Forests. In May of 1958, the FON chastised the Department of Lands and Forests for having no clearly established policy or master plans for parks. In response, MacDougall instructed the Pembroke District forester, D.M. Omand, to produce a preliminary master plan for Algonquin. The emphasis in the plan was to be the location of future logging roads, road access, preservation of wilderness areas and recreational facilities. Omand's plan was met with withering criticism by the Department of Lands and Forests because no attempt was made to specify what sections of the Park were to be zoned roadless, in other words as no-logging zones. Only after such zoning was established could meaningful studies be done on remaining areas. Omand had been unfairly asked to carry out an impossible assignment because he was given neither the theoretical framework nor the resources to do the job.

Logging in Algonquin and Quetico, and Other Issues in the 1960s

Within the Department of Lands and Forests there was considerable conflict between the powerful Timber Branch, which was backed by the woods industries, and the developing Parks Branch. The Parks Branch insisted that when a park was established park uses became paramount and timber operations became secondary. Instructions from the Parks Branch stated that trees near public areas had to be marked and roads were to be temporary and unobtrusive, as well as gated. Cutting closer than 400 feet to lakeshores and 200 feet to portages was to be selectively done, if allowed at all. However, these measures had no teeth in them to insure compliance by logging companies. For example, in Quetico cooperation of the lumber companies was not forthcoming, and their flagrant disregard of these instructions irritated the Parks Branch and infuriated the recreationists.

In Algonquin, Walter Plonski of the Timber Management Branch recommended that forestry be the first priority in the Park. In 1962, he sought to have the 1944 Wilderness Research Area set aside to allow for timbering of the entire area. He wanted all park management and operating plans to be prepared by the Timber Branch. There was obviously a wide chasm in thinking between the Parks Branch and the Timber Branch.

Some at the higher levels of Department of Lands and Forest and a few individuals in the general public knew that a tempest was brewing. For one, Deputy Minister MacDougall sensed that Algonquin Park was about to be hit by a storm of criticism from all sides; so, in 1966, MacDougall took William Hueston from his position as development supervisor at Parks Branch, and made him Algonquin Park Superintendent, saying "He's my best guy—a hands-on guy." One of Hueston's tasks was to design and write a land use plan for Algonquin Park. But like Omand, he didn't have the necessary resources at hand to carry out the task satisfactorily. MacDougall was forced to retire on reaching 70 years of age in 1966 with many problems unresolved.

The Preservationist Movement and Douglas Pimlott

A major catalyst for change in park management was Douglas Pimlott, a professor at the University of Toronto and chairman of the Canadian Audubon Society's park policy committee. Pimlott, a trained forester, had a well-earned academic reputation for his wolf research in Algonquin, from the days when he was employed by the Department of Lands and Forests. He knew the interior of Algonquin intimately from his wolf research fieldwork and was appalled that so little consideration was being paid to the protection of the natural environment. One major problem was that there was no adequate framework for preservationist policy in legislation. Part of the problem, he knew, stemmed from the fact that most senior positions within the Department of Lands and Forests were held by industrial "foresters, who had been taught to think in utilitarian terms."[251]

Pimlott stated, "It took me years to shake the idea that a piece of wood left to decompose on the ground was being wasted; to shake the idea that it was sinful to let even an occasional over mature stand disintegrate and change naturally."[252]

From his sound academic background and further studies as a biologist and researcher, Pimlott launched a crusade for preservation on the natural landscape. He criticized the *Wilderness Areas Act* as being ineffective, since the limitation of the size of the areas that could be set aside to 640 acres made it impossible to create primitive-class parks. He was chagrined when he found by chance that the Wildlife Research Area in Algonquin no longer had official status because the necessary Order-in-Council had lapsed. He discovered also that a mature stand of rare red spruce in Algonquin Park had been placed under a timber licence and was open to harvesting.

Pimlott aimed for a classification policy from the government. "For every park and nature reserve in the system, he expected a master plan to be fashioned which would identify recreational, natural and cultural features, state the primary purpose for which the area was set aside, and specify mandatory management guidelines."[253]

People of like mind supported Pimlott, such as J. Bruce Falls, a biologist at the University of Toronto and chairman of the FON's Parks and Reserves Committee, and Bruce Littlejohn, the author of "Quetico-Superior Country" Wilderness Highway to Wilderness Recreation.[254] In 1965, the National and Provincial Parks Association of Canada (NPPAC) was established with Gavin Henderson, formerly of the Conservation Council of Canada, as executive director. This organization also backed Pimlott.

Events in the United States also energized thoughts and activities on wilderness parks and ecological problems in Ontario. The 1950s and 1960s were explosive decades in terms of ecological thought and action. In the United States, John F. Kennedy and Lyndon B. Johnson enacted legislation in support of nature and wildlands. The public's ecological awareness was heightened by Rachel Carson's *Silent Spring*.[255] In Canada, Fred Bodsworth's *The Last of the Curlews* and the CBC's *Air of Death* in 1967 had national impact. The cap to these ideas—one that was global in scope—was the American publication in 1972 of *The Limits to Growth, A Report for the Club of Rome's Project on the Predicament of Mankind*.[256]

The Algonquin Wildlands League

On a cold January evening in 1968, just inside Algonquin Park, a group discussed the problems of planning and managing the park. They were protectionists in philosophy. Gathered together were Doug Pimlott, Russ Rutter, a seasonal Park naturalist and canoeing partner of Abbott Conway, Bill Swift of Camp Pathfinder and later owner of Algonquin Outfitters, along with several others. At the conclusion of the evening, Pimlott said that he was going to put together an organization dedicated to changing park usage.

In March of 1968, Abbot Conway argued before the Standing Committee on Tourism and Recreation of the Ontario legislature that timbering interests in Algonquin Park should be phased out and that 50 per cent of Algonquin be zoned as a primitive area. Timbering was recognized as a use and included in a multiple-use zone on the eastern side of the Park (Map 16). Not realizing the large backing that Conway had for the ideas he presented, the committee essentially dismissed what they heard. They were very wrong, for the battle between the conservationists, the status quo elements in the Department of Lands and Forests, and the timber industry was to be joined in full fury and in full view of the public. An organization was needed to drive forward the protectionist agenda and it was soon to blossom.

On June 1, 1968, the Algonquin Wildlands League was chartered at a meeting in Huntsville. Abbott Conway, a former tannery executive from Huntsville, Ontario became president.[257] Other members of the executive were Douglas Pimlott, first vice-president, and Patrick Hardy, secretary. Present also were Fred Bodsworth, Walter Gray, a Toronto public relations professional, and Jack O'Dette of the Ontario Federation of Anglers and Hunters. Spokespersons for the league were always well-schooled, and they were usually one step ahead of their opposition. The league constantly used news conferences and other media means. They were never short on hyperbole, nor were they caught out on their facts.

On July 10 1968, the Wildlands League held its first well-publicized news conference in Toronto proclaiming the first "Algonquin

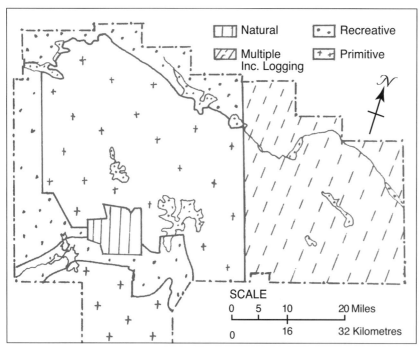

Map 16 Recommendations for Zoning in Algonquin Park: Algonquin Wildlands League, March 1968

Interestingly, the Algonquin Wildlands League map included a multiple-use zone where logging would be permitted. Conway suggested a primitive area that would occupy some 50 per cent of the Park. The Algonquin Wildlands League later withdrew the logging area from its suggested land use in Algonquin.

Source: Gerald Killan, *Protected Places: A History of Ontario's Provincial Park System*, Toronto: Ministry of Natural Resources, 1993, 171. Used with permission of Ontario Ministry of Natural Resources.

Alert." The league objected to "the expansion of logging activities and the spread of lumber roads, the rapid increase of water, air and noise pollution, the presence of various forms of mechanized transport in the interior of the wilderness, and the intrusion of permanent construction into the centre of the park."[258]

The media lapped it up, both in print and on the airwaves.

Such alerts kept coming. There was the reporting of a road crossing a busy Otterslide Creek portage—complete with a stop sign! Then there was the notice of the nine-year renewal of a timber licence for Weyerhaeuser Canada Ltd. covering an area of over three townships before zoning in the Park had been settled.

Hueston's Provisional Master Plan: Algonquin Provincial Park, 1968

In November 1968, William Hueston's Provisional Master Plan was made public in Huntsville, despite receiving severe internal criticism previously from the Department of Lands and Forests.

Recreational Planning Supervisor James Keenan said the plan was lacking in the imagination and initiative required to chart a new course for this important area. "Surely our economy can afford to preserve a more meaningful proportion of Algonquin Park than has been recommended," he stated. Timber management policies, however, suggested only a long-term continuation of the status quo, something preservationists had already deemed inappropriate. The proposed policy on outboard motors also raised Keenan's ire: "Surely they can be prohibited on at least some of the lakes outside of the Primitive Zone," an area which included less than 5 per cent of the Park. Interpretation Supervisor Grant Tayler shared these concerns and added that "there appears to be little thought given to the need for preservation of large ecosystems."[259]

The reaction to Hueston's plan was predictable. Gerald Killan, in his book *Protected Places*, wrote that the plan

> rested on shaky foundations and was not supported by sufficient interdisciplinary research and analysis. Comprehensive inventories of the earth and life science features of the park had not been made; consequently, the zoning had been based on incomplete ecological data. No economic impact studies had been conducted to establish the cost-benefit ratios of various zoning options. Without such research and analysis, no one could possibly assess the accuracy of the contradictory claims in the logging versus wilderness controversy. Only with such data could park officials and politicians make rational decisions about the size of the primitive zone and the future of logging. The provisional master plan also suffered from serious gaps in basic information about visitors to the park. Further, it contained no estimates of user demand in the future. No carrying capacity studies had been undertaken to sort out the problem of distributing canoe trippers to reduce overcrowded conditions in the park interior and to minimize conflicts between

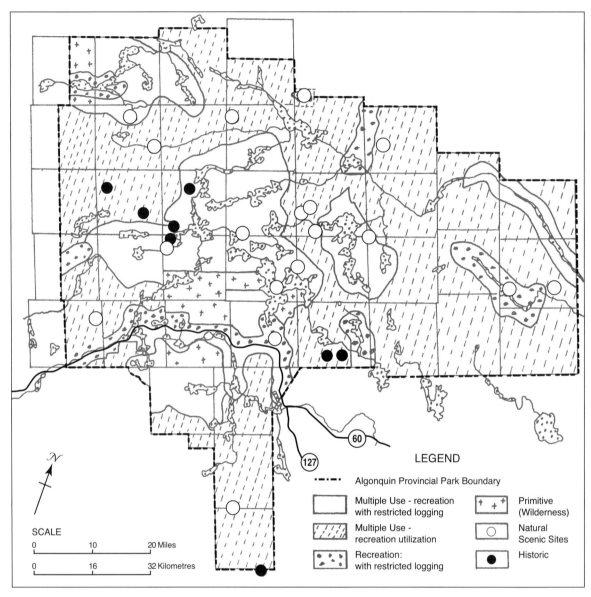

Map 17 **Algonquin Provincial Park Provisional Master Plan (William Hueston, 1968)**
Source: Algonquin Provincial Park Archives

canoeists and motor boat fishermen. Unless the Department of Lands and Forests addresses these matters, concluded James Keenan, park planners did not stand a chance of establishing a politically acceptable rationale as to the regional and provincial significance of Algonquin, from which would evolve clear statements of purpose, objectives, and policies, and a zoning plan for the park.[260]

Chapter 9: Reconciling Timber Harvesting

In 1969–1970, timber licences, held by some 20 companies (Map 18) covered much of Algonquin Provincial Park. Some of these were held by the McRae Lumber Company. In 1973, the timber from Algonquin Park supplied wood to 12 sawmills, 4 veneer mills, 1 pole plant, 1 pulp mill and 1 splint (matchstick) plant. Park-based companies employed a

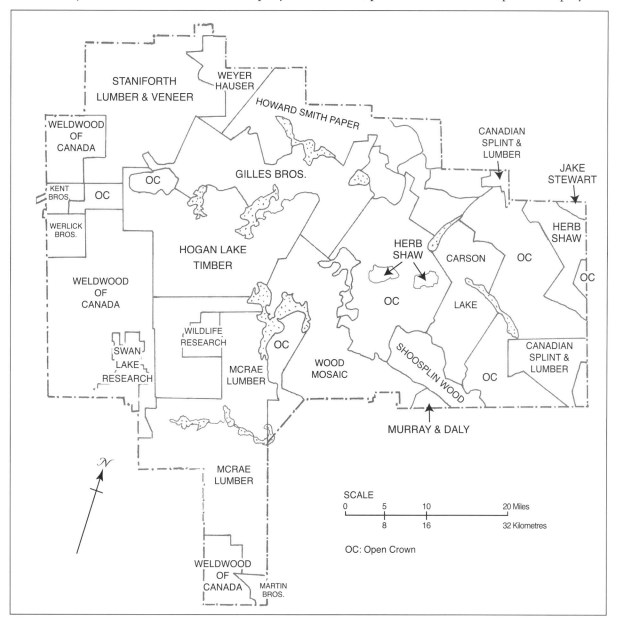

Map 18 Timber Licences: Algonquin Provincial Park, 1969–1970
Source: Lands Surveys Branch, Department of Lands and Forests, 1969

significant percentage of the local labour force and produced a valuable portion of Ontario's wood products. As such, the mills dependent upon Park wood were economically significant, as they were commonly the major employers in the small towns surrounding the Park. At the upper economic level, mill owners had their political contacts, while at the worker level, votes in the small lumber-mill towns were very significant to local MPPs.

Urban-based MPPs, however, were sensitive to the interests of their very mobile, recreation-seeking constituents. These contrasting positions made for interesting times in the Ontario legislature and within the Conservative governments of Leslie Frost and John Robarts, before the situation was quieted in the early days of the premiership of William Davis.

Leslie Frost resigned as premier of the province in 1961, and finding solutions to the land use issues in parks fell to Premier John Robarts and his government. The Robarts government was, at the same time, heavily embroiled in the process of dealing with several other difficult issues, including regional government, tax reform and mandatory school board reorganization. The park issues were not appreciated but were finally met head-on, and the needed resources were allocated.

In December 1968, the Department of Lands and Forests created a special Algonquin Park Task Force (APATAF), composed of senior administrators from every branch of the department, to develop interim policies and to produce a final Master Plan. Rene Brunelle, Minister of the Department of Lands and Forests, issued a series of new restrictions on logging in Algonquin as an interim measure in April of 1969.

It's the evidences of logging such as the noise of trucks, bulldozers, and power saws, as well as roads and bridges in certain areas, which have proven to be the most offensive to the recreationists. Through new restrictions, the minister hoped to minimize "these irritations to the canoeists." Along 1,175 kilometres of designated canoe routes, a 1,500-foot (457-m) reservation was established, in which only marked trees could be cut. No logging was to be allowed within 100 feet (30 m) of shorelines or within 200 feet (60 m) of portages. The department also created a two-mile-wide "sound buffer zone," [later reduced to 1.5 miles] within which no cutting would be permitted. Road construction was prohibited in all reservations, and all road locations and construction specifications required the approval of the superintendent. The only concession made to the 23 companies operating in the park was the permission to cut during the summer months; however, the hauling of logs would be permitted only from 8:00 a.m. to 5:00 p.m. Monday through Friday. No hauling would be allowed on weekends or holidays, and no mechanical equipment was to be used at night.[261]

The government hoped that these measures would give it some peace for a while, but such was not the case. An Algonquin Wildlands Alert publicized a serious water pollution problem at the Lake of Two Rivers campground. Here, a crush of recreational campers had been located on the abandoned airfield; they were serviced by hastily dug earth toilets located too close to the water. While the problem was dealt with quickly, the publicity was not good, and the pressure remained on the government.

To cope with the controversy, the Robarts government resorted to establishing a special study to develop an understanding of the situation and to make appropriate recommendations. This was the Algonquin Park Advisory Committee. Chaired by former premier Leslie Frost, its 15 members were representative of all three parties, plus conservation groups, the logging industry, park leaseholders and the Ontario Camping Association. Frost determined that the committee would work *in camera*. With these measures, the Algonquin controversy quieted down. But the Algonquin Wildlands League continued to plague the government.

In Killarney Provincial Park, a large new logging road was discovered. The Algonquin Wildlands League demanded that the park be reclassified as Primitive, thus preventing logging. The government said that it would remain a Natural Environment park, but 53 per cent would be zoned Primitive, with no logging. It was then found that the government had permitted construction of the road before any zoning plan had been formulated, and the conservationists asked why they had not been consulted in the planning of the park. After allowing the conservationists a seat in the planning process for Algonquin, the government found they could not deny them the same position relative to other provincial parks.

In October 1969, Chief of Parks Peter Addison, acting on this realization, informed his superiors that henceforth his branch would develop a standard procedure in master-planning large parks like Killarney, Lake Superior, and Quetico. Public submissions would be invited at the outset. A task force of experts from all branches of the Department of Lands and Forests would be appointed to co-ordinate research and analysis and to prepare a preliminary planning document. Before final authorization of the plan, the public would once again be consulted.[262]

The issue of timbering again arose with reference to both Quetico and Killarney parks. Both ended up being designated as Primitive-class wilderness parks; no further logging was permitted in them. The Algonquin Wildlands League and other conservation groups had carried the day in these parks.

Lake Superior Provincial Park then made news in 1970. Here, a licensed timber operation was being permitted to harvest an area of ecological significance. Alerted to this situation, the Wildlands League sent in a qualified team to investigate the situation. Their report was sufficient for the Ministry to send in Shan Walshe, an outstanding provincial botanist/naturalist, and he confirmed "that the area in question was worthy of designation as a natural zone."[263] Logging was stopped and the site saved by the intervention of protectionists. Again the Department of Lands and Forests and its management, planning capability and control over timber operations were made to look inadequate.

On December 8, 1970, Premier John Robarts announced his resignation from politics. Three months later, on March 1, 1971, William Davis became premier. On May 11, 1971, it was announced in the legislature that there would be no further commercial logging in Quetico Provincial Park. The Algonquin Wildlands League and other conservation groups had carried the day again.

One day later, Rene Brunelle released the report of the Frost Advisory Committee. The protectionists, led by the Algonquin Wildlands League, did not achieve the elimination of logging in Algonquin Park. Algonquin retained the designation, Natural Environment, with multiple land uses allowed. As such, Algonquin was and is the only provincial park to keep MacDougall's philosophy alive.

Why was this decision made?

> For social and economic reasons, the report indicated, commercial logging could not be removed from the Park. "Recognizing the strategic importance of forest resources in the economy of the region and the Province, the forests of the Park will be managed to ensure a continuing supply of wood products while safeguarding the recreational values involved."[264]

How did the preservationists react?

The decision caused a split in the ranks of the preservationists. Abbott Conway of the AWL termed the result, "the death knell for Algonquin ... the most arrogant sell-out of the people and their heritage we've come across yet."[265] But the NPPAC, under Gavin Henderson, accepted the decision and the protests collapsed. With the Master Plan released in the fall, the government moved on to other matters.

The Algonquin Park Master Plan, 1974

In October 1974, the Algonquin Provincial Park Master Plan was released. Bill Calvert, then of the Parks Branch, wrote the document with Leslie Frost. Arthur Herridge (then head of the Timber Branch) and Walter Giles (the Assistant Deputy Minister at the Ministry of Natural Resources) served as a steering committee. While neither the timber interests nor the Algonquin Wildlands League or other conservationist/protectionist groups liked it, the Davis government accepted it, and, as such, it was tabled and passed by the legislature.[266] The Master Plan was the compromise.

> The aims of the Master Plan were: 1) to maintain the economic base for local communities and to continue to provide Ontario residents with a diversity of recreational opportunities; 2) to provide continuing opportunities for a diversity of low-intensity recreational experiences within the constraint of the contribution of the Park to the economic life of the region.[267]

Two points should be recognized. While they didn't achieve the elimination of timbering in Algonquin, the pressure generated by the Algonquin Wildlands League and other conservation/protectionist groups was considered by some foresters to be essential in forcing better silvicultural practices on the timber industry, including such practices as partial cutting, which is employed in both the selection and uniform shelterwood systems. Another essential contribution that is seldom stated is that the pressure mounted by the recreationist/conservation movement enabled the Parks Branch of the Ministry of Natural Resources to better face the very powerful Timber Branch. Later, the creation of Ontario Parks in 1996 out of the former Parks Branch strengthened the position of Parks within the MNR.

Description of Zones in Algonquin Provincial Park, 1974

Algonquin Park is designated as a Natural Environment park. It is zoned and managed in accordance with policies for this class. Parks in this class provide a wide variety of non-intensive recreational opportunities within an environment of educational, recreational and scientific significance.

Through zoning, long range programs ensure three principal objectives: preservation, visitor use and forest production supply.

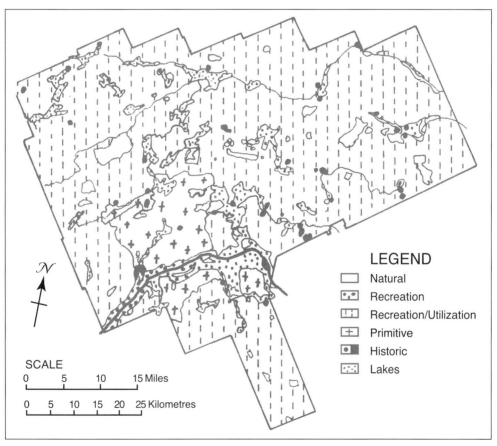

Map 19 **Algonquin Provincial Park Master Plan, 1974: Zoning**
Source: Adapted from Ontario Ministry of Natural Resources, *The Management Plan, Algonquin Park*, Toronto: Queen's Printer for Ontario, 1974.

Algonquin is a park—and the Master Plan was the compromise between the protectionist and timber interests. Is there any significance in the prioritizing of the land use categories?

Establishment of natural, historic and primitive zones was given priority over access, development and recreational zones. In turn, all of these zone types were given priority over recreational/utilization zones within the limit determined by the current contribution of Park forest products to the region's economy.[268]

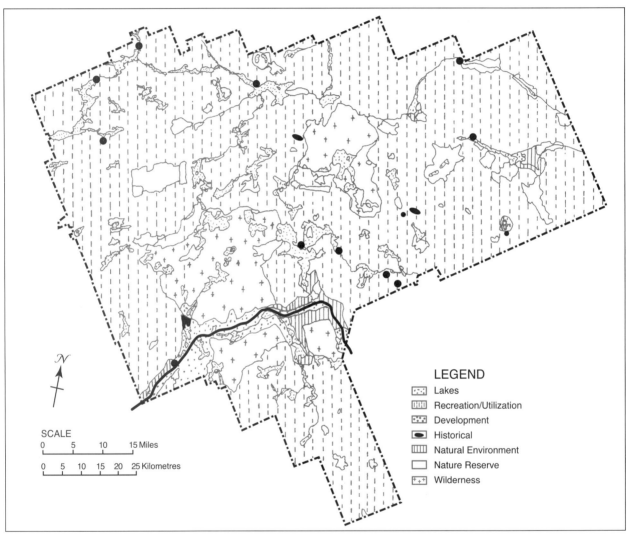

Map 20 **Land Use Zones in Algonquin Provincial Park: Management Plan, 1998**

The major change between the zoning map in the Master Plan of 1974 and this map from the Management Plan of 1998 is the addition of the Lavieille/Dickson wilderness area.

Source: Ontario Ministry of Natural Resources, *Management Plan: Algonquin Park*, Toronto: Queen's Printer for Ontario, 1998.

Legend:

Natural Zones: 63 natural zones totalling 71,730 acres have been established after five years of intensive scientific field investigation. This will be expanded to about 80,000 acres (32,390 ha).

Historic Zones: Historic and archaeological sites are included in the 48 sites.

Primitive Zones: These areas include the best natural wilderness areas existing in southern Ontario because evidence of technological and industrial impact are almost non-existent since 1940. These areas represent the best natural landscapes in an ecological sense and include numerous large lakes with a large canoe-camping capacity.

Recreation Zones: These zones include campgrounds, interpretive areas and concessions. They also include land used for walking trails, opportunities for day canoeing, and other low intensity recreation sites.

Access Zones: From these areas visitors may disperse into the Park interior in a controlled manner.

Recreational/Utilization Zone: This large zone of 1,527,000 acres (618, 244 ha) provides low intensity canoe-camping opportunities (81 per cent of the Park). While forest harvesting takes place in this area, its presence does not impact negatively on the wilderness experience for most people as zoning and the separation by sight and sound separates recreationists from timber operation.

Source: Ontario Ministy of Natural Resources, *Algonquin Provincial Park Master Plan*, Toronto: Queen's Printer for Ontario, 1974.

Over two decades later, The Management Plan, Algonquin Park was published in 1998. It brought up-to-date the changes that had taken place over the years.

Legend:

Nature Reserve Zones: Some eighty-eight zones include significant landform types or rare floral and faunal species. Field notes by specialists that examined them are stored in the Park archives at the Visitor Centre.

Wilderness Zones: Four areas of this class have been designated. These zones protect entire landscapes and show little evidence of technological and industrial impact.

Historic Zones: Forty-eight areas include any significant historical resources that require management distinct from surrounding areas: for example, logging is excluded.

Development and Natural Environment Zone: These areas provide the main access to the Park. Facilities and services provide a wide range of day use and camping facilities including the interpretive facilities of the Park.

Recreational/Utilization Zone: This zone contains the commercial forestland in the Park. Few signs of such use are seen by those who use the backpacking trails or paddle the rivers and lakes of the area.

Source: *Management Plan: Algonquin Park*, 1998.

Table 9 **Land Use Zones in Algonquin Provincial Park, 1998** (Includes water areas)

Zone Type	Area (ha)	Percentage of Park Area
Nature Reserve	39,250	5.1
Wilderness	90,475	11.9
Natural Environment	13,765	1.8
Historical	1,680	0.2
Development	22,545	3.0
Access	735	0.1
Recreation/Utilization	594,860	77.9
Total	**763,310**	**100.0**

Source: *Management Plan: Algonquin Park*, 1998.

Changes to the Land Base of the Timber Cutting Area

In 1990, some 25,000 hectares of the Recreational/Utilization zone were incorporated into the Lavieille-Dickson Wilderness Area.[269] In 1993, the McRae Addition of 6,312-hectares was added to the Park and included into the Recreational/Utilization zone.[270] This area comprised the "English Lands" in Eyre Township that J.S.L. McRae had bought for taxes back in the 1930s and later sold back to the province. As of 1998, the Recreational/Utilization zone totals 594,860 ha or 77.9 per cent of Algonquin Provincial Park.

CHAPTER 10

The Algonquin Forestry Authority: Making a New Management System Work

The major promise by the government to the mill owners in the Master Plan that cancelled all timber leases in Algonquin Park was that they would get the logs necessary to keep them in business. In a little over two months the legal framework was set up and the Algonquin Forestry Authority (AFA) came into existence in January 1975.

The Algonquin Forestry Authority is an Ontario Crown agency that reports to the Minister of Natural Resources. The AFA was subject to the *Crown Timber Act* and conditions detailed in the Master Plan. Today it follows the *Crown Forest Sustainability Act*.

On January 2, 1975, the *Algonquin Park Forestry Authority Act* was passed by the legislature. It contained the following objectives:

a) subject to the *Crown Timber Act*, to harvest Crown timber and produce logs therefrom and to sort, sell, supply and deliver;
b) to perform, to undertake and carry out such forestry, land management, and other programs and projects as the Minister may authorize, and to advise the Minister on forestry and land management programs and projects of general advantage to Ontario.[271]

To establish the Algonquin Forestry Authority as a functioning entity, a team led by D.J. Vance worked with the former mill licence holders to determine their requirements. A consulting team for the implementation of the Algonquin Forestry Authority was also created.

In their final report, the Consulting Team for the Implementation of the Algonquin Forestry Authority recommended that "the Authority should operate as a normal commercial operation, responsible to the Minister for the operation of its licence, which is the total area of Algonquin Park."[272]

The committee spelled out in detail the organization structure, the necessary start-up program, primary tasks, estimates of the costs involved and how the organization's finances were to be managed. They recommended that "initially the Authority should utilize existing operators to the greatest extent possible" for harvesting operations. A major concern of the Authority would be maintaining jobs in communities that had historically depended on park wood for their livelihood.[273]

The first board of the directors of the Algonquin Forestry Authority proved to be a capable, hard-working, diverse group that was determined to make the AFA work well. Vidar Nordin, Dean of the University of Toronto's Faculty of Forestry, was the first chairman. "He was respected, able and acted with great style."[274]

Photo 79 **First Board of Directors: Algonquin Forestry Authority**
Back Row: Frank K. Roberts, George B. Priddle, Gordon Godwin, George B. Aballah, Harry J. Searson, Arthur Lock
Front Row: Bernard B. Reynolds, J. Wes McNutt, V.J. Nordin, Ward Smith
Source: AFA Archives

One of the first acts of the board of the AFA was to find a general manager. At a meeting of the search committee one member withdrew from the interview room saying that he knew I.D. (Joe) Bird and it would not be correct if he stayed. But on leaving he said, "If we don't hire Joe Bird, we are damn fools."[275]

Joe Bird was a graduate of forestry from the University of New Brunswick. Before coming to Algonquin he had achieved the position of woodland manager for the Quebec North Shore Paper Company in Baie Comeau. This was the corporate world at a significant level.

He was also well connected with the Ontario Forestry Industries Association (OFIA). Bird was well-known before taking over the AFA, both in the industry and in professional forestry circles. While other qualified foresters shied away from the position as being an "impossible one," Bird accepted the challenge.

As a manager, Joe Bird had a marvellous way of dealing with individuals and with groups. On May 1, 1976, a general meeting was held in Huntsville between timber operators, industry clients, the AFA Board of Directors and AFA personnel. Tensions were high. Joe Bird spoke to the group. While they may still have had doubts, by the end of the meeting the skeptics were shaking Bird's hand.[276]

He was a respected and shrewd businessman. In difficult discussions, he was a low-key negotiator, letting others make their points. Then he would give his position and he usually came away with some type of agreement. He not only understood industry and the pressures that CEOs were under, but he also had a deep understanding of labour.

Bird needed all of these qualities, for he had to deal with client mills whose owners, resentful at having their cutting licences taken away, were being forced into dependency on an organization they didn't think could succeed. For the mill owners, this was hard to swallow—they were no longer independent. In addition, things were being done differently and more expensively, for the AFA charged the mills for its services and the logs it delivered, while mills outside of the Park continued to receive services from the MNR without charge. The former licence holders/mill owners had no choice—it was a case of adjusting or folding up their businesses.

The number and complexities of the tasks facing the board and Bird were monumental:

1. To hire a staff from scratch that had the strength and resiliency to cope with the problems that he knew the AFA would face;

2. To operate the AFA as a normal business. It was made very clear by the government that the AFA was not to cost the treasury money;

3. To supply the logs required to individual mills that had diverse needs;

4. To achieve the efficient use of wood from the forest—formerly licencees cut their licences to satisfy their specific needs rather than for best use;

5. To assess the resources of the Algonquin hardwood forest for which there was no directly useable inventory method;

6. To adjust to new cutting regulations mandated in Master Plan tree marking, the selection system in maple and uniform shelterwood in white pine, all supervised by the MNR [Don George];

7. To cope with the fact that the Algonquin forest under mandated cutting procedures was yielding a large volume of low value pulpwood for which there was initially a very limited market;

8. To deal with an MNR field staff that at best was dubious of the AFA, believing that the AFA's prime motive was satisfying "the bottom line", for which the AFA would if necessary sacrifice forest and recreational values that the Park (MNR) was charged with protecting;

9. To effectively rationalize a road network developed previously by each individual mill into a system of roads that would efficiently service the total licence area while at the same time satisfy the protectionist/concervationist/recreationalist concerns. Roads are still a point of controversy.

Joe Bird and the Early Years of the Algonquin Forestry Authority: 1975–1985

A major key to the AFA's success was Bird's remarkable eye for picking talented professionals who also had people skills. His initial hirings, Ray Townsend, Bill Brown, Brent

Photo 80 **Four Key Algonquin Forestry Men**
Bill Brown, Brent Connelly, I.D. (Joe) Bird, Ray Townsend
Source: Algonquin Forestry Authority

Connelly, Carl Corbett and Bob Pick, formed the core of the AFA team that delivered the promised logs to a very skeptical group of mill owners. These men, along with others, went on to put in place cutting and silvicultural methods for the Algonquin forest that are now showing signs of producing more and better logs and have won the respect of mill owners and even some conservationists.

In 1979, Joe Bird wrote,

> The Authority is incorporated without share capital, and thus is entirely debt financed. Working capital is borrowed from the provincial treasury at prevailing rates of interest. The loss of $500,000 experienced in the start-up year, 1975, was effectively written off through a government grant following a voluntary commitment by the Authority to be self-sustaining by 1979.

Results in 1976 improved to the extent that the exercise ended in the black with a profit of $9,000 on sales of $3.6 million. The government was repaid $300,000 of term loan principle and $42,000 of interest. Operations in 1978 are expected to break even on sales of $4.5 million and repayment to the government of the remaining $200.000 of term loan principle.[277]

The AFA achieved self-sufficiency one year ahead of schedule. Expenses were recovered by the sale of products to client mills. This success was duly noted by the Ministry of Natural Resources and the politicians.[278]

In its first year of operation, the AFA had the former licensees carry out the cut on their former licences. The trees had been marked by the MNR. By the following year, the AFA had second-party agreements with mill owners, who continued to cut trees marked by the MNR.

One of the first major tasks facing the AFA

Chapter 10: The Algonquin Forestry Authority: Making a New Management System Work

Table 10 **AFA Allocations of Wood to Former Licensees, 1979**

Company	Hardwood Sawlogs	Softwood Sawlogs	Veneer Logs	Pine Poles	Pulp Wood
Weldwood Canada	6.00 M	0.40 M			
McRae Ltd.	6.00 M	1.00 M			
Canada Veneers	4.50 M	1.75 M	3.24 M		
Consolidated Bathurst		9.00 M			
Goodman Stanforth	4.50 M	2.00 M			
Carson Lake		4.00 M			
Canada Splint	2.00 M**				
Shaw Bros.		2.50 M		1.50 M	
Murray Brothers	3.50 M	3.50 M			
Pembroke Lumber		1.50 M			
Kent Brothers		0.50 M			
G.W. Martin	1.50 M				5.00 M
Sklar	2.00 M				

** Poplar Splint (match wood)
M Million board feet
Source: Consulting Team for the Implementation of the Algonquin Forestry Authority, "Report," 1979.

was that of assessing a hardwood forest for which there was no uniform inventory. While the previously mentioned Forest Resources Inventory created by the province was a great advance and resource, it lacked the detail necessary to formulate cutting plans in the hardwoods of Algonquin. Researchers at Swan Lake had previously encountered this problem, and Mac McLean and Don George along with biomathematician "Doc" Raymond devised a prototype cruise design for hardwoods for cutting under the selection system in 1968.

Bill Brown of the AFA, with Dr. J.A. Mervart of the MNR and staff from the Timber Branch, also played a leading role in the design and data analysis of a new management plan. They "used information in the Forest Resources Inventory (FRI) cruise data to develop the first volume estimates for Algonquin Park, by adapting the cruising method developed by Mac McLean and Don George."[279] This work was applied to the uniform shelterwood (pine) as well as to the selection system (hardwoods) mandated by the Master Plan.

The forest inventory of the recreation-utilization zone in Algonquin Park is the database used by the AFA in the preparation of the Forest Management Plan. From this inventory, the total allowable annual cut is calculated and from that the share of wood for each mill that previously had a licence is allocated. The annual cut is maintained at less than the estimated growth of the growing stock in one year.²⁸⁰ This information is critical for the development of the following plans.

An Operating Plan Cruise (OPC) was undertaken to provide the data to write the Five-Year Operating Plan and also to generate information for the annual harvesting program. This information was vital when making up logging cost appraisals for contractors and was even used to set the cut and skid rates. The amount of cruising (averaged over 20 years) has amounted to 12,990 hectares (32,100 acres or 50 square miles) per year.²⁸¹

Bill Brown wrote the first 20-year Forest Management Plan for all of Algonquin Park for the period between April 1, 1980 and March 31, 2000. This plan developed the mechanics of allowable cut calculations for both the uniform shelterwood (mainly eastern hemlock and white pine) and the selection system as mandated by the Master Plan. Volume estimates for the Park are now calculated by using FRI data. Methods developed in this plan are still in use today and have been adapted for use by other areas in the province.²⁸²

A Five-Year Operating Plan, again a first of its kind, was submitted in April 1981. Along with 1980 data from a planning cruise, it gave the allocations of the allowable cut and calculations of product volumes. On this basis, wood allocations to individual mills were made. The mills received essentially the same amount of wood from Algonquin as they had received previous to 1975 (Table 10).²⁸³

Within the civil service, some senior personnel and many members of the MNR field staff distrusted the AFA, while the AFA didn't like being told how to do forestry. For the AFA personnel, the poor relationship simply heightened their desire to make it work. Bird was the leader. Without him it might well have failed. It didn't, and his organization, the AFA, has continued not only to deliver logs to the mills, it has accumulated a highly respected history of service to the forest industry.

In 1974, the Master Plan mandated tree marking and the application of selection and uniform cutting procedures, which in large part replaced the long established cutting criteria of diameter limits, or high-grading. This change in cutting procedure was protested by many of the mill owners, as they were only interested in good logs. For a time, the AFA had great difficulty in meeting its commitments to the mills and the mills had difficulty in meeting their production requirements.

When the AFA took over the management of cutting operations, it was subject to the new cutting rules as interpreted by the MNR and its tree markers. There were difficulties initially matching what was marked for cutting by the MNR with what was cut under AFA direction. John Simpson, who was the Algonquin Park superintendent between 1975 and 1980 reflected that "the AFA, at that time at least, appeared to us to be more interested in the bottom line than it was in good forest management."²⁸⁴

But the MNR hierarchy was determined to have the AFA succeed, and Don George was constantly being told to get along with the AFA. The problems were essentially ended when the Interim Forest Management Understanding (IFMU) was signed between the AFA and the MNR in 1983. This agreement was made permanent in 1985. Don was not pleased with this change in policy because he did not trust the AFA to carry out the prescribed silvicultural procedures in the agreement. The AFA to Don was simply big business. He believed that if big business executives had to make a choice between a procedure that was good forestry and one that was good for the bottom line, the latter would win out every time. Don George commented, at the time the IFMU was passed in 1983, "We have made the goat the gardener."[285]

In time, the number of infractions by AFA personnel or their jobbers diminished in number as experience in the new system took hold.

In April of 1985, Joe Bird, resigned from his position of general manager of the AFA and took up the position of president of the Ontario Forest Industries Association. On leaving the AFA, Bird said to Bernard Reynolds, then chairman of the board of the AFA,

> I can say without reservation that my ten years with the Authority have been the most fulfilling of my career. The job of starting up an organization from scratch in a climate of controversy has been exciting and satisfying. It has been made possible by the people involved and I will always treasure the association I have enjoyed with the Chairman, directors and staff. I regret leaving these associations which have been enriching both professionally and personally.[286]

Asked to comment on Joe Bird, people who knew him, people who worked for him, and even those who didn't like the AFA, inevitably start with thoughts such as: "a good family man," "old school politeness," and "respected, likeable, ethical."

Carl Corbett, the present general manager of the AFA, fondly remembers Bird as a mentor, and Carl now passes along to his younger staff advice that he received from Bird. "You will be involved in difficult situations—be professional and don't take issues personally. Bird would not accept work of poor quality."[287]

Ray Townsend, retired operations superintendent of the AFA, said of Joe Bird, "Joe was a good forester, and the best manager I ever worked for: diplomatic, knowledgeable and fun loving."[288]

John Simpson recalled that "Joe Bird and I had a number of disagreements but to both our credits, we remained on a friendly basis."[289]

By the 1990s, working relations between the AFA and MNR field staff had smoothed out considerably. Ray Townsend wrote,

> Today there exists a feeling of mutual respect (between the AFA and MNR) found lacking in the earlier years and working together to solve problems is a common occurrence.[290]

AFA Relations with the Ministry of Natural Resources

The Interim Forest Management Understanding (IFMU) of 1983 between the AFA and the Ministry of Natural Resources brought a major change to forestry operations in the Park. This agreement gave the AFA full control of both cutting and forestry operations in the Park. The MNR retained a "right of approval" role.

At that point in time, the AFA had satisfied senior MNR officials that it could carry out the total forestry program in Algonquin Park. The MNR retained overall supervision and its expenses were slashed dramatically, while the government retained much of its income. The Interim Forest Agreement of 1983 in many ways foreshadowed the *Crown Forest Sustainability Act* that was passed in 1994. This act created Sustainable Forest Licences, which are run as private businesses and are owned by the timber industries of their areas. SFLs look after planning and forest management, and monitor forest operations. The MNR monitors and audits all activities and the government receives funds from stumpage.

Since 1983, the AFA has had control over both harvesting and management (silviculture). However, safeguards were put in place. An independent audit of AFA operations is carried out every five years.

The advent of the IFMU proved a very difficult time for many MNR employees in Algonquin. Some were let go, while others were forced to relocate. For example, the large MNR east-side Petawawa Management Unit at Achray was essentially closed down. Gone were the road construction crews, forestry personnel, cooks, tree markers and scalers. Contract workers replaced them, when they were needed, along with three AFA supervisors. As one commentator said, "The MNR presence went from being steak to baloney overnight."[291]

Critics vary in their comments on this agreement. Some 20 years later, Bob McRae stated that the "AFA's work in the bush, both in harvesting and in silviculture, was a considerable improvement over that previously done by the Ministry of Natural Resources, however at a higher cost to the company."[292]

New Ways of Doing Business: The Interim Forest Management Undertaking

The 1980s saw the creation of the first of a series of acts by the province that, by the turn of the century, transferred many of the expenses of maintaining Crown forests directly to the forest industries, while allowing the government to maintain ultimate control.

Interim Forest Management, 1985

In 1976, the influential Armson report, *Forest Management in Ontario*, noted that good forest management involved both harvesting and regeneration and that harvesting should be the first step in forest renewal. When established in 1975, the AFA was responsible only for harvesting and planning, but in 1981 the AFA took over responsibility for tree marking, road maintenance, silviculture and scaling from the MNR, which retained an administrative and

Chapter 10: The Algonquin Forestry Authority: Making a New Management System Work

regulatory role. This Interim Forest Management Undertaking Agreement came into effect in Algonquin Provincial Park in July 1983. It was signed by then-Chairman Bernard Reynolds for the AFA. Negotiations leading to this agreement were chaired by Joe Bird, then general manager of the AFA, and Tim Millard, superintendent of Algonquin Provincial Park for the MNR. This interim agreement was called an "Agreement between the Minister of Natural Resources and the Algonquin Forestry Authority to Undertake Forest Management Activities in Algonquin Park." This agreement was made permanent in June 1985 by Chairman Bernard Reynolds, Frank Parrott, Bill Brown, then general manager of the AFA, and Mike Harris, who was then Minister of Natural Resources in the government of William Davis.[293]

This was a historic first in the province, for forest management had previously been included in the MNR budget presented to the legislature within the Consolidated Revenue Fund. Because this was considered a budget item, forest management funds were not always specifically allocated to forestry. At the bush level, this often caused awkward delays and sometimes even cancellation of pending tree-tending projects that were needed in the woods. With the passage of the new agreement, normal silvicultural work would be funded by the retention of some of the normal stumpage dues charged to the mills supplied by the AFA. The government, however, maintained both legal control, through monitoring of the forest operations, and flow of income from stumpage fees to the provincial treasury.

Environmental Assessment Act, 1988

This bill required the compliance of the MNR in all its activities. In 1988, the Environmental Assessment (EA) Board began an assessment of the Crown land timber management planning and activities on productive forest land, known as the Areas of Undertaking. Crown lands total 34.1 million hectares, of which 26.0 million hectares is productive forest. This legislation was passed by the William Davis government, Vince Kerio, minister of natural resources and George Tough, deputy minister.

Six years later the board released its findings in the *Class Environmental Assessment by the Ministry of Natural Resources for Timber Management on Crown Lands in Ontario.* The EA Board approved the MNR's planning process with 115 terms and conditions.

The Crown Forest Sustainability Act, 1994

The *Crown Forest Sustainability Act* of 1994 made sustainable forestry the law on Crown land in the Areas of Undertaking. These areas were divided into 68 management units as of April 1, 1999. Two types of licences are issued under the *Crown Forest Sustainability Act:* the forest resource licence, and the far more common sustainable forest licence. The board's objective was "to ensure long term health of the forest so that all the benefits of Crown forests are available to future generations."[294]

Forest industries located within former MNR management units were forced to form companies in what were then termed forest management units (FMUs) and qualify to hold a sustainable forest licence. This licence is renewable, and is adjusted every 5 years up to

197

a period of 20 years, provided the licensee has met all terms and conditions—including an independent forest audit. This licence permits the licensee to continue harvesting trees on Crown land. Annual provincial audits and regular five-year independent agreement reviews were pioneered in Algonquin by the AFA and MNR, and are now standard practice across the province under the *Crown Forest Sustainability A*ct of 1994.

The FMUs are required to carry out, at their expense, silvicultural operations formerly covered by the MNR. Money is also paid into the forest renewal trust fund from each management unit and these monies cover silvicultural expenses in Crown forests destroyed or damaged by fire or natural causes. This legislation was passed by the Bob Rae government, Howard Hampton, minister of natural resources and Bob Milton, deputy minister.

In 1999–2000, the Ontario government collected revenue of $249.6 million for wood harvested; $102.7 million went directly to maintenance and renewal of the forest. Currently, monies from the forest futures trust are being used to help control, and it is hoped, eliminate the Asian long-horned beetle that has infected an area in northwestern Toronto. Maple trees have been the major target. If this infestation spreads into the commercial maple forest including Algonquin, it would be a monstrous disaster. The bulk of the money for this project is coming from the federal government. Another destructive pest is the emerald ash borer. This insect has cut a wide swath in southwestern Ontario and is still spreading despite efforts to control it.

While the ultimate control of Crown forests still legally rests with the Ministry of Natural Resources, the industry has become essentially self-monitoring, albeit with MNR inspections and non-compliance reports, as well as independent audits every five years. As a result of these acts, the Ministry has been able to cut many staff positions by passing along to the industry responsibilities that it formerly covered. The public still pays for them—previously through taxes and now through higher costs passed along to the consumer.

The Timber Management Planning Process

In response to *the Environmental Assessment Act*, the MNR developed the Timber Management Planning Manual in 1986. Carl Corbett, of the AFA, authored the first Timber Management Plan for Algonquin Park using the new Timber Management Planning Manual. The 1990–2010 plan outlines the harvest areas, primary and secondary road construction locations, areas of supplementary silvicultural work, areas sensitive to forest management activities, and management systems to be used within Algonquin Park. Approval of the plan was given in March 1990 following public participation and review.[295]

The Algonquin Park Independent Forest Audit: 1997–2002

Prepared by KBM Forestry Consultants of Thunder Bay, Ontario, this audit report presents the most current analysis of forest operations in the Park. This type of audit is now also required of the sustainable forest units that cover

Chapter 10: The Algonquin Forestry Authority: Making a New Management System Work

all Crown forest lands of the province. A summary table of recommendations, suggestions and best practices is given in Appendix VI.

The audit was tabled in the legislature in March 2003 and is there for all to see and comment upon. Basically, it was considered to be a "good report" for the AFA. Yet there are always questions of levels of inquiry and application of standards. Where is the line drawn between acceptable and not acceptable? You can't satisfy everyone all of the time.

Algonquin is a provincial park, and the superintendent has on staff a professional forester whose job it is to monitor the field operations of the AFA. This task is impossible for one person to carry out satisfactorily. The MNR replied to this criticism by KBM by providing funding to Algonquin Park for one full-time senior compliance technician and a winter forest compliance technician.

Cultural information for the Park covers archaeological sites as well as early sites connected with lumber activities. This information is variable in quality and coverage. At times, personnel have not known of the existence of information that has been available. An effort is being made to incorporate what is available on the computerized Natural Resources Value Inventory System (NRVIS) and then to make this available during forest management planning. Training of field personnel and bush workers has been undertaken. For example, there has been training of skidder operators in the recognition of cultural features that they might encounter in the field, such as when woods operators near Clancy Lake discovered what turned out to be a depot farm of John Egan. When piles of stone were found and reported, the operation was stopped in the area and investigated, but unfortunately by the wrong office.

Subsequent investigations found that the site had been known, properly reported and filed in the archives of the Visitor Centre, but the existence of the report was unknown to the Park East Gate headquarters staff. That staff thought the rock piles might possibly be an aboriginal burial ground, investigated through channels appropriate to that assumption and were told that it was not such a site. Forest operations were resumed and minor disturbance of the site occurred. It is to be hoped such an event will not reoccur, but kudos must be given to the skidder operator who noticed something different in the woods and reported it. The use of the NRVIS inventory at the planning stage of a forest operation may prevent the repeat of a situation similar to that described above. However, the NRVIS inventory requires continuous updating and monitoring. It will be a difficult, never-ending task to maintain it.

The audit asked for a Cultural Resources Management Plan, as was called for in the Algonquin Provincial Park Management Plan of 1998. This would be a huge effort for which the money is presently unavailable.

There was also a suggestion in the audit that prescribed burns be used when costs and benefits are reasonable. This relates to jack pine regeneration. However, it has been found that where scarification has exposed sandy soils, on hot sunny days the ground becomes warm enough to have jack pine cones release their seeds. Satisfactory regeneration then takes place. Expensive and potentially dangerous, prescribed burns are apparently not necessary.

At the conclusion of their report, KBM stated that

> the finding of the audit team is that, with the exceptions noted, management of the Algonquin Park Forest conformed to program direction and legislation during the 1997–2002 period, and that the Algonquin Park Forest is being managed effectively and in a manner consistent with accepted criteria of forest sustainability. In addition, the audit team finds that the Algonquin Forest Authority has complied with the provisions of the Algonquin Park Forestry Agreement and recommends that the term of the Agreement be extended for a further five years.[296]

The road network in the interior of Algonquin continues to attract attention. It serves the AFA and forest industry, as well as the MNR when it is carrying out needed maintenance functions. Despite concerns expressed by some, the roads appear to have had little impact on the natural environment in the Park. Biological impact studies have revealed next to no significant changes. The most notable change has been in the increase in the population of the American toad. This creature apparently likes the increase in open spaces in the forest.

However, the use of these roads by unauthorized persons is a major concern. Dealing with this issue involves confronting difficult social situations. At the present time, there is a marked reluctance to do so.

Twenty-Five Years of Change in Harvesting the Algonquin Forest

By direction of the Master Plan, mills and lumber camps were to be removed from the Park as soon as possible.

In 1973, the United Oil Products (formerly Staniforth Lumber and Veneer Limited) mill at Kiosk burned. The residences in the lumber mill village of Kiosk were gradually removed by the MNR by 1996. In 1975, two sawmills remained in the Park. The Pembroke Lumber Company mill on Lake Travers ceased operation in 1976, while the McRae mill on Whitefish/Rock Lake closed in 1979, with its operations moving to Whitney. A portable mill operated for a brief period in 1986 at Odenback on Radiant Lake and then was removed. An inventory of logging camps taken in the same year indicated the presence of 16 camps within the boundaries of Algonquin Park. The Master Plan also called for their removal. Bush camps, now unpopular with the men (and their wives), were not needed because the road system was good enough to permit daily commuting.[297]

Nonetheless, the burning of the old camps evoked nostalgic memories of camp life for older men. Some buildings were left, however: "On one occasion, two AFA supervisors left a fine "eight-holer" still standing. It had originally been fitted with a large box stove and electricity."[298]

Table 11 **Percentage of Tree Species Harvested: 1975–1976 and 2003–2004**

	1975/76	2003/04
Maple	31.0	32.0
Yellow Birch	7.0	2.0
White Birch	1.0	4.0
Beech	0.5	4.0
Oak	0.1	1.0
Poplar	3.0	19.0
White Pine	24.0	19.0
Red Pine	12.0	10.0
Hemlock	8.0	2.0
Other	13.4	7.0

Source: AFA 2005.

The maple harvest has remained constant while that of yellow birch has declined. This is because the impact of birch dieback, first noticed in the mid-1940s, is still showing. The results of silvicultural treatments and tree marking since the 1970s have yet to show in the harvest of this tree. White birch harvesting has increased, but only some 15 per cent of this species is graded for sawlogs because many of these trees are both small in diameter and frequently crooked. Most of these stems end up as pulp. Beech and poplar, whose volumes have increased, also go mainly for pulp, reflecting the development in that market. The white pine numbers indicate a decrease but this is deceiving; the volume produced in both time periods was approximately the same. Hemlock harvesting has decreased. This reflects a change in the marking prescriptions that have been made in an effort to restore this species to the same percentage of the forest that it previously had. When this is accomplished, winter deer yards in the Park will be restored to the levels existing before the heavy cutting in the 1950s. Scarification aids in establishing young trees, but unfortunately young hemlock are a favourite browse of moose and heavy damage can occur.[299]

The forest industry had to adapt to what the forest could provide. Some lumber company managers early on understood what was taking place. John McRae made the following comment in 1986:

> The lumber business is something like farming. It's the land base that to a degree determines its success or failure, as well as how you manage that land base. Our limits weren't really superior in the past, and today the same would hold, though we're improving them. But it's going to take a generation to do it.[300]

It was foreseen that the first 20 years after 1975 would be the most difficult for the MNR, the AFA and the mills. Benefits are now starting to show. Managed softwoods have shown a response to tree marking and selective harvesting. Red and white pine shelterwood-managed areas indicate increasing volume growth in the 40 to 130 year age ranges.

In 1985, an optimistic opinion stated that

> reports from one mill operator, now starting to receive logs from areas cut twenty years ago [under tree marking], indicate that they are sawing as high as forty-seven per cent number one common and better from the managed hardwood stands. Ten years ago, the same mills were experienc-

ing number one common and better yields of only twenty–five per cent ... Also, the proportion of saw logs to total hardwood harvested has increased to fifty per cent from the traditional thirty-five to forty per cent range in these managed areas.[301]

Another authority cautioned, however, that 40 years might be a more realistic time period to realize a significant upgrading of harvested stems received at the mill. As of 2005, not all of the recreation/utilization zone had received silvicultural treatment, and, in consequence, mills are still receiving high volumes of "cull" wood. Cull wood, before tree marking, would have been left standing in the bush. Now this poor wood is cut and ends up at the mills. It does not yield lumber and therefore is chipped. When all cutting is done on land tree-marked, harvested and then rested for at least 20 years, better yields of superior timber should result. However, this will obviously take time.[302]

Between 1975 and 1995, the average harvested volume was 394, 000 m³ per year. The Algonquin forest could supply close to 620,000 m³ and maintain sustainability.[303]

Overall man-day production increased steadily from 8 m³ per man-day to 15 m³ per man-day between 1975 and 1995 as better equipment developed and logging techniques were refined. Before 1975, two-man crews, a feller and a skidder operator, were common, while log making at the stump was still taking place. By 1995, some 85 per cent of the harvest was tree-length. Grenville Martin began hauling tree lengths out of the Park as early as 1975–1976 and McRae began hauling tree-length stems to its Whitney site in 1979; the AFA has since adopted this method in concert with other mills.[304]

Operations in the woods have undergone continual modifications. In 2003-2004, approximately 60 skidders operated by independent contractors were required to handle the timber cut. In the summer, from the spring breakup through to mid-September, if weather permits, skidder numbers are currently reduced to just over 50 to lessen noise. Skidders are not used in wet weather because of safety concerns and because these heavy machines can cause significant damage to the ground when it is wet and soft. Since 2000, the feller-buncher has come into use (Photo 120). The introduction of this machine has cut down on manpower and greatly improves safety as there are fewer men to worry about and that the operator is in a big machine and not running out of the path of a falling tree. Because the feller-buncher grabs the trees and cuts them, then places them on the ground, the trees don't hit things as they fall and other trees are spared the occasional hit when a tree would be felled by a traditional chainsawman, thus lessoning the footprint from logging in the forest.

Logging is permitted in only one zone— the recreation/utilization zone—to lessen conflict with recreationists. This zone is divided into summer and winter logging areas. Within summer operational areas, sound zones and other restrictions apply.

The Impact of the AFA on the McRae Lumber Company

Bob McRae reflected on the pressures that his father faced before he retired.

> The company went through the Algonquin Wildlands League campaign to end logging in the Park. This was very distracting for Dad and took some of the focus off running the business. He wasn't sure that logging would survive. In 1975, the Algonquin Forestry Authority was created and took over all logging in the Park. For the company, the AFA complicated life.[305]

No matter how frugally the AFA was managed, the cost of the timber it supplied to the mills was greater than it had been under the previous arrangements. This was inevitable since costs formerly borne by the MNR now became those of the AFA. These costs were passed on in the price of the wood delivered to the receiving mills, while mills outside of Algonquin continued under lower cost arrangements. This was a competitive disadvantage for mills receiving Algonquin wood.

The AFA was unsettling for the workers as well as for the mill owners. Their jobs were being threatened. Gary Cannon, the McRae bush supervisor, commenting in 1986, stated,

> Mind you, in 1975 we lost about 150 miles of gravel roads to the AFA in Bruton and Clyde townships. The AFA came in and took all the licences from McRae, Murray, Weldwood, and so on. And the AFA took over all the logging in Algonquin Park. So when they took Clyde Township, which I was working in at that time, and I had about two years of roads ahead already—they came in and took them just like that, and said, "No, we're going to log in there; you can log in some other area." We lost a hell of a pile of roads. In fact, one road from Rock Lake around to Whitney here we had it built in 1970–1974, I think was 46 miles of mainline. When I was building that, Donald McRae gave me to use five bulldozers, seven gravel trucks, two 950s. It was just like a construction company building those roads. The AFA took over cutting on those roads. So the work I done to get two years ahead, and the planning of the cutting on those roads which wasn't very good for us … It hurt the company pretty bad. And there was no compensation. McRaes never got five cents for all the roads they built on Crown lands and in the Park.[306]

The Impact of Harvesting Changes at the Mills

In the mid-1970s the Algonquin forest had a high pulpwood component since the forest had been previously high-graded. Mills had to adapt to the large quantity of low-quality wood coming out of the woods as a result of silvicultural-based tree marking and cutting.

Switching from cutting only the best trees to a tree-marking system that left many of the best trees standing was a radical change for the timber cutters and the mills.

The mills themselves, particularly the mills sawing hardwood, have almost all undergone profound changes. They have adapted to handle what the forest was growing. An example of this is the McRae Whitney mill yard, which consists of two sawmills—a large log mill processing approximately 400 logs per day, and a small scragg mill that saws up to 2,400 pieces per day. Tons of chips are produced daily from these operations. Some of this chip volume is the result of cull trees (not containing merchantable volumes of sawlogs or pulp material) being cut and delivered to the mill.[307]

The table below shows the remarkable increase in pulpwood cut and sold to local mills. This wood had not been cut previously; instead, it was left standing in the bush where it took up growing space. At the mills now, this pulpwood is usually chipped. Fortunately, the market for this product grew dramatically during the 1980s, with new paper plants opening in Portage du Fort in Quebec, and Temiskaming.

The Forester and Silviculture

Carl Corbett, the current general manager of the AFA, commenting in 2005 on the conflict between the desire for profitability and the importance of environmental and social responsibility, made reference to the fact that all AFA foresters are members of the Ontario Professional Foresters Association. As such, they are committed to:

- The sound management of Ontario's forests;
- The highest professional standards of practice;
- Public accountability for their actions;
- The principle of stewardship and sustainability;
- Adherence to a code of ethics.

Applying these precepts to silvicultural practices in the Algonquin forests requires knowledge, skill and an element of art, or the feel of what a section of land can best produce. This is what AFA foresters are expected to do.

Considerable information is needed about a potential area before it is cut. Today, the lands harvested in Algonquin are first photographed and mapped to assess the species and size of the trees present; they are then sample-cruised on the land to verify the previ-

Table 12 **Pulpwood as a Percentage of Algonquin Forest Timber Cut**

Year	Total Cut	Pulpwood	Per Cent
1975–1976	401,550 m^3	38,270 m^3	9.5
1989–1990	342,965 m^3	132,678 m^3	38.7
2002–2003	543,584 m^3	294,204 m^3	54.1

Source: AFA Annual Reports for each year

ous interpretation. Following a prescription for an area, provincially certified tree markers, supervised by the AFA and audited by the MNR, mark the trees to be cut. Every five years, the AFA operation is audited at the office and in the field. The roots of these procedures go back to the work of researchers at Swan Lake and Don George.

Several MNR publications are based on Swan Lake research:

- In 1973, a document entitled *Management of Tolerant Hardwoods in Algonquin Provincial Park* was published. This report was updated in 1983 by a committee comprised of Mac McLean, Harvey Anderson, Dave Wray, Jim Scott and Bill Brown;
- In 1987, Don George asked Harvey Anderson to design a scientific quality assurance sampling scheme to monitor the AFA tree-marking program. The results were published in a 1996 report entitled *Quality Assurance in Hardwood Tree marking: A Case Study*;
- In 1990, a provincial guideline called *A Silvicultural Guide for the Tolerant Hardwoods Working Group in Ontario*, written by a committee of provincial management foresters chaired by Harvey Anderson was published. In 1993, a companion document entitled *A Tree marking Guide for the Tolerant Hardwood Working Group in Ontario* was produced. Both documents have been recently updated. Much of the science background in these reports was derived from the research work at Swan Lake and other Park locations.[308]

The above publications have each led to improved tree-marking procedures.

Harvey Anderson credits his mentor, colleague and friend Mac McLean with being the driving force behind the

> research involved in the use of tree marking and partial cutting techniques to improve uneven-aged sugar maple-beech stands. Intensive analysis of stands of variable cutting history provided an insight into the dynamics of stand growth, structure, regeneration, and tree quality. Provisional structural, stocking, and cutting-cycle targets for the single-tree selection system were then tested, starting in 1966, in research cutting trials at Swan Lake and elsewhere, many of which continue to this day.
>
> His [Mac McLean's] ability to see the forest, and not just the trees, perhaps was rooted in his early experiences with the holistic approach espoused by Angus Hills. [Hills was a pioneer of land use studies in Ontario.] In a way this set him apart from many of his peers, and seemed to provide him with an intuitive understanding of forest dynamics … Many others may walk in Mac's footsteps, but very few will ever fill his shoes.[309]

Tree Marking

Tree marking is considered to be the most critical activity of the AFA's silvicultural operations. Done well, tree marking and harvesting lead to a more productive, higher quality, efficient, "industrial" forest. All trees to be cut are individually marked by trained and certified tree markers.[310] An annual task is to keep tree-marking some 9,000 hectares ahead of planned logging operations. Ray Townsend wrote,

> Tree markers work independently of the tree harvest as a whole. What is marked is the product of environmentally-sound forest management practices: the blending of silvics, species diversity, fish and wildlife habitat and recreational values rather than the economics of the harvest. The basic requirements of individual trees like shade tolerance, seed establishment, soil type and water regime are considered along with topography, site class, stocking levels, growth and recruitment. Tree markers base their decisions on a much wider range of factors and variables involving the incorporation of wildlife values into the marking process.[311]

During the course of a day, tree markers must consider

> some or maybe all of the following: tree species diversity, moose calving sites, feeding and nesting cavities, stick nests, fruit and nut trees for birds and mammals, heron rookeries. Additional considerations include wintering requirements for deer and moose, fish habitat, slope percentages, siltation, as well as where canoe routes and campsites are located.[312]

Tree marking is not easy work. Jamie McRae, of McRae Lumber, reflected on his experience.

> It was a really difficult job, walking all day long with a heavy pack on your back full of paint, over and under fallen trees, through more spider webs than I had imagined existed, while all the while trying to figure out which trees to mark … It did give me a good appreciation of the quality of trees that come out of the bush. I learned first-hand how stand improvement works and all the details of the selection system.[313]

Tree-marking prescriptions are still evolving, based on new science. Items recently added to protected status are habitats suitable to such birds as the red-shouldered hawk and great blue herons, as well as the pine marten. Early tree-marking prescriptions called for the removal of cull trees, but now a number of cavity-nesting trees are left on each hectare of marked area. Frequently these cavities have been made by pileated woodpeckers, but these trees are now left to provide nesting places for owls and wood ducks.

Silvicultural Management Systems

Two main forest management systems are used in the Park. They are the selection and the uniform-shelterwood management systems, both of which maintain forest cover on the land at all times. Clear-cutting, with standards, is employed on a small fraction of the area harvested.[314]

The selection system, based on partial cutting, is used on about 70 per cent of the area in the tolerant hardwoods common on the west

Chapter 10: The Algonquin Forestry Authority: Making a New Management System Work

Figure 33 **Basal Area of Sugar Maple Trees**
Growth of sugar maple is maximized when trees are spaced so that the total cross-section area at breast height of trees greater than 25 cm in diameter (sum of shaded areas) is equal to 12.3 square metres per hectare of forest (60 square feet per acre).

Source: Dan Strickland, *Trees of Algonquin Provincial Park*, Whitney [ON]: Friends of Algonquin Park, 1987, 36. Used by permission of MNR.

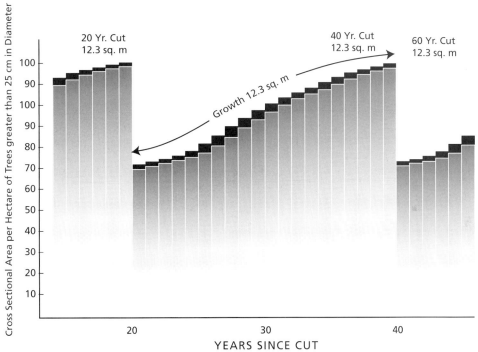

Figure 34 **Growth of Sugar Maple under the Selection System: Theoretical Stocking Development Projection**
The diagram is a conceptual model illustrating the theoretical stocking development for an ideal stand that has been harvested to 20 m²/ha/year, the net periodic growth (accretion) for the following 20 year period would be approximately 10m²/ha. A volume of wood equal to the accretion may then be harvested and the process repeats itself indefinitely. This model assumes minimum damage during harvest, favourable site conditions, and optimal residual stocking and structure. Actual values in previously unmanaged stands often range from 0.3 to 0.4m²/ha/year.

Source: Adapted from Ontario Ministry of Natural Resources, *Ontario Tree Marking Guide*, Technical Series, Toronto: Queen's Printer for Ontario, 1983, 145.

side of the Park. In this system, marked trees are removed singly or in small groups, and regeneration is usually natural. The aim is to have a distribution of age classes in the forest at all times. These forests are commonly referred to as "uneven-aged" or "continuous forest." In this system, many better stems are left to provide good quality seeds. Poorer quality stems, many of which are low, pulpwood quality, are cut. After two or three cuts, 20 years or so apart, with possible tending, high-quality stems will become more common and the value of the forest upgraded. Sugar maple is the dominant tree, while red maple, American beech and yellow birch are also common.[315]

In the selection system, sugar maple stands are cut about every twenty years leaving the density of trees (12.3 square metres of trunk cross sectional area per hectare) where growth will be at a maximum. In this way the land always remains forested and the volume of wood approximately doubles before the next cut.[316]

The system described above was worked out at the Swan Lake Research Station in Algonquin Park.

The uniform shelterwood system is applied to conifer stands and in some tolerant hardwoods, most commonly on the east side of the Park. Here, trees marked are left. The others are harvested in a series of two or more cuts while natural regeneration occurs under the shelter of residual trees. White pine, hemlock, white and black spruce, balsam, fir, red oak and some sugar maple are harvested under this system.[317]

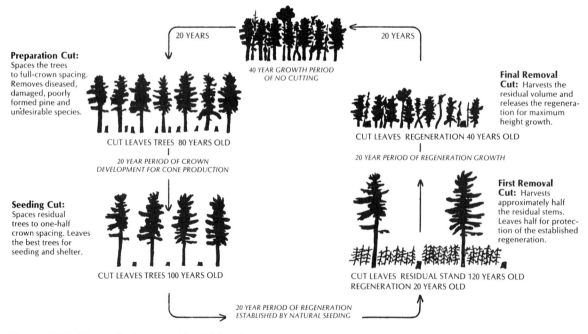

Figure 35 **Uniform Shelterwood in White Pine**
Source: Strickland, Trees of Algonquin Provincial Park, 36. Used by permission of MNR.

The problem of providing just the right amount of sunlight also occurs with white pine. Too much light promotes disease and too little leads to less than optimum growth. Experimentation led to the uniform shelterwood system.

> A mature pine stand is removed in a series of four cuts at 20 year intervals. Although these cuts thin out the stand more and more, the crowns of the remaining trees expand after each cut and tend to keep the sunlight on the forest floor below in the desirable 40-50 per cent range. This makes for maximum growth while guarding against white pine weevil problems. By the time of the fourth cut, when the last of the overhead trees are removed, 60 years have elapsed since the first cut and the new, even-aged stand of pine was established in the shelter of the old trees is ready to stand on its own and grow for another 40 years before the cycle starts over again.[318]

The term "clear-cutting" evokes an extremely negative response in many people. But clear-cutting is necessary in Algonquin as seed-bed preparation for red pine, poplar, white birch and jack pine because abundant sunlight is needed for the regeneration of these species, and the AFA is mandated to keep the forest diversity that was present in 1974.

Clear-cut areas are small in Algonquin, usually less than five hectares in extent, and some mature trees are maintained to provide seed. This procedure is needed, as white pine seedlings will out-compete red pine where shade is present. Clear-cutting is necessary to ensure red pine regeneration.[319]

The Local Citizens Committee

Every forest district in Ontario is now required to have a Local Citizens Committee (LCC). These committees are composed of people who have interest in forest uses and are representative of the local area. The Algonquin LCC, whose chairperson has been John (Bud) Doering since June 1998, has reviewed and provided comment and input on many topics, including the following: the Forest Management Plan Terms of Reference, public consultation sessions, values maps, areas of concern prescriptions, cultural heritage protection and annual work schedules.

Since this committee has members that are not foresters or mill owners, the questions raised can be different than what one might expect.

One item charged to the AFA was a bill for shotgun shells submitted relative to the collection of seed cones from an old growth white pine stand. Behind this curious purchase lie questions raised at an LCC meeting. "Does white pine growing on the drier, sandy-soil lowlands of the eastern part of Algonquin grow at a different rate than pine growing on the loamy, textured, more moist soils of the western uplands of the Park?" Another question asked was, "Is there a genetic difference between the pines growing on the lowland as opposed to those growing on the upland areas of the Park? If so, should upland stock seedlings be used in replanting upland areas?"

Figure 36 **Supplies for Old-Growth White Pine Seed Collection**
Source: Algonquin Forestry Authority

Could these questions be answered?

If such were the case, "Would seed collected from old growth upland pine grow at a different rate than seed collected from pine from the lowland area?" The shotgun shells were used to shoot seed cones from an old growth pine stand that was over 200 years old. The cones, harvested in September 2000 by Keith Fletcher and Dave Peters of the AFA from the Lizz Lake area south of Hogan Lake, were shipped to the Ontario Seed Plant in Angus, Ontario where the seed was extracted and stored. The seed was then grown by the Millson Forestry Service of Timmins, Ontario. Fred Pinto of the MNR science and technology unit provided advice

and an experimental planting plan. Twenty rows of 25 tagged seedlings grown in containers were planted north of the Hogan Lake Road in May 2003 in an upland area. Normal and old-growth trees were randomly assigned to each row.

In June 2005, the plantation was visited and all the young trees were found to be doing very well, having reached a height of over two feet. It will be a further 20 years before different rates of growth may appear. Genetic studies have yet to be done.

What progress have trees in the plantation made since being planted? Keith Fletcher, the area forester for the Algonquin Forestry Authority, said,

> The trees have been measured three times already [previous to June 2005]. A sample of each type was measured as they were planted so that we could test to see that they were about the same size when they started out. After the first and second growing seasons in the field [spring of 2004 and 2005] every tree was measured (height, diameter, crown diameter, vigour, etc.). They are next due for remeasurement in year 5. No detailed analysis of the data has been done as of yet—it will be more informative to do it when there is a longer term information to look at.[320]

The trees will likely have to be ten or more feet in height before a growth differential, if any, becomes apparent. Only further time will tell if the growth rates of old-growth upland pine seedlings are superior to the usual nursery stock in this upland location. In addition, it is also hoped that the genetic makeup of the old-growth pine will be determined and compared to the lowland white pine. Results could influence restocking procedures. Realistically, the planting of old-growth seedling pine across the Algonquin Highlands would be prohibitively expensive. But this may not be necessary.

Most Algonquin pine today comes from the sandy lowlands of the east side, where it dominates other species. Similar sandy soil conditions favoured by white pine are spotty across the hilly and rocky Great Lakes forest region. Yet, white pine grows best in the upland areas where sugar maple now dominates. Why has the upland pine not regenerated there over the past 100 years? Early loggers sought these pine out and eliminated all but a few of these giant trees, which grew to over 135 feet (40 metres) and 45 inches in diameter (1.2 metres). These upland trees sometimes grew in small groves but commonly they were scattered across the hardwood forest. Research has shown that many of these trees started their lives by seeding after a forest fire. These flames removed the thick leaf litter deposited by the sugar maple and other deciduous trees, and opened up the canopy by killing some of the trees. This enabled light to reach the ground and provided the 40 to 50 per cent direct sunlight favoured by the pine, enabling them to become established over a growth period of 15 to 20 years. Today, forest fires are quickly extinguished. Apart from the shorelines of lakes, there are now few areas with ideal growing conditions for upland pine.[321, 322]

Photo 81 Tagged White Pine Plantation: Hogan Lake Road, June 2005
Left to right: Bud Doering, Chairman, Local Citizens Committee; Arlene Hyde, Pembroke MNR office; Keith Fletcher, AFA, Huntsville; Joe Yaraskavitch, MNR, Pembroke.
Source: D. Lloyd

A recent observation by Carl Corbett is also very interesting. Where harvesting operations in sugar maple-dominated sites have taken place, incidental scarification during skidding operations have apparently led to considerable natural seeding by upland stock white pine in some areas. Perhaps, decades from the present, magnificent white pine will again tower in significant numbers above the hardwoods of the western uplands of Algonquin Park.

The AFA and Relations with the Public

The AFA has contributed to the public's enjoyment of Algonquin through the years in many ways. The preservation of sites that foster an appreciation of timbering history has been a consistent agenda item of the AFA. Some of the AFA's projects are given below:

- In 1984, the AFA led by Operations Supervisor Ray Townsend, restored the "The Office" at Basin Depot. This building is said to be the oldest structure in the Park and dates back to the early square-timber days in the 1800s (Photo 6);
- The Kitty Lake Ranger cabin received a new roof and related repairs. It is one of the cabins in the Park that is available for rent;
- With the aid of the Friends of Algonquin Park, two footbridges on the trail system were constructed in 1991;
- The video of current forest management practice now shown at the Logging Museum was paid for and produced by Forestry Canada. W.J. Brown, then manager of the AFA, arranged a donation of $10,000 on behalf of its clients. *Algonquin Forest – Maintaining the Balance* was produced in 1987–1988;
- In 1989, J.R. Booth's Ottawa, Arnprior and Parry Sound Railway bridge at Whitefish Lake was given a completely new deck. This work involved welding, which was done by the McRae Lumber Company, with material supplied by the AFA. Part of this railroad bed is used now as a bicycle trail between the Rock Lake access point and the Mew Lake campground;
- Two donkey engines are now on display at the Logging Museum.

What are donkey engines and how did they get there?

> On August 2, 1990, two steam donkey engines were located [by E. Ray Townsend, AFA] between Opeongo and Proux Lakes, where they had done their last day's work pulling loaded sleighs of pine logs up to the height of land. This allowed the teams to proceed downhill from there with the loaded sleighs to Opeongo Lake from which point the spring drive would move the logs down the Opeongo River. This Booth operation took place in the late 1930s.[323]

In the fall of 1992, the donkey engines were skidded out to the nearest bush road, loaded onto a low-bed float and transported to the museum. The following spring, McRae Lumber Company sandblasted and painted the two engines and returned them to the museum, where they are now on display.

In December 1990, E. Ray Townsend provided drawings and a cost estimate for the log dam and chute at the Logging Museum. An AFA crew headed by Dave Barras and supervised by E. Ray Townsend built the structure under contract to the MNR.[324]

As acknowledged on the entrance sign at the Logging Museum, the Algonquin Forestry Authority is sponsor, and the Friends of

Chapter 10: The Algonquin Forestry Authority: Making a New Management System Work

Photo 82 **The Opening of the New Logging Museum, Algonquin Provincial Park, August 15, 1992**
Left to right: Carl Corbett, AFA; Brent Connelly, AFA; Bill Brown, AFA; Gary Cannon, McRae Lumber; Bob Pick, AFA; Rob Keen, chair of AFA; George Garland, chair, Friends of Algonquin Park; and Ernie Martelle, superintendent, Algonquin Provincial Park.
Source: Algonquin Forestry Authority

Algonquin Park [FOA] is a co-sponsor of the facility, with Ontario Parks. These three organizations sponsor Loggers Day at the Logging Museum in July each year.

On this day, volunteers in period costume are stationed at the exhibit stations along the museum trail and they explain how the various pieces of equipment and facilities were used in days past. A nominal fee is charged along with an additional charge if one partakes of the "Logger's Lunch." This day is a highlight of the summer season in Algonquin Park, and the large parking lot fills to overflowing for this very popular event.

It is fair to say that the AFA, being a Crown agency, does not operate under quite the same constraints as private business relative to the "bottom line." For example, when ice flows damaged the important road crossing of the Petawawa River near Lake Travers, the bridge was replaced by a single span structure that was considerably more expensive to build than repairing the damaged pier in the middle of the river would have been. Aware that this crossing was on a popular canoe route, the less intrusive, but more expensive, structure was built.

In his book, *Algonquin Forestry Authority: A*

Photo 83 **Ribbon-Cutting Ceremony at the Official Opening of the Algonquin Provincial Park Visitor Centre, August 1993**
Left to right: Dan Waters, MPP for Muskoka-Parry Sound; Dr. George Garland vice-chairman, FOA; Bill Brown, general manager, Algonquin Forestry Authority; Ralph Bice and Adele Ebbs, FOA Directors Award recipients; Sean Conway, MPP for Renfrew-North; Marjorie McRae MacGregor and John McRae, McRae Lumber Company, generous contributors to Logging Museum exhibits; Ethel LaValley, Reeve of Airy Township.
Source: Courtesy of Dr. Don Beauprie and the Friends of Algonquin Park.

Twenty Year History. 1975-1995, E. Ray Townsend wrote,

> Twenty years has done much to dispel some of the apprehension as to whether the Authority could deliver on its mandate. I think it's fair to state that there has been steady improvement in this relationship from poor to better, and quite good on many occasions. One has to remember that the AFA and its clients are perpetually in adversarial roles when sales contracts come up for renewal at least once a year, and big bucks are on the table. Allocations, once a major concern of all former licensees, are much less of a concern than they used to be.[325]

Over a history of almost three decades, the AFA has proven to be a success. Arthur Herridge, a right hand man to Frank MacDougall said in 2001,

> The AFA has prospered under all three political parties—each Minister became a staunch supporter of the multiple-use concept and the Authority's role in carrying out government policy as laid out in the Master Plan. The Ministry, regardless of who has been in power, has had the wit to put people into positions that have had a record of accomplishment in the face of challenge.[326]

The Quieting of Turbulent Times

It has been three decades since the Master Plan for Algonquin Provincial Park was made public and the Algonquin Forestry Authority established. The time has come to pass out credits to people who contributed significantly to their particular cause and ultimately to the solution of problems and thus to the lessening of conflict between adversaries.

Frank MacDougall as superintendent of Algonquin, saw the conflict between logger and conservationist/recreationist, and started the process of separating the two groups in order to lessen the conflict between them. MacDougall was an advocate of multiple use policy, which is still in effect in Algonquin Park. Prof. J.R. Dymond, as first chairman of the Federation of Ontario Naturalists, introduced to Ontario the philosophy of protectionism, and started interpretive programs that evolved into Ministry programs that serve

Chapter 10: The Algonquin Forestry Authority: Making a New Management System Work

and inform visitors to the Park today. While their Park plans were highly criticized, D.M. Omand and William Hueston started the Park on the way to orderly planning. The protectionists, led by the Algonquin Wildlands League under Abbott Conway and Douglas Pimlott, supplied great energy and passion. While they managed to exclude logging from all other provincial parks, they did not win in Algonquin, because in this area there were no other forested lands that could feasibly supply wood to the local mills. The government could not see putting firms out of business and destroying the livelihood of the small towns surrounding Algonquin Park. Multiple land use would continue in Algonquin under the Master Plan. The AFA, under Joe Bird's direction, delivered the logs and kept the mills going. Not generally recognized was the impact of the pressure brought by the various protectionist groups that helped promote the adoption of the better forestry methods devised by the Swan Lake researchers—Don Burton, Mac McLean and Harvey Anderson with hardwoods, and Al Gordon with conifers. They contributed greatly to the knowledge of how to grow our trees better. Don George, "the Dirt Forester," took this information, along with his experience, and set up partial harvesting operations across the Park. Donald McRae fought two wars, one for all of us, and the other for his company. He was on the winning side on both. Along with others, mainly smaller mill owners, Donald McRae first gave the new forestry a chance of succeeding. Credit also has to be extended to the governments and politicians of the day. The compromise brokered by the government, the Master Plan, provided zoning to satisfy some of the protectionists' requests and provided rules for the timbering that continued under the AFA. William Calvert and Leslie Frost, the retired premier who chaired the Advisory Committee, wrote the Master Plan, aided by Arthur Herridge and Warren Giles. Premiers Frost and Robarts fuelled the process of finding a solution, which William Davis concluded. Finally, Gavin Henderson in effect decided that, "enough is enough," and withdrew the National and Provincial Parks Association of Canada from the fight.

Perhaps the most significant reflection on the turbulent times is that in 2006, the conflict between logger and recreationist in Algonquin, while still present, simmers well beneath the general public's concern. The Master Plan for Algonquin Provincial Park (and now the Park Management Plan of 1998), and the Algonquin Forestry Authority serve as outstanding examples that compromise can be reached and made acceptable to most people. Most importantly, Algonquin continues to work its magic on many people.

CHAPTER 11
Whitney, Its People and Algonquin Park During and After World War II

Whitney and its immediate area have a population of slightly over 800 people in 2006, and a history of dating back just over 100 years. Whitney, in 2006, is still a very small place. It would be considered by most to be isolated. The major employer in town, as it has been for over 80 years, is the McRae Lumber Company. The MNR, with its headquarters in nearby Algonquin Park, is the second largest employer, followed by the tourist industry. Communications have greatly improved—the highways are now paved and well maintained. Yet the main intersection in town at Highway 60 and Post Street has no traffic light. Here can be found the bank and post office, along with the general store, a supermarket and three gas stations. Schooling goes to the end of Grade 8, and then busing is required for high school in either Barry's Bay or Bancroft. Yet satellite TV and the internet bring in the world—too close at times.

In its time, Whitney has housed some exceptional characters and it continues to produce citizens who respond to the needs of their community. This chapter touches on some of them.

Tragedy in Whitney Woods: The Billings/Stringer Deaths

On January 9, 1926, John Billings, a Department of Lands and Forests game warden, and Joseph Stringer, a guide from Whitney, set out in search of two trappers suspected of breaking fish and game laws in northern Haliburton. They travelled south from Whitney some 25 km (15.6 miles) measured in a direct line, and stopped for lunch in the office cabin of the abandoned Mickle, Dyment bush camp at Birch (Little Hay) Lake. Sand Lake, the destination of the men, was some 25 km (15.6 miles) farther by direct line to the south (Map 10).[327] Both men had been sergeants in the army, with Billings much decorated and nominated for the Victoria Cross.

The bush camp at Birch Lake was located on a high hill to the north of the lake. About 8 p.m., S. LaValley saw flames from his farm on Hay Lake. Woodsmen farther down the lake noticed the reflection of flames around 9 p.m.

When the men didn't return, a search party was sent out that subsequently discovered the remains of the men in the burned-out cabin near Birch Lake. Initial evidence from the scene suggested that the two men had been murdered and the fire set to cover up the crime. Two observations at the scene led to this view: the bottom logs of the cabin were entirely consumed by the fire, suggesting that they had been uprooted and added to the fire; and the bones of the two men appeared to have been raked up, as they were found in a very mixed-up condition. In consequence, foul play was suspected.[328]

Suspicion fell on John Parks, a trapper from the Whitney area who was also a war veteran. When informed by his father that he was under suspicion, Parks turned himself in to the authorities. Parks was able to substantiate his presence in Whitney on the day and evening when the death of the two men took place. The coroner at the inquest decreed that there was

> no evidence to show that he [Parks] was in any way connected with the deaths of Sergeant Jack Billings, game warden of Barry's Bay, and Joe Stringer, guide and bushman of Whitney. An open verdict attaching no responsibility for the double fatality was returned and charges against Parks were at once withdrawn.[329]

This was a popular decision with the people of Whitney, who were convinced of Park's innocence. Such was not the case in Barry's Bay, where Billings had been the local hero.

George Garland, in *Names of Algonquin*, suggests that the scene of the tragedy was Billing Lake, then called Sand Lake. That name "came into use after the 1926 incident and was officially adopted in 1957."[330]

Stringer Lake was not named after Joseph Stringer of the above story, but a relative, Park Ranger George Stringer, who had cut the trail from North Grace to the lake about 1920. George Stringer, "who later drowned in Rock Lake, was also a member of the family that has had many Park connections. His cousin, Jack Stringer of Canoe Lake, also was a Ranger for many years. Of Jack's numerous sons, Wam and Jim were guides, Dan was a Ranger, and Omer was a teacher, a famous canoeist and a builder of canoes."[331]

The Post Family and the Red Cross Outpost Hospital: 1928–1978

The Post family served this area for three generations and they are marvelous examples of dedication to one's community.

O.E. Post had come to Whitney as the last manager of the Dennis Canadian operation, and when the mill moved out, he stayed on and took over the company store that had been originally established by St. Anthony. Post had an entrepreneurial side to him and proposed to the provincial government that Ontario "lease" him the town. A deal was struck for $1,100 a year, and Post assumed the omnipotent position last assumed by Edwin Whitney. Post, fortunately, was far more benevolent. Houses were rented for $2.50 a month, $5 for the larger ones down along the tracks. He even sent one of his own children, Gilbert, off to medical school on the understanding that "Gib" would come back and serve the poor logging families, which he did, setting up a small Red Cross outpost in 1928 and staying on as the only doctor in the area until his death more than thirty years later. O.E., landlord and storekeeper, dealt in vouchers and often traded in goods—chickens for flour, venison for yeast—and for the next two decades the little town of Whitney was entirely his concern.[332]

In 1942, O.E. Post, the government landlord and storekeeper, died as did his son Albert.

Post's other son, Dr. Gilbert Post, ran the Whitney Red Cross Outpost Hospital from 1928 until he died in 1965. Among the women of the area, he had a great reputation for delivering babies. His son, Geoffrey, was the very popular principal of the Madawaska Valley District High School in Barry's Bay for a long time.[333]

Helen McRorie served as a Red Cross nurse in the hospital before World War II. She returned to Whitney after the war and married Donald McRae.

> When I was nursing here [before the War], you saw people who were pretty well shut off. You made regular visits to people. I think it was twice yearly school inspection, unless there was some reason for other things. And then we were sort of on call. I guess we were the first line in helping in sickness. Sometimes you were left with a home delivery of a baby or something. Yes, you delivered babies when you had to. The nearest doctor was at Madawaska. Actually, Whitney wasn't all that isolated, except when the doctor visited the lumber camps. He'd go in for three or four days and then he was really gone. That was Dr. Post. But otherwise he was in Madawaska, and he made regular trips up once or twice a week at least to see his patients. I think Dr. Post came in about 1941; it was after I'd left. He was in Renfrew then for some time, and then he became ill; he had a lung condition, and he was in the sanitarium and wasn't back until about '46. Then he was here in Whitney until he died at the end of '65 or in '66.[334]

The Red Cross dinner in support of the hospital was the social event of the summer—it brought the people of Whitney together. Janet McRae Webber remembered:

> Many were involved in the Red Cross dinner. The dinners were mammoth undertakings, and were essentially run by the women. The men would set up, but it was the women who did the meal and the organization. The Red Cross was the big organization. The Red Cross was very inclusive. The doctor's wife was active in all these things as well. Mother [Helen McRae], Marj and Granny [Janet McRae], and Mrs. Munro. There were lots of other people involved, such as Mrs. Phil Roche, and two war brides, Edna Nicholas and Mrs. Stubbs. I remember one year it was cancelled, because it was such a bad forest fire year. There were only skeleton crews on at the mill—they were basically fire protection. They [MNR] drafted the remainder of the mill men and bush crews to fight the fire.[335]

After Dr. Post died, other physicians ran their practices from the Red Cross house which had been given to the town. The Red Cross Hospital closed in 1978, when it was replaced by the new medical centre which opened in September of the same year.

Chapter 11: Whitney, Its People and Algonquin Park During and After World War II

Photo 84 **Red Cross Outpost Hospital** Source: APM 609

"The Automobile Road" from Whitney to Lake Opeongo

Long before provincial highways reached the village, there was a road that ran some 22.4 km (14 miles) northward to Sproule Bay on Lake Opeongo in Algonquin Park (Map 10). Why was it there?

The St Anthony Company built a railway in 1903 to haul pine logs from Lake Opeongo to its mill at Whitney. The Munn Company, which had bought the St. Anthony mill, was ordered, in its 1910 settlement with the government, to remove the rails from the line, but apparently they didn't. When the rails were removed as salvage is not clear, but it was likely before 1916.[336]

At that time, the fishing was good in the lakes adjacent to Whitney, but some fishermen always want to go where the fishing is great. Opeongo had great fishing and those wanting to fish the lake kept the railroad bed passable as a cart trail.

In 1916, Jack Whitton, a butcher in Whitney and father of Charlotte Whitton, later mayor of Ottawa, was awarded a commercial licence to net lake trout and whitefish in Opeongo. This enterprise lasted two years. To bring out the fish, Whitton used a horse-drawn wagon and the railroad bed, taking four bumpy hours to make a one-way trip from Sproule Bay, Lake Opeongo to Whitney because the railroad ties were still present (Map 10).[337]

To have a car in the twenties was something quite unusual; most people certainly didn't. But to have a car in Whitney in the twenties was very unusual, since there were no highways into the village. Some people have a compelling urge to go somewhere—just anywhere, and a road made that possible. In Whitney, the easiest place to develop a road was along the abandoned railway bed and cart trail to Lake Opeongo. With a road to go on, cars appeared.

In 1925, "Sandy" Haggart of Whitney was issued a licence of occupation on Sproule Bay, Lake Opeongo by the Department of Lands and Forests for a tourist business known as Opeongo Lodge.[338] The ties, however, had not then been removed from the abandoned railway bed. To help Haggart remove the ties, a young George Holmberg, then working as a blacksmith for Booth Lumber, made two large iron hooks that were dragged behind a wagon. These hooks gradually removed the ties from the line. The wagon road was then improved by dragging ties to smooth out the road.[339] With the road improved, Haggart obtained a Model A Ford and used it on the "automobile road" in 1924. The car must have been shipped to Whitney on the railroad because there were no roads into the village until 1930.

Other locals also used cars on the road. Janet McRae told the story of one excursion.

> We had a little bit of an old Ford car with no top on it. The trees were only so high—they had to take the top off it or they couldn't get through. One time we thought we would go towards the Opeongo to pick berries. Donald [13 years] and Marjorie, who was about six, and Beatrice Shalla and I heard the "thing" grunt at me, there in the blueberry bush. So I let a shout out of me, and told the kids to run to the car. I got them into the car all right. And I turned around and Donald was out of the car with stones. Him and Marj were throwing stones at the bear. The bear disappeared into the gully. Well now, if that bear had had any cubs, it would have turned on them! I was screaming! I was hanging onto Marj and Beatrice—they were only little things.[340]

The development at the Sproule Bay end of the road has had an interesting history. In 1930, Haggart purchased the bunkhouse built by the St. Anthony Company known as "Red Row." Pine board salvaged from this building was used to build two additional cabins at the Opeongo Lodge site.[341]

Improvements on the road continued:

> In the fall of 1930, $750 was allocated for "hazard disposal" (i.e., road improvements and brush clearing) along the 14 miles of road from Whitney. This was the beginning of "The Dirty Thirties" and half of these funds came from the provincial government's special allotment for the relief of unemployment.[342]

However, the level of water in Lake Opeongo interfered with the road in 1930. The J.R. Booth Company helped pay for the costs of fixing the road as their dam at Annie Bay was causing the problem.[343] The Opeongo Road, much shortened, was then connected to Highway 60.

Soon, more than fishermen and tourists were using the road. By 1935, scientists beginning fish research studies were using the road. In 1936, these studies were expanded with the establishment of the Opeongo fish lab. The Department of Lands and Forests rangers and fire wardens had also built accommodations at Sproule Bay. McRae Lumber used the road to service its bush camp on Sproule Bay and to haul logs from 1946–1949.

The road still received a great deal of tourism-related traffic. In 1936, Joseph Avery bought Opeongo Lodge and leased the site under a licence of occupation. In 1946, he installed an in-ground British-American gas pump and tank.[344] The tank was used, but later, during renovations, it was covered over and forgotten, only to be discovered some 57 years later. During excavation for an extension to the outfitting store, the buried gas tank was found when a front-end loader hit it and created a flash explosion. The contents of the forgotten tank were then carefully removed along with the surrounding contaminated soil. Construction was then completed.[345]

In 1976, Svend Migland won the bid for the concession from the MNR. Later, Bill Swift took over the concession and operated it under the name Opeongo Algonquin. The Avery's re-established their outfitting business in Whitney under the name Opeongo Outfitters while they continued to operate their water-taxi service on Opeongo for campers and fishermen.

Whitney in the 1930s

The decade-long Great Depression that began in 1929 created great hardship for all. Whitney survived with resourcefulness and often good humour, but after 30 years of wear and tear, ten of which were under government ownership, Whitney was not an attractive place.

Three roads broke Whitney's isolation and total dependency on the railroad: a road eastward to Barry's Bay, one to the south through Sabine Township to Maynooth, and Highway 60 through Algonquin Park. But there was a price. The following story may bring a smile now, but the passing of the cows was a sad day for Mrs. Cenzura.

> We had cows for a long time, then the police came here and stopped that … there was always cows on the highway. The policeman said if something happened somebody could run into that cow of yours or someone else's, you're going to pay for all the damages to the car and then if somebody gets hurt they gonna pay for that too. So everybody got rid of them.[346]

The roads may have been opened, but their condition was something else, being gravel and corduroy. Janet McRae remembered her first experience of driving to Whitney.

> They started to build the road from Barry's Bay the year after I came up [1930]. I'll never forget the first trip I had up here from Eganville. It was in our Willys-Knight. It took us all day. We left at seven in the morning and never got

here [Whitney] till eight o'clock at night. Straight driving! It was the mud holes. The roads were made with old logs and then covered with sand. Bump, bump. Corduroy roads. And the hills! Oh, it wasn't like going up a hill. It was like going up a mountain! You'd just make a rip at it and half the time back down. I didn't make that trip often. I came in and out by train.[347]

The people coped by tending their gardens, keeping a cow, fishing and hunting, and working at anything that became available. Mrs. Alex Cenzura remembered:

> They put us on relief—$7 for the four of us. We had to live on those $7 a month! They'd go out on Thursday and get enough fish for Friday. We couldn't eat meat on Friday in them days. Now you'd have to fish all week to get enough fish. Fishing was very, very good when we come here. When times were hard, we lived on lots of fish—and lots of wild meat. Ha, Ha! There were lots of deer, and wild ducks and rabbits. Years ago you could get away with hunting out of season and poaching in the Park. Now, they don't even dare try it. If one neighbour got a venison, they'd have to share it because it wouldn't keep. So everybody got a snack. Everybody would share; it was good that way.[348]

Donald McRae and Felix Shalla used to go up Mud Bay to hunt deer in the 1930s. They would shoot a buck and bring it down to Paul Shalla who would gut it and distribute the

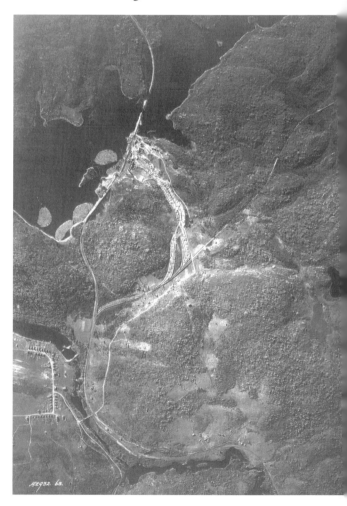

Photo 85 **Aerial Photo of Whitney and Airy, 1930**
Airy lies to the top. Crossing the entrance to Mud Bay is the CN railroad, which cuts through the centre of the picture and follows the north shore of the Madawaska River off to the east. Connecting Airy to Whitney is Algonquin Street, which crosses the Madawaska River. A short distance to the west of this bridge, the approach to the railroad bridge that once joined the St. Anthony/Dennis Canadian Mill to the CNR can be seen. Outlines of the railroad siding attached to that mill still show. The dam across the river is upstream from that point.
Source: Aerial photo A2932.63, September 3, 1930. National Air Photo Library.

meat to the families around Airy. Bob McRae remembers his father telling him how he and Felix shot a doe once and Paul Shalla just about shot them.[349]

Airy was a very small, close-knit community in the 1930s, comprised of the Cenzuras, Shallas, LaVallys, Chenskies, Leveans, Kuiacks and Micowskis.

The women had to contend with their men being in the bush over long stretches of time—say three months to the Christmas break and a similar period starting in the New Year. Mrs. Henry Taylor, of Bancroft, commented on life in these days. Living conditions in the Bancroft area would have been similar to those in Whitney. Her husband worked for McRae from 1927 to 1942.

> If you didn't have your husband there, you have to have some person who was a responsible individual to stand to your back. Some of the women, I presume, would be pregnant and that would make a difficulty too, in feeding the cattle and so on. But I presume they would make arrangements with some, maybe some of their relatives to come in to do the heavier work for the time being.
>
> I always honour the women of this locality because they must have been very brave and strong-minded to endure some of it, because it was back before the tap, the push button lights and the whole bit. The whole thing rested actually on those women, carrying through that length of time, but it was needful because of their incomes. That few hundred—I don't know how much they would come home with, but it meant so much to their particular situations.
>
> Now, on a farm, understandably, they would have cattle to sell. But that was a spasmodic thing. They would have two or three head to sell now, and maybe in the fall they would have more, but their income was a hit-and-miss type of thing because they could take eggs to the store and get what they wanted. I say what they wanted, but rather what they could buy at that time. You can say what you like but the men were strong, healthy men in a general way. But back of them stood a crowd of really brave people. Because, as I say, the women was there with the children and, of course, generally had to have someone in there to stay with them, but that wasn't always the case. A great many of them could manage very well on their own, but it meant trailing in water, and filling the wood box.
>
> Of course their children were taught to do that. It was just as common as breathing when we came from school, and I didn't grow up in this area. We had to fill the wood box and to see about the water supply, feed the hens and all these sorts of things that we could do. Everybody contributed. Everybody was part of the getting-along situation. The children in this area, naturally they would have wood boxes to fill and all that sort of thing.[350]

Phil Roche remembers his arrival in Whitney:

> I arrived in Whitney on February 25, 1935 on a CN train consisting of a steam locomotive pulling two freight cars and a combination baggage car-passenger car

and caboose. I had been hired by O.E. Post, general merchant in Whitney.

After crossing the bridge on the Madawaska River, we headed to the hamlet several hundred yards away. This was the only bridge for vehicular traffic at this time and was the main artery through to the village. The highway bridge was not built 'til the early forties. There were a few cars belonging to American tourists who left them in the care of Mr. Post when not in use. These old stables were behind the present property. On the same side of the road was an icehouse belonging to the store, and nearby the blacksmith shop owned by Henry Fuller.

My first glimpse of the village proper was shattering to say the least. On top of the fairly high hill on the left was a row of unpainted wooden houses with wooden shingles. This was Church Street and its length was between the Roman Catholic church and rectory at one end and continued past the combined St Anthony and St Andrew's Anglican and United Church and met with Paradise Road. From Church Street down the hill as far as the general store was called Post Street and it was lined on both sides with larger houses, also unpainted and with very weathered wooden shingles on the roofs.

From the store south to Pine and Mr. Stubbs' property was Ottawa Street, in the same dilapidated condition except for the Red Cross outpost, which stood out like a sore thumb with a coat of gleaming white paint. These three streets made up the hamlet as originally built by the St.

Photo 86 **Whitney Store**
Gasoline for cars was available in the village but a horse harnessed to a sleigh was a common means of transportation in the 1930s during a Whitney winter.
Source: APM 04310

Anthony Lumber Company; [the town] had subsequently been taken over by the Ontario government, and leased back to Mr. O.E. Post for $1,100 annually. He in turn rented out the houses on Church Street for $2.50 monthly and those on Post Street and Ottawa Street, for $5 monthly. No other business or building was allowed in this area. The two-storey store had a warehouse attached, and across the road were a butcher shop and another ice house. Two Imperial Oil gas pumps stood beside the store.

Ottawa Street ran right through to what was called Hay Creek Road; there was a settlement there. When you went up Church Street to the top, there was Paradise Road. When you came down the hill, you turned right where the municipal building is, and before you crossed the bridge there was a road that went down along the river, which was Jerusalem.

Chapter 11: Whitney, Its People and Algonquin Park During and After World War II

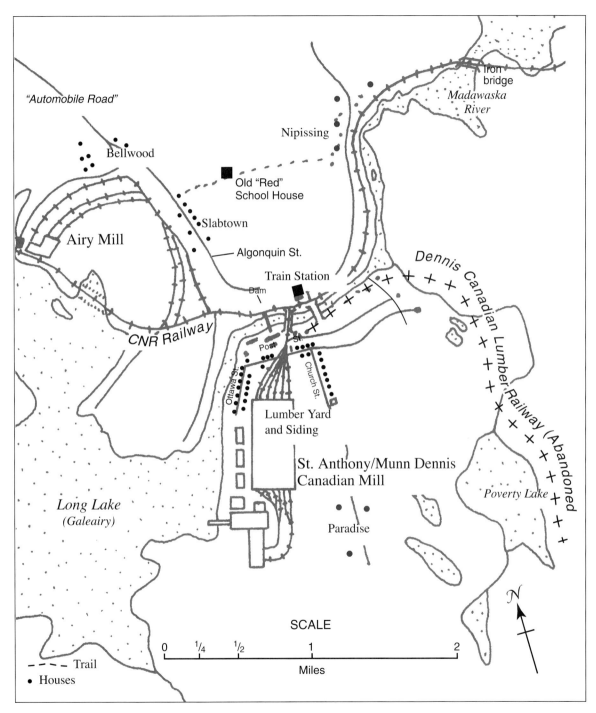

Map 21 **The Village of Whitney and Vicinity, Early 1930s**
Source: Jamie McRae/Mervin Dunn.

When you crossed the bridge, there was a road that went down along the other side of the river, which was Nipissing. When you went a little farther towards the Park near the police station—it was called Slabtown. There was a railway spur there where McRaes piled their slabs—some of which were "retrieved" and used for siding on houses. Of course, Airy itself was a settlement, where the McRaes and some of their employees lived.[351]

Whitney was not a pretty place in the thirties. Bernie Stubbs Sr. remembered:

> The boiler [of the St. Anthony/Dennis Canadian mill] couldn't burn all the sawdust generated by the mill. Great quantities of sawdust were dumped in Galeairy Lake and all over Whitney, as far down as Bernie's house and up toward the current location of McRae's mill. Deposits were always catching on fire.[352]

This dumping took place despite the 1873 law "prohibiting the throwing of sawdust, slabs, edgings, bark or refuse into any part of a navigable stream."[353] Bernie Stubbs continued:

> Pine logs filled Galeairy Lake. They had a series of stone piers that ran from the mill to Post's Island. At times, log booms would break and the river wasn't passable by boat. There were two cedar mills making shingles on Galeairy Lake. They dumped their waste directly into the lake, as well as depositing waste on nearby land.[354]

In the Whitney area, the Depression years from 1929 on through the 1930s were difficult times. The Airy mill was shut down in 1930 because there was no market for lumber. Almost everyone in Whitney and Airy was unemployed. There was no unemployment insurance, and welfare payments were very small.

Until 1929, the McRae family lived in Eganville and summered at Barry's Bay. Then the Great Depression forced the family to move to Airy, which was then considered a part of Whitney. Whitney was effectively an island in the forest with the Canadian National Railway providing the only link to the outside world. There were no roads into the village.

"A large white building still stands on Mud Bay of Galeairy Lake. It was the original boarding house, post office, and office of the McRae Lumber Co. at Airy."[355]

Janet McRae gave up her fine home in Eganville and came to live in this building.

> When the Depression came in 1929, why, things were bad. And so I came up there to live then. But the bad times came along. We had this business here,

Photo 87 Railway Trestle with Airy in Background
Source: APM 05613

and, well, it was either lose our business, or sell the Eganville house and me come up here. So what could I do? I had to come. So we ended up in Whitney.[356]

The McRaes may have been mill owners and better off than most, but the Depression required sacrifices of them as well. Janet McRae remembered Whitney in the thirties.

> Whitney looked pretty bad. The houses weren't painted or anything—an awful

Photo 91 Henry Fuller Walking to His Blacksmith Shop
Source: Larry Fuller

Photo 88 Two Young Girls on Post Street
The expression "A photograph is worth a thousand words" certainly sums up the condition of Whitney well into the 1930s.
Source: Mrs. Bissonette. APM 05715

> looking place. The government owned the houses. Nobody would fix them up. Everybody was waiting to get them. The foundations were good and there's lots of good lumber in them because they built those things out of pine.[357]

Photo 90 The Stirling Bank of Canada and St. Anthony's Horse Stables
The Stirling Bank was on Post Street. The building behind the bank is the horse stable. The lady is Mattie Johnson.
Source: McRae Lumber Company

Janet McRae remembered water being an issue in the early days in Whitney.

> When I came up here, there were people living in this house in Airy—the McBrides. Ben McBride had been manager of the office for the company that we bought it from. It had been the office house. They all slept here that worked in the office.[358]
>
> We had our own electricity, but there was no water in the place. They used to bring barrels of water on stoneboats. Did you ever hear of a stoneboat? Well, a stoneboat is a flat thing that a horse will pull; it's flat down to the ground. There were no wheels on it. They used it for drawing stone. When I came here there was no bathroom, so I got them to build a place outside. I had a bath put in, and you could let the water go. They used to have to bring me two barrels of water every day outside of this bathhouse. They left a big window and you just dipped the pail down through the window and pulled it up and warmed some water. And poor Mr. [Paul] Shalla used to say, "Oh, Jan, you use a lot of water. My God, what youse doin' with all that water? Two barrels a day, that's really bad!"[359]

In a 2003 interview, Janet McRae Webber later reflected on what her grandmother's feelings could have been on coming to Whitney and her position in the village.

> I think she was quite a person—I think she was into that vision of what a lady in the small town should be and played the part. She was conscious of manners and what was proper behaviour. For her, coming up to Whitney was the end of the world. She had grown up in a middle-class, small town in the Ottawa Valley—Eganville. Her family was very affluent in her childhood—I think at the end of the square-timber logging era. She had a couple of brothers who went to university. She had come from a very comfortable family, but one that had a sense of their-own middle class respectability—she was a MacGregor! She could remember when Queen Victoria died—she was very much of that era. She had been surrounded by family and friends when young—a very comfortable social circle. Going to Whitney was going into a totally different environment.[360]

Government Work in the Depression

Highway 60 was begun through the Park in 1933 and the gravel road was completed by 1936. The highway was one of Conservative R.B. Bennett's relief-work projects, as was the clearing of an emergency airstrip at Lake of Two Rivers in 1935.[361]

One labour camp was located very close to McRae's new mill. Soon the relief workers were bartering their quality army clothing for cigarettes or moonshine, while the mill men began to look like soldiers.

Lorne Boldt said, "From those airport guys I got a lot of clothing. Boots—brand new boots—for a dollar. Jeez! Good mule skin mitts for workin' in the yard—ten cents. Pants, shorts, mackinaw coats. The fellows were only getting 20 cents a day."[362]

The roads meant jobs. The wages were very low, yet the men lined up to get them. Mrs. Alex Cenzura commented, "They was making a road. There was no road from Sabine to Lake St. Peter or to Whitney. So they were making one. One month with a team of horses, he got $25—a team of horses and a wagon and him! A bushel of oats was 75 cents. And the team, they was working hard for only $1.50 a day."[363]

Frank MacDougall: Superintendent, Algonquin Provincial Park, 1931–1941

In 1930, needing to make a change in Park administration, Finlayson, Minister of the Department of Lands and Forests, appointed J.H. MacDonald, a university-trained district forester in Pembroke, as Algonquin Park superintendent. Unfortunately, McDonald died four months into the job. The minister then turned to Frank MacDougall, a war veteran and also a university-trained forester to fill the position. The MacDougalls lived at Cache Lake in Algonquin Park.

MacDougall saw Algonquin through the crushing years of the Depression. Then in 1941 he left for Toronto and the position of deputy minister of Lands and Forests. He served in that position until 1967 and built that department into one that was capable of managing all of Ontario's forests.

Frank MacDougall was born in Toronto but grew up in Carleton Place, Ontario. While enrolled at university, he enlisted in the artillery of the Canadian Army, serving in World War I. At Vimy Ridge, he was gassed and suffered lung damage. His war experience showed throughout his career in the very warm spot he reserved for ex-service men.[364]

MacDougall spent his first year after his army service working for the Forest Commission, and then he entered the University of Toronto Forestry School. During two summers, while at university, MacDougall worked surveying the forests of northern Ontario and the James Bay area. The use of flying boats captivated MacDougall, and this led to his being taught to fly by George H.R. Phillips and in him gaining his commercial pilot's licence in 1930.[365] His interest in flying was lifelong. He was an active supporter of the Provincial Air Service and he enthusiastically supported the development and use of the famous de Havilland Beaver, Otter and Turbo-Beaver aircraft. He was three times awarded the McKee Trans-Canada Trophy for contributions to Canadian aviation, and in 1974 he was elected to the Canadian Aviation Hall of Fame.[366]

After graduating from the University of Toronto with a degree in forestry in 1923, MacDougall joined the Department of Lands and Forests as an assistant forester in the Pembroke District. In 1924, he was transferred as forest inspector to Sault Ste. Marie and became district forester in 1925. He started the Kirkwood Forest, the first management unit in the province, where some 630 square miles of land that had been made a virtual desert because of heavy logging and fires were reforested. The unit became an important centre for research and experiment.[367]

Before he accepted the position in

Algonquin Park, he insisted on making an administrative change. He amalgamated the year-round park rangers and the formerly seasonally-employed fire wardens. Tom McCormick, the chief fire ranger, became chief ranger of Algonquin Park, based at Whitney.[368] This move saved money, as the retired older staff that could no longer do the job efficiently. MacDougall later subdivided the Park into administrative units with deputy headquarters at Brent, Achray and Whitney.

Socially, the MacDougalls mixed with both lumber interests and with lodge and cottage leaseholders in the Park. Among those who knew him as a friend were J.S.L. McRae and his wife, Janet. In 1986, Janet recalled, "His wife was a friend of mine; they lived up at the lake where the hotel was, above Two Rivers. She used to come down to Whitney often."[369]

Janet noted that her brother Bill "was a war veteran, a captain in the British army's Black Watch Regiment. He and Mr. MacDougall were very great friends—and Bill used to always go up to see Mr. MacDougall."[370]

MacDougall's direct contributions to the Park were numerous and long lasting. As superintendent of Algonquin Park, it was his duty to look after the administration of health, game, fish, timber and aviation matters. In a floatplane, a Fairchild KR-34, he made near-daily patrols, looking for fires and poachers, and transporting personnel about the Park. He was also very active in improving radio communications.

The construction of Highway 60 was perhaps the single most important event during his superintendence and it didn't come without problems. The road, initially a dirt and gravel road, was begun in 1933 and opened during the summer months in 1936 to through traffic. The highway was first opened for winter travel in 1940, paved in 1948, and improved and repaved in the 1960s and 1990s.

Department of Lands and Forests Minister William Finlayson pushed the highway as a Depression work project in 1933. MacDougall saw the highway increasing the recreational potential of the Park and looked to improving hotel and lodge accommodation.[371]

The Cache Lake community, however, objected to the idea; they were joined by the Ontario Federation of Anglers, who claimed that the road would have a detrimental effect on fish stocks due to an increase in fishermen using the road. A short while later, the same leaseholders backed the highway when their beloved railway talked of closing down the line, which was then their only means of accessing Cache Lake.

To keep the road aesthetically pleasing, MacDougall saw to it that there was no cutting for lumber along the highway. Tourist accommodations were improved, as Killarney Lodge was built on a private lease on Lake of Two Rivers in 1935. A 5.7 hectare (14 acre) campground was opened at Lake of Two Rivers to accommodate highway campers, followed by two others at Tea Lake in 1938.[372]

In 1934, MacDougall sought "to emphasize scenic and wildlife protection, recreation, and visitor education, rather than the exploitation of natural resources."[373] This was a concept borrowed from the National Parks Service of the United States. The idea of setting up nature sanctuaries as ecological benchmarks was quite radical for the day.

He learned of the sanctuary concept from J.R. Dymond, a biologist at the University of Toronto and a founder of the newly-formed Federation of Ontario Naturalists. Dymond, a leaseholder on Smoke Lake, became a friend whom MacDougall frequently consulted. The first pamphlet of the FON, entitled "Sanctuaries and Preservation of Wildlife in Ontario" (1934), stated,

> In most civilized countries today, sanctuaries are being set aside for the preservation of representative samples of the natural conditions, including the plants and animals, characteristic of those countries. This movement for the preservation of nature as a whole has important points of difference from the conservation movement as ordinarily understood.[374]

Killan, quoting from the FON pamphlet, continued by saying,

> Since the nineteenth century, that movement has been "frankly based on utilitarian motives." The wise-use impulse was still "deserving of support," argued the FON, "but there is real danger that overemphasis upon the conservation of a particular form may prove detrimental to the preservation of nature as a whole … The need for nature preservation does not depend only on the economic value of the natural life usually considered worthy of consideration."[375]

Research took on a prominent position in Algonquin Park with the establishment of a field laboratory at Cache Lake by Professor J.R. Dymond and Dr. E.B. Ide in 1933. The fish laboratory was built at Sproule Bay, Lake Opeongo in 1936. This facility has since been renamed the Harkness Laboratory in memory of its founding director. MacDougall also started limited research initiatives aimed at improving general resource management. Topics that MacDougall may have encouraged included a report on birds in 1938, a census of beaver lodges in 1939–1940, and a survey of moose distribution in 1941.[376]

During MacDougall's time, Park regulations controlling logging began to change. MacDougall recommended that

> as timber licences came up for renewal, clauses be inserted to establish no-cut reserves along shorelines and portages of major canoe routes. Such a policy would give priority to the recreationist, not to the timber operators. The licence holders, added the superintendent, should be reimbursed for the areas withdrawn from logging.[377]

MacDougall did not seek a ban on logging in the Park, for he believed in multiple use of parkland: that logging, recreation, and nature conservation could coexist in an area the size of Algonquin. But he was frustrated by reports of canoeists bumping into timber operations. Incidents of encroachment by timber operators infuriated him. For MacDougall, "there are very few accidental trespasses. It is quite discouraging and wasted effort to work for such reservations and then … have a 100-year setback to the scenic attraction of a Park."[378]

In a foreshadowing of what was to come from recreationists and protectionists, the Federation of Ontario Naturalists was quoted in *The Globe & Mail* of November 3, 1938 as saying that Algonquin Park "was being ruined for all time by present lumbering activities."[379]

In an attempt to avoid such situations, MacDougall sought to separate the two user groups:

> All timber licensees were informed, in June 1939, of a new standard shoreline protection policy, designed "to establish a just and fair balance between the interests of the lumberman and the tourist." Henceforth, scenic preservation would be given priority over timber extraction as the first principle of multiple land use in Algonquin Park. Park regulations now decreed that no timber could be cut "within 300 feet of any lake or highway or within 150 feet of any river or portage" unless special dispensation was obtained from the superintendent. Significantly, the new regulation had teeth. "In future," declared then Deputy Minister Cain, "any trespass on a reserved area shall be subject to a minimum charge of five times the regular price of the timber illegally extracted."[380]

In May 1940, the first nature reserves in Algonquin Park—two small virgin pine areas containing a mere 100 trees—were established for both educational and scientific reasons out of J.R. Booth and Gillies Brothers leases. Both companies complained, much to MacDougall's chagrin.[381]

In his time as superintendent, MacDougall was unsuccessful in establishing a position of full-time forester-biologist, which held back the accumulation of basic knowledge of wildlife, and, in consequence, the scientific management of wildlife lagged. Wolf control was altered from the intent to exterminate these animals to one of control.[382] Illegal trapping or poaching became less of a problem because of enforcement measures and the fact that the harsh conditions of the Depression eased with the onset of World War II.

MacDougall left the Park pleased that Algonquin was the first large area in the province to be administered under the concept of multiple use. He took up his appointment as deputy minister of the Department of Lands and Forests in 1941. MacDougall then set out to build and reform the department both in administration and policy.

John Stanley Lothian McRae

J.S.L. McRae dominated Whitney from the early 1920s to 1969. The Post family was justly beloved—but Whitney survived because of J.S.L. and the people knew it.

Observations from the men that he worked with paint a picture of a commanding man.

J.S.L. knew first-hand the hard work he was asking of his men, according to Frank Shalla.

> Before his wife was livin' there at Airy he used to come to our place after supper every night; he'd be up there and talk to Dad [Paul Shalla] at night. Dad called him

Chapter 11: Whitney, Its People and Algonquin Park During and After World War II

Photo 91 McRae Alligator, 1922
J.S.L. McRae is the gentleman on the left wearing the white hat and holding his "picaroon" or pike pole. This was an early model of an alligator that J.S.L. may have owned before he purchased the Airy mill.
Source: McRae Lumber Company

Jack, but I called him Mr. McRae. Dad used to say when they went up to Hay Lake, you wouldn't know he was the boss at all. He just put on old clothes and he'd drive them logs. He'd be working with a bunch of men and they wouldn't know him. He'd be out there with a picaroon [pike pole], drivin'. That's what Dad said.[383]

J.S.L. knew his town. Alex Cenzura said,

McRae liked talking with the men. He liked to know what was going on. He used to come to Greg LaRochelle for that, at the Shell Station [Algonquin Lunch Bar]. So he'd be sitting there and the young lads would be coming in from all over the place. And he finds out one

Photo 92 J.S.L. McRae
Source: McRae Family

thing from one and another thing from the other. He knew everything, everything that was going on.[384]

With the purchase of the Airy Mill, McRae had to become knowledgeable about its operation. He didn't have to know every job in the mill but he had to know when things were not right. Then there was no question that something was going to happen. J.S.L. was not a quiet man, as his daughter Marjorie attested.

> If the production started dropping in the mill, for example, or the grade started dropping, he'd get pretty angry. He'd end up at the mill there. I remember when Shereck was there he broke three saws in a week and that was the end of Shereck [Shereck was the head sawyer at one time]. The last saw he broke, he was drinking too much. He'd been drinking all night then tried to come and start sawing. Dad walked into the mill the last morning he broke the third saw and chucked him out. He never came back.[385]

J.S.L. had a reputation of being gruff at best. But the men knew him, and knew that if they did their job as required, he would back them if they needed help. His brother-in-law, Duncan MacGregor, said, "John was tight and miserable, but if you came after him for money, he'd give it to you. I guess he'd get it back, but lots of people wouldn't give it to you. Lots of fellas building a house'd need help, and he'd give it to you."[386]

Whitney was J.S.L.'s town. He was the major provider of work. The hamlet was home to many of his men and most of them were Roman Catholic—not his faith. Yet when the parish built a new Catholic church after the war, McRae made a sizeable donation according to Duncan. "When they were buildin' the church, why I'll bet you McRae gave a lot of money, over $50,000."[387]

While the mill was at Lake of Two Rivers, J.S.L. had a serious accident that kept him away from the office and the mill for the better part of a year. In 1933, the trestles on the railway line between Lake of Two Rivers and Cache Lake were condemned and through traffic on the line was discontinued. Rail service continued to the mill, but on a need basis. Since Highway 60 through the Park had not then been constructed, a method of moving men to and from the mill was needed. A gas car owned by the CNR was used for this purpose. The mill had a small gas jigger of its own, suitable for up to four men. Neither of these vehicles had brakes—you simply coasted to a stop! When the smaller jigger was used, a phone call was typically made to make certain the line was clear. On the day of the accident in 1934, the cautionary call from the office to the mill was not made and a crash occurred. J.S.L.'s wife, Janet, remembered it vividly.

> You see, they stopped bringing the train up here. They had a great big gas car for passengers—to bring the men up and down from the mill. There was no road then. They always called the line to find out if the gas car was coming down. But this time the fella never called to see if the gas car was coming, and they got right down, and that thing wouldn't stop. It just smashed into them, threw them all off. He [JSL] was thrown on the other

side where the rocks were. He broke his leg in two places and his arm in two places and hurt his back. He was laid up a whole year and never walked a step to the next December. He was in a cast for four or five months, and then he was in a wheelchair and he had to take therapy down at the hospital.

He'd phone every night to see if the truck was in. Well, he'd look after everything. It was too hard on him. It was enough to kill anybody. They got along just the same without him. I mean, he was payin' those men to do that, why should he have to do it himself? But he was just like that. He'd check on everything.[388]

While J.S.L. was in hospital, Duncan MacGregor kept him in touch with operations.

When John was laid up in the hospital in Ottawa, I'd go down every two weeks or so, and he'd tell you what to do—write it on a piece of paper. If he didn't like the way somethin' was done, boy he'd straighten it up! He was so goddamn used to boss'n' everybody, he'd boss the nurses in the hospital and everybody! He was always the boss, you know.[389]

Remembered by Gary Cannon, "J.S.L. had a bad accident with the gas car once, and one leg stayed stiff. He couldn't bend his knee."[390]

J.S.L. was somewhat of a terror when behind the wheel of a car. Fortunately, traffic in and around Whitney was very light. According to Gary Cannon,

He always had a car with a standard transmission in it. He'd put it in low gear, step on the gas, rev her right up and just let out the clutch. He'd dig a hole 15 feet in the gravel and away he'd go. He got by 30 years like that.

Finally the last car he had, a '65 Pontiac, somebody talked him into buying an automatic; he'd always forget to put it into park. Right in front of John's house [then J.S.L.'s house at Airy], the road was kinda tilted a bit forwards, and then there was a wire fence across the front yard. He'd get out of the car and go into the house, and his car it'd be up against the fence. It was all scratched from that fence.

J.S.L. always wore a grey felt hat. He'd put his seat back as far as he could. You could just see the top of his head stickin' through the spokes of the steering wheel, and he was over six feet. He was quite a lad, I'll tell ya.[391]

Nephew Roy McGregor wrote, "J.S.L. was a driven workaholic who had no ability whatsoever to relax. His rare breaks he took by drinking and he had a cardinal rule of refusing to do business whenever he was on a binge. 'That was his holiday,' Duncan often said."[392]

Apart from business trips and his annual attendance at the Lumbermen's convention, J.S.L. didn't travel far. He did travel on business, often with a driver.

However, J.S.L. was not all business; he enjoyed spending time with his wife and both of them greatly enjoyed spending time with their family, in particular the grandchildren. His daughter, Marjorie, remembered:

Small things pleased him, and he loved children and animals. When a dog came here and chewed into the television, chewed it all out. I don't know how much it cost to get it fixed, but he never said anything.

I remember the first time we went to Florida, Mother wanted Dad to go. He thought about it a bit, and then said, "Oh no." He wouldn't go. He loved to watch sports. He'd love to go and see the ballgames. The one thing he said he would have to take off was his long underwear. That was his excuse anyway.[393]

Whitney and World War II

Whitney survived the Depression only to become involved with World War II. In 1941, the census showed Canada had a population of some 11.6 million. Montreal was the largest city with 890,000; Toronto had 657,800 inhabitants and Vancouver had 272,000. Canada was still 46 per cent rural.[394] Sons joined up to fight, despite their father's 1918 victory in "the war to end all wars."

Early setbacks and casualties ended any sense of romanticism about the war. Initial losses were suffered by the navy and air force; pessimism and grief spread across the country. Newspapers periodically printed casualty lists—they took up pages. The army was largely formed from locally-based units. When ground forces went into action, whole communities were devastated by losses. Defeats at Hong Kong and Dieppe were portents of what was to come.

The record of those who served from Whitney and the local area is preserved in church records. Despite a population of less

Photo 93 Felix Shalla and Donald McRae at Airy, 1941
Felix Shalla was a casualty on D- Day plus 2.
Source: McRae Family

than 900, 95 men and 4 women from the area served their King and country. Six men from the Whitney area were killed in action.

Donald McRae joined the Air Force in 1940. Then a second McRae went overseas.

Marjorie (McRae) McGregor

Asked why her Aunt Marjorie went to England with the Red Cross, Janet McRae Webber ventured, "She had just graduated from high school, she was from a small town and her brother was overseas." However, proud as she was of her service in the war, Marjorie's mother, Janet, was also concerned for her daughter. "Granny would have been very much for 'King and Country,' yet on the other hand [it was] 'Her daughter'!"[395]

Chapter 11: Whitney, Its People and Algonquin Park During and After World War II

World War II Servicemen and Women from the Whitney Area

Affiliated with the Anglican and United Churches

Annie Bowers	Fred Bowers	Dick Callighan	Harry Cheeseman
Hiram Conroy	Maxwell Conroy	Fred Doolittle	Gordon Dufresne
Arthur Eady	Charles Eady	Eldon Eady	Lloyd Eady
Allan Fox	Howard Fox	Murray Fox	Guy Heggart
Jack Heintzman*	Clarence Holmberg	Hartley Holmberg	Dennis Hyland
Mike Lundy	Irvine McCormick	Edelore McGuey*	Donald McRae
Marjorie McRae	Kenneth Shields*	Norman Stubbs	Walter Stubbs
Gerald Walters			

Affiliated with St. Martin's Roman Catholic Church

Frank August	Harvey Bissonnette	Leo Bissonnette	Emanuel Bowers
Henry Bowers	Norman Bowers	Cyril Coghlan*	Eddie Coghlan
Jimmy Coghlan	Roger Coghlan	Aurel Cousineau	Francis Cousineau
Ray Cousineau	Joseph Dargus	Stanley Dargus	Leo Duchaine
Lorne Duchaine	Alfred Dumas	Aurel Dumas	Godfrey Dumas
Albert Caverly	Leonard Caverly	Mabel Caverly	Ted Cenzura
Tom Chapeskie	Lloyd Easto	Edward Frederick	John Gorgerat
Julius Gorgerat	Bennie Kubisewsky	Joe Kubisewsky	John Kubisewsky
Peter Kubisewsky	Dominic Kuiack	Maxie Kuiack	George Labarge
Leslie Labarge	Gerald Laginski	Bud Lalonde	Mike LaVally
Alex Luckasavitch	Bernard Lynch	Earl Lynch*	John Lynch
Sylvester Lynch	Ted Lynch	Bill Lystiuk	Fred Lystiuk
John Lystiuk	Archie Martin	Grant Mask	John McMeekin
Alex Mochulla	Fred Mochulla	Charlie Nicholas	Joseph Nicholas
Paul Nicholas	Elvin Olmstead	Gordon Palbiski	Wallace Palbiski
Vivian Pigeon	Felix Shalla*	Isadore (Bob) Shalla	John Shalla
Anthony Spielleck	Michael Warrankie	Stephen Warrankie	Edward Werroneau
Stephen Werroneau			

* Killed in Action

Photo 94 Marjorie McRae Overseas with Her Ambulance
Source: McRae Family

In a taped interview, nieces Janet McRae Webber and Cathy McRae Freeman commented on the work that their Aunt Marjorie and other women contributed to life in the village.

Janet: When we were young, it was Marj [Aunt Marjorie, Donald McRae's sister] who was doing the more active stuff. Marj wasn't working then and didn't have a lot to do in terms of daily obligations, so she got involved. In a town like that you need people to do such things—to run the Cubs and the Scouts for the boys, and the Brownies and Guides for the girls.

Cathy: She taught Sunday school when we were there. Marj was involved in the Ladies' Aid, teas and that kind of thing.

Rachel McRae, in a later conversation, remembered that Marjorie

> fundraised for the Red Cross Hospital and was a staunch supporter of St. Andrew's United Church, the Scout organization and the Red Cross swimming program. At various times in her career, she taught Sunday school, supply-taught at the local Airy school and worked at the mill office. She was honoured with the "Most Valuable Citizen Award" in the Whitney 25th Anniversary celebrations.[396] Later in life, Marjorie married Alex McGregor, a widowed distant cousin who was clerk and manager of the Rock Lake mill.[397]

Donald McRae

Donald McRae was born in Barry's Bay in 1915 and went to elementary school there and in Eganville. In 1932, he went to Albert College in Belleville for five years, and this was followed by Ontario Business College. Donald McRae joined the business fulltime in 1937, beginning by clerking out of the Lake of Two Rivers mill. He served later as manager of the mill. World War II changed Donald's life as it did that of many others.

Don McRae enlisted in the air force in 1940. Donald was a top man in an essential war industry. He could have been excused from active duty.

> Donald had fought with his father, J.S.L., over a minor incident, and had set off to join the air force without permission. He got Dunc to come with him to Toronto, where they got gloriously drunk in the old Ford Hotel the night before Donald went into the service, leaving Duncan to explain to his furious brother-in-law why he had let his son go. It was made loud and clear to Duncan [by J.S.L.] that the mill could not afford to lose a second top man.[398]

Bob McRae taped conversations with his father in which Donald recounted many of his wartime experiences:

> After training, I was flying coastal patrol in the Maritimes. We flew Hudsons—down to Boston and up to Gander in Newfoundland and also in the St. Lawrence. If the convoys were going out, we flew out with them on patrol. We saw a sub once but we never got to attack it. I put in one tour with Coastal Command. The U.S. was still neutral at this time. The RCAF then asked for a gang to go across.
>
> In February 1942, I went to Britain on the Dutch ship Volendam. On the trip over, the destroyers were running around depth-charging and a sub came up. We were with one of the anti-aircraft guns, but we couldn't depress the gun enough—we should have had it.
>
> We landed at Glasgow and went to Coastal Command, flying Halifaxes from Yorkshire. We flew out towards Iceland. Twice we went to Gibraltar, when they were taking Spitfires to Malta. Once, coming back, we crept up on a sub and got it. Batch Elder, our pilot, said they found it out of the pen in France at the end of the war. The pen was the sub's concrete bomb-proof base. It was damaged enough that it was unusable.
>
> The RAF said they would give me more rank if I transferred to them. I ended up as gunnery officer of the 102 RAF Squadron. As an officer, you were a Flight Lieutenant. I got the pay, but not the rank until after we were shot down. I made four missions in Halifax bombers.
>
> On the last mission we were going to Cologne. They called half of us back and we went to Essen and bang—a night fighter got us. We were shot down outside of Vermeer, one of the German towns near the Dutch frontier. It was June 16, 1942.
>
> I shouted, "We are on fire!" Batch, the pilot said, "For Christ sake, get out of there, come forward." He meant instead of dropping out if you turned the turret around. One kid was badly shot in the leg and the other lad, from the top turret, was about the same. We spent a bit of time with them doing first aid. The other two kids were fine. Oh Christ, all I can remember was one kid pulled his chute. I guess he got excited. We had a hell of a time getting him out. We had to push him out. One lad's name was Peacock and the other was an Irishman from Liverpool. They both were repatriated through Sweden at the time Don Morrison, ex-head of the Canadian POW Association, who lost a leg flying at Dieppe, was sent home. One lad died a year or two after. The other one was not too bad. Our navigator was an English kid from Chile, where his dad was a banker. The kid had gone to school in England.
>
> Then the rest of us dove out. Batch said he got out about 1500 feet and said I wasn't too long hitting the ground. Jesus, it was a windy night, I remember. I hit a fence and it turned an ankle—my ankle has never been good since. I was on the run for three to four days heading to Holland.

I was looking for a way across the Rhine River and was trying for an unguarded bridge when one of the land police saw me dive into a ditch. He searched me for a gun, then put me on the back of his motorcycle and took me to his home in Wiseburn. There he and his wife fed me breakfast before taking me to one of their squadrons. One of the lads there said, "I shot you down."[399]

Shortly thereafter, Donald's parents received the following stark notification.

They kept me for a couple of weeks at the interrogation centre at Dulag Luft near Frankfurt, and then they sent me to Sagan in the East camp. Kommandant von Lindeiner was in charge. He was old school Prussian—an officer in World War I. His duty was to keep his POWs, or Kriegsgefangen, safe behind the barbed wire of the camp. The "Kriegies" had the duty to escape. It was a serious battle of wits.[400]

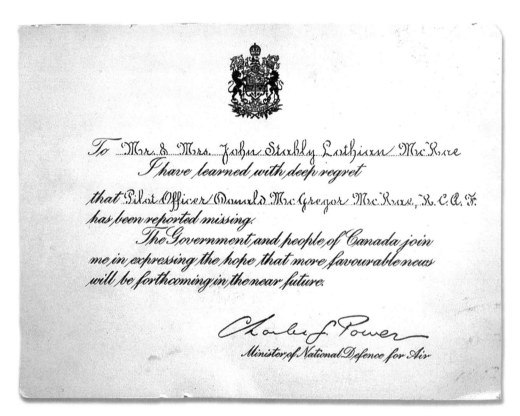

Figure 37 **Missing in Action Notification**
Many such notifications were followed by ones, weeks later, stating that that the missing person had been killed in action. Donald McRae was lucky, but thirty-three months spent in a prisoner of war camp were not fun.
Source: Library and Archives Canada, Personnel Records Services Branch.

Chapter 11: Whitney, Its People and Algonquin Park During and After World War II

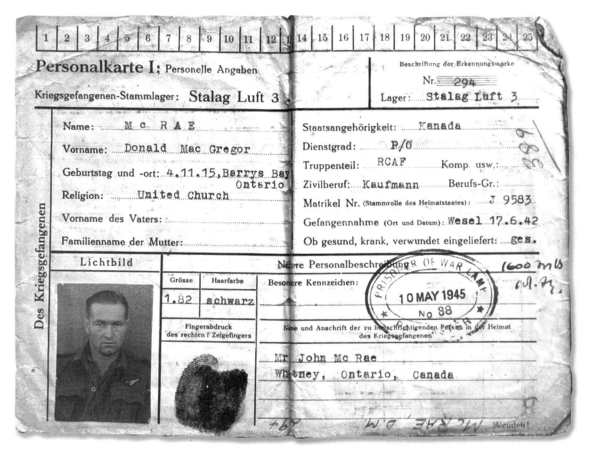

Figure 38 **Personalkarte I: Donald McRae's Prisoner of War Identity Card**
Source: Donald McRae

Sagan Stalag Luft III, famous for the "Wooden Horse" and "Great Escape," was designed by the Luftwaffe as a major camp for air force prisoners. Five compounds were eventually built: East was completed first; it was followed by North, Centre, then South, and, some distance away, Belaria. The buildings were of frame construction and double-floored, but were below the standard for a 1900 Canadian bush camp. The North camp was run by the British Royal Air Force (RAF). It had 15 one-storey barrack huts enclosed by barbed wire. A corridor ran down the middle of a hut, with ten rooms on each side besides the kitchen. Usually there were six officers to a room, which had two windows. Three double bunks, one on each of the three sides, and a couple of tables and stools made up the furniture.[401]

The senior British officer at Stalag Luft III was Wing Commander Harry "Wings" Day, who had served in the Royal Navy in World War I. Then in his late forties, he earned the reputation of being a really good officer while in a German prison camp. He got along with everybody and he was very reasonable. Day had been shot down on Friday, October 13, 1939, as he flew his first combat mission into the German Ruhr Valley to reconnoitre the railway lines. His twin-engine Bristol Blenheim was no match for the Messerschmitts.[402]

Every prisoner reporting in to the camp was questioned by the RAF camp command.

> They made goddamn sure you were who you said you were, as the Germans tried to insert spies. They checked you out with other people from your squadron who had been shot down. I knew one lad that had been with our squadron, but had gone to 419. He was one of the observers and he knew me. The Americans were with us for a while in the East camp. We had two Americans in our room. One had played first base for the Toronto Maple Leafs of the International League.
>
> A lot of the Poles got out of Poland before the Russians and Germans took over, through Hungary and the Middle East. Some of them were experienced. They had a Polish block, most of the camps did, along with a few Czechs and everything else. A few French got out to Britain. I started in the East Camp. Then they took some of the old ones who had been there awhile and put us in the North Camp. A committee or command, comprised of exceptional men under Day, controlled the camp. They designed operations or plans for escaping. Squadron Leader Roger Bushell was always in on interrogations.[403]
>
> Bushell was a trained criminal lawyer: he was a hard man, resolute, tough and coldly determined to escape. He was Big X, the head of the Escape Committee. Three long and deep tunnels were started from North Camp: "Tom," "Dick" and "Harry." Three Polish flyers, who also were engineers, Gotowski, Minskewitz and Kolanowski, designed the ingeniously hidden entrances. Lt. Commander Peter "Hornblower" Fanshawe, of the Fleet Air Arm, was in charge of dispersing sand. George Harsh had been in the chain gang in Georgia for murder. In 1940, he enlisted in the RCAF as an air gunner, and was commissioned as a Flying Officer. He was then attached to 102 Squadron of Four Group as squadron gunnery officer. In October 1942 he was shot down and ended up at Sagan. Harsh looked after security in the North camp. He was pretty good at it. We all had jobs to do. In the camp they ran a first class intelligence operation; they knew what was going on.[404]

The daily routine in the prison camp was very repetitive.

> We got up about 9:30 a.m. and formed up in a hollow square for the first roll call and count; the second came around 5 p.m. in the afternoon. Time was spent in conversations concerning politics, home and the war, while a phonograph and some records helped. We played bridge, euchre, 45, poker, a little of everything. We played softball—the field was too small for hardball. I played hockey one year. There were no boards—just snow. There were no pros—the players were just like me. Other nations found that

Photo 95 Canadian Hockey Team at Luft Stalag III
Hockey—the Canadian game—was played at Stalag III. Red Noble, bottom left, played the Germans every way and at every chance he had. The skates and sticks were supplied via the Red Cross.
Source: Donald McRae.

the Canadians played a rough game. Jimmy Kanakackis had a real good garden—onions, carrots, cucumbers and tomatoes. The sandy soil wasn't very good —it didn't seem to support big trees.

Some of the men found the time pretty long. Oh yeah, it was long."[405]

The men in each room did their own cooking. With only one small stove in the hut, each room of men got 15 minutes to cook, or time could be pooled for 30 minutes between two rooms. Because of the time shortage, there was too much fried food. German black bread was toasted because it was too hard and heavy to eat any other way. Raisin pie was common, with the crust being made from crushed crackers and margarine. The Red Cross parcels were most important, as all meat came from them— bully beef, canned pork chops and sometimes bacon in cans.[406]

Canadians in Sagan comprised at least 10, maybe 15 per cent of the prisoners. Roommates were a living geography lesson of the country. "Quite a few lads were from the West. They went across on cattle boats like my friend Clancy, who was from Seamans, Saskatchewan. His family owned a chain of drugstores in the West and he was a pharmacist. After the war, he went into Diefenbaker's government."

Another in Flying Officer Don McRae's room in January of 1942 was Pilot Officer Flight Lieutenant Don Morrison [Morrison became head of the Canadian Prisoner of War Association]. "He lost a leg flying his Spitfire over Dieppe in August of 1942. He bailed out at a high altitude and the cold air froze his leg and slowed down the bleeding. The Germans got him to hospital in time to save him. He was out of Fergie's squadron."[407]

Photo 96 POWs from Ontario at Stalag III
Top row: Larry Sommers, Simcoe; Harry Crease, Windsor; Wally Floody, Toronto; John Buchham, Toronto; George Seveanos, Port Hope; Glen Gardiner, Merlin; Ross Gillespie, Kitchener; Jim Lago, Timmins; John Fry, Jordan.
Bottom row: Butch Adams, Smithville; Henry Kadslon, Ottawa; [First name unknown] Higgeunet, Ottawa; Skeets Ogilvie, Ottawa; Keith Brackenburgh, Aylmer; Donald McRae, Whitney.
Source: Donald McRae

The East camp was "primary school" for the POWs learning escape methods. The first tunnels were crude and shallow, and frequently the sandy earth caved in. They found they couldn't tunnel without shoring.

> I can remember we had to pull Scruffy Weir out from under our block in the East Camp. He got stuck. That was the start of shoring up the tunnels. They [the POWs] had to go deep enough to get out. At least 50 tunnels were lost or abandoned at East Camp. They learned a lot—they got all the bed boards and everything they had to have.[408]

One ingenious escape used a gymnastic wooden horse. The horse was placed in exactly the same place every day. While teams of men vaulted over it, men hidden in the box dug a tunnel under the fence. Three or four men escaped and two hit "a home run" by making it back to England.[409]

As the war progressed, more prisoners appeared at Stalag III. Many Canadians ended up with English and Polish flyers in the North Camp. It was from here that the deep, long tunnels, Tom, Dick and Harry, were excavated. This was dirty, dangerous work. "In the East camp, Wally Floody was next door to us for a while, before being transferred to the North camp. Wally Floody had experience in mining in northern Ontario. He knew how to shore everything up."[410]

Despite shoring up the ceiling and walls, cave-ins happened. Excavated soil had to be disposed of secretly. In the beginning, soil was carefully spread under the raised huts but

Photo 97 East Camp, Stalag III, Spring 1943
This photo shows Eric Williams, author of *The Wooden Horse*, in the front row, second from the right, just prior to his escape from the prison camp. Donald McRae is on the far right of the row standing.
Source: Donald McRae

when no more could be put there, it was deposited under the raised seats of the prison camp's theatre. The Germans were suspicious and surprise inspection raids were common.

The air in the tunnels was so starved of oxygen that the wick-grease lamps would just go out. The prisoners devised a ventilation system by constructing an air duct using empty Klim tins. Klim, a form of powdered milk, derived its name from the word "milk" spelled backwards. The tins, with their bottoms removed, were then pushed together and the joins were sealed with tarred paper. This pipeline, with vents along the line, was buried in the tunnel. A bellows, which required very strong arms, was used to move the air. Playing down his contribution, Don McRae said, "All I ever did was work a little bit on the air pump."[411]

Security for the tunnels was very tight, not just because of the digging. There was also a sending radio in the tunnel. The Germans listened for this type of communication throughout all the lands that they controlled and they were capable of quickly determining the location of a transmitting radio. Repercussions would have been very harsh if the sending radio had been found, as possession of a transmitting set would have been treated as espionage. Only a select group of POWs knew about it and they were sworn to secrecy. Donald McRae said, "I knew it and I didn't want to know it."[412]

One of the prisoners was a top radio expert. Howard Cudall was a civilian. He should not have been in the camp—he should not have been flying since his expertise with electronics made him very valuable. He had flown as a lark—just to see what a raid was like. The Germans always counted the men in the planes that they shot down. Fortunately, in this case, the second pilot managed to escape successfully to Spain, so the "correct" number of men was accounted for and the expert was not discovered. He did, however, have to be covered up by a few Kriegies who knew who he was, because he couldn't pass himself off as a flyer to other prisoners!

Because of the danger of having transmissions intercepted, the use of the sending radio was problematic. Stalag III camp history states that the sending set was never used. Instead, coded information from prison camps was sent by mail. By this method, POWs in the various camps in Germany were able to send valuable information on German developments in radar, jet planes, naval and army

Photo 98 Flight Lieutenant Ferguson and Donald McRae

Fergie, a close friend of Donald McRae, was one of the tunnellers and had dirt left on him that the German ferrets saw. Ferrets were unarmed Germans that wandered around inside the camp looking for signs of escape attempts. "Fergie dug all the time and he ran the radio. We got the British broadcasts. Ours was one of the spare ones and it was built into our stove. Fergie made another stove that we could cook on."[413]

Source: Donald McRae

movements. This information was a very valuable check against information gathered from messages in the German code "Ultra," which had been broken. Cudall, however, is credited with communicating valuable information concerning German night fighter radar. Cudall's biography does state that he used the set to transmit directly.[414]

Bribing the German guards was not easy. If caught, the punishment for the guard might well have been a posting to the Russian front. Nonetheless, obtaining items such as cigarettes tempted some guards.

> I had lots of cigarettes. Plunkett, who was an American buyer of McRae lumber, was sending them to me. We stored them in the wall or the Germans would steal them. We had to get a tube for our radio that was hidden in the stove. Towards the end, you could bribe the guards. I kept throwing cigarettes to a German that had been in America. The German stuck the tube into the third base bag. During the game, one player slid into the base and I thought that's it for the tube, but it survived.[415]

A brash attempt to escape was made a month before the tunnel was used. Delousing was a regular occurrence and took place in a facility outside the camp. One day early in 1945, "A gang went down to get showered and so forth to the delousing unit with two or three dressed up as Germans. They took blankets. They went so far down the road—then they took to the bush."[416] This was the first mass escape attempt from Sagan.

When a second group of prisoners appeared at the gate to go for delousing, the Germans sounded the alarm. All the prisoners were lined up and checked against their photos. On the way through the line, Red Noble stole a handful of the identity cards and gave half of them to Donald McRae. To cover the theft, an impromptu diversionary fight was staged by the other prisoners, while Noble and McRae buried the cards in the sand. On restoring order, the Germans continued with their check of the prisoners only to find they had more Kriegies than cards! Don McRae still has his card, which was among the ones stolen.

As punishment, the Germans made all the prisoners stand on the assembly ground through a cold winter day until midnight. The Americans, in the adjoining compound, found out what had happened. They assembled their band and played "God Save the King." Remembering that moment, Donald McRae commented, "I'll never forget that!"[417]

Red Noble was a character. Donald McRae tells why Noble had that reputation.

> Red Noble came from some place up near Owen Sound. He was different than most of us. Red Noble—well he was the worst thief, and he gave me half of what he stole! He stole a handful of the identity cards of the group after the delousing escape and gave me some to bury. I remember another time the Germans were rigging a system of loudspeakers. Red Noble grabbed the wire when the man was up a pole and then he got a gang to start a fight as a cover while he ran and stashed the wire. I heard they shot the two lads that were stringing the wire when they found the wire in the tunnel. The wire was used to install lights in the tunnel that plugged into German camp electricity. The lights replaced the smoky grease lamps the prisoners had made. The electric system worked well unless there was a nearby air raid. Then the Germans cut off all the camp power.[418]

Coming into the spring of 1945, the Germans knew that the escape season had arrived. Kommandant von Lindeiner pleaded with "Wings" Day not to attempt escape. The Kommandant knew that the wrath of Hitler, through the Gestapo, would come down on escapees in full force. The digging of the tunnels and escape plans went ahead.

> About a month before the Great Escape, about 20 of us were then moved out to Belaria, which was about six miles away. George Harsh, Bob Tuck, Peter Fanshawe and Wally Floody were all members of the escape committee. Fergie and I got heaved and Red Noble got kicked out too. Red Noble and I got bounced because of Red stealing that bunch of identity cards. The Germans may have suspected it, but they didn't see it. Fergie was a digger and didn't get a chance to get his underwear cleaned.[419]

Being transferred to Belaria could well have saved the life of Don McRae and the others who were moved. If they had stayed, they would have been amongst those eligible to have their names drawn by lot to attempt the escape. Yet the move to Belaria was no doubt a bitter one for these men, for they had all contributed to the digging of the tunnels. Roger Bushnell, the escape committee leader, was so deceptive in practising for a lead role in a camp theatre production that the Germans didn't include him in the group they sent to Belaria.

> We were marched to Belaria. It had been a Russian camp and had lots of bed bugs. Beatie went over and told the German major that we would burn it down if they didn't do something. So they fumigated it and we had to stay for a time in other rooms—our building was the worst.[420]

As the Russians advanced, the prisoners were moved west.

> We were then moved to Luckenwalde. Fergie built a sleigh that took some of our stuff, including 100 cartons of cigarettes. The Germans gave us a Red Cross package apiece. That's all we had. We were better off than some of the English. Most of us had our flight jackets. Some of the English didn't.[421]

In 1945, the Russians were close to Luckenwalde:

> The Germans left the camp two or three days before the Russians came and I took over policing the camp along with an American by the name of Turner. Major Turner was a big guy who had played football at the University of California. He had been captured at Anzio in Italy with the 69th. We met an American—he was following the Russians. He said the Americans were across the Elbe so we walked out—four of us. Fergie wouldn't come. He said the hell with it. We had a hell of a time getting across the Elbe. The Russians wouldn't let us use the bridges. We picked up arms from dead Germans and went from farm to farm, and eventually we had nine Germans in tow at gun point until they finally got us a boat to cross the river. We met some Americans in jeeps and they took us back to where we had crossed. That's where we hit

Turner's 69th Division. Jesus, white bread tasted like cake.

We then went downtown and got into a house. With a wagon we got some coal, heated up water and washed ourselves and our clothes. Both were lousy. We had been unable to wash and these biting insects infected our head hair as lice and bodies as cooties.

They flew us out from the Elbe to Brussels and then to Britain. We flew over Aachen. My God, there was nothing left, and it was a fair-sized city. It was totally destroyed. They must have fought through it, because it was really bad, one of the worst I had seen.[422]

The Great Escape

For those prisoners who remained in Stalag III, North Camp, tension and anticipation of escape heightened. Two tunnels, Tom and Dick, were discovered by the Germans. Harry, some 340 feet in length, remained. The Great Escape took place on the night of March 24–25, 1945. Eighty-six men exited the tunnel, their order determined by lot. Three men managed to "hit a home run" and make it back to freedom.

The Germans sounded a *Grossfahndung*—a national alert. Hitler ordered that all recaptured prisoners be shot. The number was reduced to 50 and the Gestapo was put in charge of capturing the escapees and carrying out Hitler's direct orders. General Artur Nebe, head of the Kripto, personally chose those who were to die, using their identity cards. The men were taken from their concentration camp cells individually, driven to deserted wooded areas and shot. Their bodies were cremated and urns containing their ashes returned to Sagan.

British Foreign Minister, Anthony Eden, in the House of Commons vowed,

> They will never cease in the efforts to collect the evidence to identify all those responsible. They are firmly resolved that these foul criminals shall be tracked down to the last man, wherever they may take refuge. When the war is over they will be brought to exemplary justice.[423]

The RAF's Special Investigations Branch went to work. Twenty-one Gestapo were convicted and executed, 11 committed suicide and 6 were killed in air raids during the last days of the war; 17 received long sentences of imprisonment; a handful were acquitted. The Kiel Gestapo chief escaped prosecution until 1968, when he was arraigned before a German court and sentenced to two years' imprisonment. His plea: he was only obeying orders.

Kommandant von Lindeiner survived a heart attack and escaped Nazi punishment; he was not charged as a war criminal, although he was detained for two years by the British while the investigation, trial and sentencing of the Gestapo personnel involved in the aftermath of the Great Escape were carried out.

A stunning memorial painting, including pictures of the 50 executed men, hangs in the Donald McRae house. Don McRae has not forgotten those times, nor should we who have had peaceful lives as a result of the sacrifices made by many.

Chapter 11: Whitney, Its People and Algonquin Park During and After World War II

Photo 99 **Joe Lavally, Wally Floody and Unknown American**

There are quite a few Lavallys in the Whitney area. Shown in this photo, taken in the spring of 1945, is "Hay Lake" Joe Lavally. He was the hero of Bernard Wickstead's delightful book, *Joe Lavally and the Paleface*, the story of a canoe trip that Wickstead took with Joe that summer.

Source: Donald McRae

Photo 101 **Ex-POWs' Hay Lake Fishing Party, 1945**
Left to Right: J.A.S. Ferguson, Donald McRae, Don Morrison, unknown American, Wally Floody
Source: Donald McRae

After the war, Donald invited some of his wartime friends on a fishing holiday. No doubt he had talked about the good fishing he had back home while in the prison camp. The photos show that they had good luck. Note that "Hay Lake" Joe Lavally guided the party. Joe died the following fall not knowing that he had been made famous by the delightful book *Joe Lavally and the Paleface* by Bernard Wickstead.

Helen Grace (McRorie) McRae

Helen Grace (McRorie) McRae was born December 1914 in Warwick Township, near Lake Huron. After graduating from Forest High School, she attended the University of Toronto School of Nursing, graduating from there in 1937. Her eight-year career was spent largely with the Canadian Red Cross, with a short stint with the Victorian Order of Nurses. She met Donald McRae when attached to the

Photo 100 **The Catch**
Left to right: J.S.L. McRae, Don Morrison, unknown American, Keith Brakenburg, J.A.S. Ferguson, Donald McRae, Wally Floody
Source: Donald McRae

Red Cross Outpost Hospital in Whitney, just before the war. She waited a long time for Donald to come home. "It was pretty upsetting when we got word that Donald was missing in action. We still have the card, an official notification to his parents. There were quite a few weeks when we didn't know anything more than that."[424]

She and Donald were married at Forest, Ontario in 1945, soon after Donald returned. They lived at Hay Lake until 1952, when they moved to the new home that they had built in Whitney. Over the years, Helen has been very active helping with the Red Cross, the Whitney Public Library and St. Andrew's United Church, as well as raising four children, encouraging them all to attain university degrees and supporting her hard-working husband.

Photo 102 **Helen McRae, Donald McRae, Marjorie McRae and Jean Bradley at Whitefish Cottage**
Source: McRae family

Algonquin Park: The War Years and Post–World War II

Recreational use of Algonquin Park was supervised by ageing rangers with some volunteer help.[425] While some canoe routes survived marginal maintenance, boardwalks over wet or marshy areas deteriorated and didn't receive sufficient attention. By the 1950s, canoe routes were in better shape and routes were expanded into new areas in the 1960s.

In 1944, the Wildlife Research Area was established with its base at Sasajewun Lake. The Swan Lake Forest Research Station was designated in 1950. While these two areas in Algonquin Provincial Park were established after MacDougall had left the Park, his encouragement and influence was behind both projects.

In 1945, MacDougall was behind the establishment of the Forest Ranger School at Dorset, in cooperation with the University of Toronto Forestry Faculty. This move was made necessary when the Department of Lands and Forests assumed responsibility for scaling, timberwork and the use of technical equipment because there was a severe shortage of technicians, such as scalers, who measured the wood cut, and others that monitored company operations.[426] In the 1970s, the name of the facility was changed to the Leslie M. Frost Natural Resources Centre. It became a very popular educational facility. In 2004, to the distress of many, the facility was closed.

In the ten or so years that followed the war, considerable development took place around the Park. In 1947, Ontario Hydro built the Des Joachims dam on the Ottawa River, and a transmission line was established across the Park to Minden. There it connected to the Ontario grid. In the following year, the formerly gravelled and corduroyed Highway 60 was paved, and the flow of tourists and canoeists increased markedly. The West Gate entrance complex was built in 1953, followed by the East Gate. Park headquarters was relocated there from Cache Lake in 1959. These moves provided a considerable boon to the economy in Whitney. Several motels and the Bear Trail Inn Resort now cater to tourists.

Across the province, recovery from World War II was rapid and led quickly to population growth, urbanization, higher domestic standards of living, increased levels of leisure time and more personal mobility, along with increased American tourism. These combined to bring about the crisis in outdoor recreation that hit Ontario beginning in the late 1940s. "They quickly strained the few existing parklands beyond acceptable limits and caused park administrators no end of headaches."[427]

In Algonquin, for example, "Superintendent George Phillips reported as early as 1946 that the campgrounds, the children's camps and resorts were 'filled to capacity during July and August. There are too many [visitors] sleeping in chairs, canoes, etc. around the Park hotels on weekends.'"[428]

In October 1947, Professor J.R. Dymond wrote to Frank MacDougall to express his concern about the litter problem that was spreading throughout the Park and the damage that was being inflicted upon natural features by recreationists insensitive to the environment and lacking in basic woodcraft skills.

George Phillips: Algonquin Park Superintendent, 1944–1958

In his time as superintendent in Algonquin, George Phillips saw many changes. Although he had but peripheral contact with many of these projects, they were completed during his watch. Contact with his friend Deputy Minister Frank MacDougall was no doubt frequent. During his time, the following events took place:

1944 Wildlife Research Area and interpretive program established;

1945 Planning for the Petawawa Forest Management Unit begun;

1947 Cutting of the electric transmission right-of-way and access road across Algonquin from Des Joachims on the Ottawa River to Minden begun;

1950 Swan Lake Forest Research Station established;

1953 West Gate complex and Park Museum at Found Lake built. Telephone line established along Hwy 60;

1954 Plan to bring Park back to its original condition with the intent to phase out all leases developed;

1956 Highland Inn, Hotel Algonquin, Opeongo Lodge, Minnesing Lodge and the Barclay Estate purchased, dismantled and burned;

1957 Wolf Research Program begun under Douglas Pimlott;

1958 Frank MacDougall instigated work that eventually led to the creation of the Master Plan for Algonquin Park in 1974;

1959   Park headquarters moved from Cache Lake to the East Gate; staff residences moved to nearby Clarke Lake;

1959   Pioneer Logging Museum opened near the East Gate.[429]

George Phillips was born to fly and did so in both World War I and World War II. He was the instructor who taught Frank MacDougall, and they remained life long friends. In her book, *Tallying the Tales of the Old Timers*, Joan Finnegan set down this story about Phillips told her by OPP officer Roy Wilston:

> One time he scared the pants right off me. He took me into the lake one time—in those days you had to, supposing somebody set fire to the bush in the park. Well, the Forestry men got paid for it, but I had to go in and investigate how much was burned and all that. And there was this little lake up near Aylen Lake and part of it is out of the park [possibly O'Neill Lake]. But this little lake from a thousand feet up looked about like a teacup and old George flips the plane like this and I said, "It's pretty small, isn't it George?" "Oh," he says, "we'll make her." There was pine trees all around the perimeter of this lake, see? He slips the thing in like this into that lake—oh, it was a beautiful landing. I go in there. He waits for me and I do what I have to do.
>
> And he'd been sizing the thing up and he says, "Roy, I've got news for you." I said, "Jeez what is it? News from you is never good." He laughed and he said, "Do you see that mountain there?" I said, "Yes." He said, "On the other side is a big lake." I said, "Is that right?" He said, "Well, you're going to walk up that." I said, "You go to hell." He said, "I can't lift off the lake with the two of us in the plane. You're going to have to walk out over the mountain, and I'll get out of here somehow, and I'll meet you with the plane over on the other side, on the big lake."
>
> The plane was sitting at the shore and George says, "Here's a piece of rope. Put it in that ring at the back of the plane and tie the other end to the tree, and I'm going to start her up, and I'm going to run her right up to the rope and, when I signal with my hand, you cut that rope with your knife." He took off—he got off the lake—and he was about 30 feet from the tops of the pine trees heading straight for them and I thought he'd had it. But at the last minute he pulled back on the stick and just skimmed over the tops of them.
>
> Well, it was a hot day in the summer and I had my gun belt on and handcuffs, and everything. I was supposed to walk through the bush and up the mountain for about two miles, no trail or anything. And I did. George Phillips was then about 65 and he lived at park headquarters on Cache Lake, and from Cache Lake to Smoke Lake where the plane used to be kept was five miles and, right up to the day he left the Forestry, he ran that five miles every morning. A fantastic man![430]

Life in Whitney was not easy for many. Work was not always available and much of that was poorly paid. Yes. There were some real characters and some young people growing up were more than restless and got into trouble. Yet when earned, there was respect between those that lived close to the line, and those that enforced the law. Roy Wilston told this story to Joan Finnegan.

> I had this amusing little incident happen when I was two-and-a-half-years in Whitney. One of my girls had finished public school and we had to move someplace where there was a high school. So, I asked the "powers-that-be" to move me. And so they said they'd move me into Pembroke, which was fine. The day before I was to leave, the old lady, Mrs. Parks, the local bootlegger—"the Queen of the Bootleggers" we called her in Whitney—she had several boys and two daughters who were the worst hellhounds in the whole area. Well, she came to the house on the afternoon before I was to leave and said, "I want you to come down to my place tonight. I'm having a corn roast for you!" "Holy jeez!" I said to the wife, "I'll never come out alive." Every one of those guests at the party I'd arrested or beat up—if it wasn't one thing or another. But she invited me down and there was going to be about fifteen of them. I said to my wife, "If I don't go I'll be a coward for life, eh?"
>
> So the Parks lived in a place that was called Death Valley—that's what it was known as in those days. No policeman was ever supposed to go there. So I went down to this little log cabin just outside of Whitney and I got there about eight-thirty at night when it was just getting dark. There was three or four cars there and I knocked on the door and Mrs. Parks says, "Come in." And, I go in and there was about a dozen or fifteen of these characters sitting in this big, huge kitchen, chairs around the wall. And each with a bottle of beer in their hand, eh? I'd arrested all of them for one thing and another in the few weeks before. So I entered into the party and I had a great time. They didn't think I'd come, eh? And they were just so pleased I went to the bootleggers' party in Death Valley!
>
> That same old lady Parks, about two years after I came to Pembroke, was brought into the hospital there. My wife was a nurse; she was working at the Civic Hospital, and Mrs. Parks was brought in. I always kind of liked the old lady. I know she was a bootlegger but I had a kind of respect for her. She was a terminal case. She recognized Marg, eh, and she said, "Do you think your husband would come and see me?" Marg said, "I'm sure he would." Mrs. Parks said, "I'd like that. You know, he arrested every one of my boys, but they were asking for it. I'd like to see him again." And I went up and took her some flowers, but the poor old lady died about a month later.
>
> If there was a Whitney girl who had gotten in trouble and was going to have a baby and the parents would kick her out—they'd do that up there, no place to go—old Mrs. Parks would take her in. And her only source of income was bootlegging. I can't remember her first name; we always just called her "Ma Parks."[431]

Bootlegger, yes, but for most of those that knew the full story, Mrs. Parks was respected.

In the aftermath of the war, there were many displaced persons seeking jobs and places to live since they didn't want to return to their homelands. Of the many Polish people who came to Canada, some came to Whitney, where they worked for the McRae Lumber Company. Three families of the village trace their roots to this time: the Sajas, the Siydocks and the Szczygiels.

Whitney was, and still is, quite isolated. In the 1940s, Huntsville was a two-hour drive and is not much less today. Highway 60 wasn't paved until 1948. Whitney only gradually acquired modern amenities—electricity in 1952 and telephone in 1953. This replaced an arrangement with the Department of Lands and Forests.[432] Phil Roche explained:

> Prior to 1961, Airy Township was one of the many small villages, mostly in northern Ontario, which due to the small population and assessment were governed by the province and therefore had no local government. The Department of Lands and Forests did the assessment and tax collecting, except for the schools, whose local school boards appointed their own tax collectors.[433]

If you wanted to drive to Toronto in 1946, Don McRae remembered, "It took six or seven hours to go to Toronto, at least. It was a day's run to Toronto always. You had to go to Belleville if you wanted to go to Toronto."[434]

In 1954, a Statute of Labour Board was established, starting the path to formal municipal government. This was followed by the creation of a Roads Board. Thereafter, each taxpayer had to contribute work or money to cover the costs of roads. The provincial government matched these monies on a 50/50 basis. Street lights and the town dump were administered by the Lions Club, until that organization folded in 1959.[435]

In 1959, the MNR transferred Algonquin Park headquarters from Cache Lake to the newly built East Gate complex.[436] This led to more MNR staff living in Whitney, which stimulated service industry businesses in the community.

After considerable preliminary local organization, the Township of Airy was incorporated in 1961. Phil Roche was acclaimed as the first reeve. On June 1, 1998, the Township of South Algonquin was formed by joining the Townships of Airy, Sabine, Murchison, Lyell and Dickens. The township seat became Madawaska.[437]

Donald and Helen McRae built their very comfortable home in Whitney in 1952 when Janet, the eldest child, entered public school. The Donald and Helen McRae home was kept very friendly to the neighbouring children. Helen and Donald McRae commented,

> We used to have lots of kids around when they were small. There were lots of kids in the neighbourhood then. When they were growing up, the yard used to be full of them. We had a sand pit out here. There always used to be six or seven around besides our own.[438]

Cathy McRae Freeman remembered, "We had one of the first TVs in town. Kids came over all the time and parents were always phoning, 'Please send so-and-so home.' Kids had a place in the basement to hang out."[439]

The village was a very quiet place. People were busy with the jobs they had and with the businesses that they were running.

Helen McRae, commenting on life in Whitney, recalled the 1950s:

> We didn't have a very busy social life at that time. I guess the people that we socialized with aren't here now: Dr. Post, Jim Munro. Jim's over there now, but he lives alone: he's retired now; he's in the log cabin. They came here right after the war. They were storekeepers. But he came because his wife had spent time here when she was young. Her father, Russell Van Meter, had been the mill manager for Dennis Canadian, and she came back with her husband, Jim Munro, and ran the store. The store has been completely rebuilt and is now the Value Mart.[440]

Janet McRae Webber recalled, "The churches, the Catholic Women's League were parts of the community."

Cathy McRae Freeman said,

> I don't think Whitney was as divided (on religious grounds) as were many of the towns in the Ottawa Valley. There was no active Orange Lodge. In Eganville there certainly was and it was evenly divided between Protestant and Catholic. Whitney was basically Catholic. There were some Protestants, people like us, the doctor, and some from the forestry.

For the young people of Whitney, public and separate schools were available to the end of Grade 10. A bus to the high school in Bancroft started in 1961. The Barry's Bay bus wasn't available until 1989. Once there was a high school bus, the Whitney schools went only to Grade 8. The McRaes, however, sent their children away to be educated. Bob and John went to Lakefield College School. Janet and Cathy went to the Ontario Ladies College at Trafalgar Castle School in Whitby. All four children of Donald and Helen McRae are university graduates.[441]

The Passing of John Stanley Lothian McRae

J.S.L. McRae, the patriarch of the family and the company that he founded, died in June 1969. Company operations stopped. In large measure the town of Whitney stopped.

Gary Cannon, a long-time McRae worker and bush boss remembered,

> Mr. McRae worked until the Thursday, and then they took him to hospital and he died—on a Tuesday, the last week of June in 1969. So the mill closed down. I remember I was loading pine out of the water up at the little lake behind the mill, and I think it was Bob came up and said, "Gary, we're not going to work today or tomorrow because Grandpa died. We're going to close everything down 'til Tuesday after the long weekend."[442]

Roy MacGregor, son of Duncan and nephew of J.S.L. wrote of J.S.L.'s passing.

> When John Stanley Lothian McRae died, it was, for Whitney, much like a state funeral. The church was overflowing with

mourners; both Protestant and Catholic priests spoke; the pallbearers were all men from the mill ... and the streets of the little village were lined with mill families, children turned out in their best clothes, as the funeral cortege wound through the streets and turned out onto Highway 60 for the long ride to Eganville and the McRae plot at the side of the Bonnechere River.[443]

J.S.L.'s granddaughters knew him well and remember him with great fondness. Granddaughter Cathy McRae Freeman said,

> Grandpa was definitely a character, and Janet, my grandmother, kept close tabs on him as best she could. When we stayed over, there were always Laura Secord candies around—Granny had a sweet tooth and liked to sleep in.
>
> Grandpa, an early riser, always cooked breakfast for himself—Schneider's maple cured bacon—and shared it with the dog, "One for you and one for me."[444]

Phil Roche knew J.S.L. very well and was a close family friend.

> The McRaes were good friends. Mr. McRae Sr., he was a very loyal employer. He was good with his employees. That's right. He had a small circle of personal friends that he wanted to sit down and talk to, very small, and I think he had a lot of business friends he knew. But he loved to talk. I used to drive him around a bit. In fact, not long before he died—about three weeks—I drove him to Peterborough. We left here in the morning; he did some business that day. I think he sold some lumber, but I know we talked on the way back—there was some property he was thinking of buying.[445]

After J.S.L. passed on, his wife, Janet, began to travel more. Daughter Marjorie recalled the following:

> I remember the first time we went to Scotland. My mother was impressed with it. I think mother was around 80. She got off the plane and of course the little customs man there ... He held up her passport and her name was Janet Agnes MacGregor McRae, and he said to her "Welcome home, love," and she thought that was just the most wonderful thing she had ever heard, because she was about three or four generations Canadian![446]

In 2005, Helen and Donald McRae celebrated their 60th wedding anniversary, and, appropriately, a gathering of the clan took place at the former office building of the Airy mill. This building, now renovated, is the home of the John McRae family. For Donald and Helen McRae, reflecting back on their years together brought great satisfaction. They had raised four children, and saw them all achieve fine academic success and develop into respected professionals. Nine grandchildren bring them continuing delight.

Chapter 11: Whitney, Its People and Algonquin Park During and After World War II

Photo 103 **Janet MacGregor McRae**
Janet McRae was the support behind J.S.L., raised Donald and Marjorie, worked hours for St. Andrew's United Church and hosted a succession of family members at her large home at Airy.
Source: McRae Family

Below: Photo 104 **McRae Family, 60th Wedding Anniversary of Helen and Donald McRae, 2005**
Back Row (L-R): Michael McRae, John McRae, David Freeman, Cathy Freeman, Jamie McRae, Jen McRae, Duncan McRae, Cam McRae, Bob McRae
Middle Row (L-R): Rachel McRae, Michael Freeman, Maggie McRae, Donald McRae, Helen McRae, Janet Webber, Val McRae, Mark Webber
Front Row (L-R): Sarah Freeman, Gillian McRae
Source: McRae Family

Whitney Today

When need arises, individuals in the town band together and get projects done. In 1967, as a Centennial project, the very active local Red Cross committee conceived the idea of establishing a small library in the town. The Red Cross donated its vacant building on Ottawa Street to the town and Jay Post, wife of the town's beloved late doctor, Gib Post, donated his large collection of books. The library was established as the Post Memorial Library. A volunteer committee started raising funds. Murray Brothers and McRae donated lumber for shelves, Lester Smith built most of the shelves and volunteers staffed the library. However, in 1969 the Township of Airy bought the library land, and in 1972 the entire building was leased to Dr. Pelletier, who later purchased it. The library was forced to close operations and store the books.

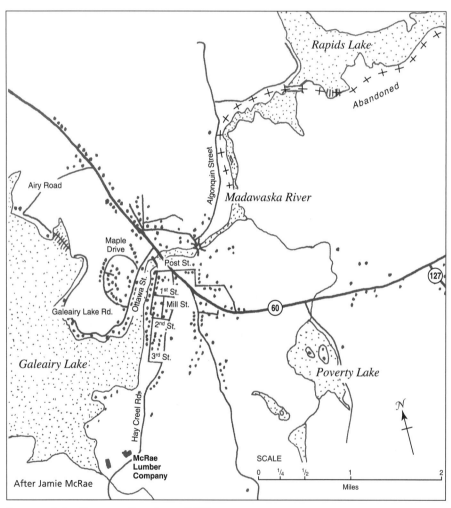

Map 22 **Whitney and the Surrounding Area, 2005**

Chapter 11: Whitney, Its People and Algonquin Park During and After World War II

Then, in 1975, a combined project for a new medical centre with a library was started. Wintario and Canada Winter Work grants were obtained and land was donated by the curling club. To match the Wintario grant, funds were raised with a great variety of activities. September 1978 saw the medical centre and a beautiful, well-lit library completed and turned over to the Whitney town council, debt free. Further grants were received, including another one from Wintario to buy Canadian-published material. Additional funds were raised and spent on more shelving and additions to the collection of books.

Community involvement in the library has continued, with staffing on a volunteer basis, and the raising of funds by the bringing to town of live stage productions, the showing of movies at the public school and several Mother's Day Teas. A trivia night was a very popular and successful fundraiser. Keeping up-to-date, the library now has internet workstations, thanks to an Industry Canada Community Access Grant and three Student Grants.

Those connected with the library have also seen fit to give back to the community by establishing two awards in English for Grade 11 Whitney students: the Theresa Kuiack Award to a student of North Hastings High School in Bancroft and the Rosemary Kukhta Award to a student of Madawaska Valley District High School in Barry's Bay. The awards honour two long-time workers for the Whitney library.

Photo 105 **Whitney: Bridge, Shell Gas Station and TD Bank** Source: Jamie McRae

Photo 106 **Whitney: East from Bridge** **Source:** Jamie McRae

Photo 107 **Whitney: Ottawa Street**
The houses in the photo are the original ones built by the St. Anthony Company.
Source: Jamie McRae

Chapter 11: Whitney, Its People and Algonquin Park During and After World War II

Photo 108 **Medical Centre and Library** **Source:** Jamie McRae

In 1998, with the amalgamation of Murchison, Dickens, Sabine, Lyell and Airy townships, the newly created Township of South Algonquin appointed a new library board to manage library services for the Madawaska branch in Barry's Bay and the Post Memorial branch in Whitney.[447]

Taken together, Whitney is greater than the sum of its parts, and perhaps this is typical of small-town Canada. People know and care about each other. When a need arises, people come forward and get things done. There is something precious in Whitney. That is space.

Bush-Wise Items

This device can be made from a portion of a hollow log sawed off squarely, about one foot long and placed on one end for holding wood while it is being split into small sticks. To prevent splitting the holder, pin a half round stick or a length of hose against the upper end, against which the axe may strike. *p. 103*

The wedge is one of mechanical powers – it has its place and is almost as indispensable among choppers as the axe. Its power to separate bodies from one another is perfectly wonderful. The power of the wedge increases as its length increases, or as the thickness of its back decreases. *p. 116*

When a tree leans, for example, from north to south, it should always be cut to fall east or west, and always if possible, at right angles to the way it leans. If cut to fall the way it leans, there is great danger that it will split at the butt. *p. 119*

Wood cut during the three months that precede the first of the year is much more valuable than if cut the three months that succeed that time. The reason of this is, probably because during the latter part of autumn, and the first part of winter, there is but little action in the sap of the tree, and therefore the wood is not filled with it, as it is after the sun runs higher and the days are longer. *p. 134*

Saw mill men must remember that the most prominent defect that lowers the grade of lumber on inspection is bad manufacture. Of course this defect can be avoided, but it is one which often costs a man more than his profits. *p. 154*

Musson's Improved Lumber and Log Book
New and Revised—1905
—Illustrated Edition

Section IV: The Recent Years of The McRae Lumber Company

The building of the modern low-grade scragg mill at the Whitney site in 1973, followed by the replacing of the Rock Lake mill by the high-grade band mill in 1979, marked a great change for McRae Lumber. The Rock Lake mill set-up was a fully integrated operation. The McRae Lumber Company at Rock Lake carried out bushwork that included road building, felling the trees and transportation of the wood to the mill. Processing included making and marketing lumber and chips. Today, the AFA delivers the logs to the mill site while the company processes and markets the products. The McRae Lumber Company is no longer directly present in logging operations within Algonquin Park.

Key personnel within the McRae Lumber Company changed. J.S.L. McRae died in 1969. His two children, Marjorie and Donald McRae took over ownership of the company, with Donald as general manager. With Marjorie's support and approval, Donald piloted the company through the turmoil of building the low-grade mill at Whitney in 1973, the cancelling of timber leases in the Algonquin Park Master Plan of 1974, as well as the setting up of the Algonquin Forestry Authority the following year. Donald McRae gradually turned over the running of the company to his sons Bob and John McRae, ceding ownership to his four children, Janet Webber, Cathy Freeman, Robert and John McRae, and retiring in 1980. In December of the same year, Duncan MacGregor was crippled by an accident at the mill and retired to Huntsville. Bob and John McRae now run the company that is owned by them and their two sisters.

At the new Whitney mills, modern equipment and developing markets led to the efficient use of the poorer quality wood coming out of the bush. While lumber remained the major income producer, residue products became the main output of the mills by weight.

In the 1990s, the privately owned Bancroft/Minden Forest Company was set up to look after silviculture in the Bancroft/Minden forest unit, subject to MNR rules and approval. Since the McRae Lumber Company had Crown timber leases in this area, it was required to be a partner in this new company.

In 2004, Bob's son Jamie became the fourth generation to work full-time for the McRae Lumber Company and the sixth generation in the family to timber in the Ottawa Valley.

Photo 109 Aerial View of the Whitney Mills of the McRae Lumber Company

This photo shows the spacious layout of the Whitney operations of the McRae Lumber Company. The village of Whitney is at the top of the photograph with Galeairy Lake to the left. Hay Creek Road leads south from Whitney to the millyard and becomes a gated, private road, proceeding south to join the Ontario Hydro access road. The high-grade mill is to the left with the garage below. The low-grade mill is in the centre of the photo with the slasher line to the right.

Source: McRae Lumber Company

Chapter 12
The Modern Mills of The McRae Lumber Company

The present mill site of McRae Lumber lies just south of the town of Whitney, at the end of Hay Creek Road, and is bordered by the waters of Galeairy Lake and Hay Creek. The entire site covers a little over 80 acres and includes space for a low-grade scragg mill, a high-grade band mill, a maintenance garage, a slasher line, a log yard, two lumber yards and a large firepond. Between 1953 and 1955, J.S.L. was buying up farms in the area. A large portion of the present mill site is comprised of the former farms of Norman Bowers and James Parks. The area now used for the large log yard was purchased in 1989.[448]

The entire operation is powered by electricity and all buildings are well-insulated to keep noise levels as low as possible. All the buildings have fire suppression sprinklers and there is a hydrant system capable of supplying a large quantity of water from the firepond.

The mill at Whitney has no bunkhouses or cookery and operates just one shift per day from Monday to Friday. The 77 employees that work at the mill come mostly from Whitney, Madawaska and the surrounding countryside. The operation is clean, efficient and modern, capable of producing up to 50,000 board feet of hardwood products a day. The company strives to be a responsible employer, a guardian of the environment and a respected member of the community.

The Low-Grade Scragg Mill, 1973

Donald McRae started development of the mill site on the south side of Whitney when he built the low-grade mill. While chipping had been started at Rock Lake, the quantity of low-grade wood coming from the forest was inundating that mill.

Bob McRae explains why his father started development at the Whitney site.

> Dad [Donald McRae] set up the small low-grade mill in Whitney in 1973 in an attempt to utilize the wood better. The idea was that the mill in Whitney would process the low-grade short wood. Also, the climate in the bush had changed con-

Photo 110 **Low-Grade Mill**
Source: McRae Lumber Company

siderably. In the late sixties, the company had moved away from diameter limit in the bush to selective harvest, where all the trees were marked. A large percentage of the better trees were left standing. The net effect on the company was that the production and grade of lumber was falling at the Rock Lake mill. Hence the rationale for the new mill.

The mill that Dad built in 1973 in Whitney consisted of a single circular carriage with a combination bull-edger/edger, Canadian-style trimmers and a boardway. The mill ran double shifts from 1974 to 1980, processing hemlock and low-grade hardwood.[449]

John Mastine remembers problems with the chipper that was associated with the low-grade mill.

> The mill is just at the edge of the bog, but the chipper is right on the bog. I laid pine down like corduroy and bulldozed sand over top. But the chipper used to shake … whaw … the lights would shake as it just pounded on that bog. So I dug a ditch all around to lower the water table, so she'd dry out. As soon as the water table was down it didn't shake anymore.[450]

Bob McRae recalled some of the later changes to the low-grade mill.

> In 1978–79, we added a new Brunette peeler, a larger 75-inch chipper and a second circular carriage line to the low-grade mill in an attempt to cope with the quality of the wood by peeling more wood and chipping more of the cull wood directly. This was an attempt to keep the poor quality cull junk wood out of the band mill.[451]

The High-Grade Band Mill, 1979

Bob McRae explains the reasoning behind building the band mill in Whitney.

> In 1979, we built a new band mill in Whitney to replace the mill at Rock Lake. The Rock Lake mill was old and the authorities wanted all mills out of the Park. The end of the Rock Lake mill coincided with the building of a new high-grade band mill on the Whitney site in 1979. Dad began to retire at this time. Alec McGregor retired from the office about one year later [1981]. Combining the mills at one location and having the office there made life simpler for John and me.
>
> It was the right economic decision. It greatly simplified the operation by locating both mills at the same site. Previously, we had to unload logs at Whitney and then ship them to Rock Lake and vice versa. Also, the mill at Rock Lake was becoming an increasingly complicated mill to run. The mill had an old boiler plant and other older equipment. It also lacked hydro, and it had 15 miles of dirt road to maintain.
>
> We built an all-steel-and-cement mill building where previously all mill buildings were made out of wood. The mill had been engineered, but it took a while to get it running properly.

The design was similar to the Rock Lake mill with a five-foot band mill and an automatic carriage with a five-foot resaw and automatic edger. The real trouble was that some of the equipment didn't work as well as it should have. In the end, we had to replace the five-foot band mill with a six-foot double-edged band saw, two resaws and one edger.

Each piece of equipment was powered by an electric motor and a gearbox. One of the major problems with building the new mill at Whitney was our hydro service. We were operating on the original power line to Whitney. When we put the new 75-inch chipper in the original mill in Whitney, we had to power it with a diesel generator. This made life more complex and expensive. It was a big step for the company to walk away from its mill in the Park without any assistance. It was the last sawmill in the Park; [it] would have been cheaper to rebuild if it hadn't been in the Park on a land-use permit, and without access to hydro and 20 miles back in the bush.

In 1982, we were fortunate to obtain hydro. The Ontario government introduced a BUILD program, and our application was accepted. The company had to put down a $100,000 deposit and the BUILD program provided the balance to upgrade the power line to Whitney to 44,000 volts. How the situation had changed! Hydro wouldn't even talk to the company about more power prior to the BUILD program. With the new program Hydro worked through the winter with all available crews to get hydro to us.[452]

Photo 111 **High-Grade Band Mill**   Source: McRae Lumber Company

A number of improvements have been made to the band mill over the years and more are planned to keep up-to-date in the technology-driven world of lumber manufacturing.

Jamie McRae commented on recent changes to the band mill.

> In 2000, through an Industrial Research Assistance Program (IRAP) project funded by the National Research Council of Canada, we began the design and construction of a new, computerized carriage and sawing system for the band saw. This new system is expected to be fully installed and operational in the fall of 2005. Along with greater sawing speeds, ease of operation and complete 3-D log scanning comes a vastly improved system of data collection and information management. Essentially, this system allows us to scan every log, know its exact dimensions and helps our sawyers make more informed decisions.[453]

The building of the high-grade mill at the Whitney site also brought with it the moving of the office from the "large white building" at Airy. While the new building, built on a cottage plan, brought more office space, which was welcomed, other aspects were missed. Merv Dunn, the company's office manager remarked, "Oh you couldn't beat the view at Airy, and on Fridays Marjorie used to come into the office with a fresh pot of percolated coffee … and that was good stuff." Having worked the past 32 years for the company, Merv is an indispensable source of information pertaining both to the administration of the mill and to the history of Whitney.

The Weigh Scale and Scaling

Tree-length stems replaced logs brought from the forest, and a new and very efficient scaling scheme based on weight was introduced in 1980.

For Algonquin Park and the AFA

> the advantages of the system modification are that it removes more labour and equipment from a physical presence in the Park with the added feature of reducing the size of roadside landings. Improved product recovery, tree utilization and operator safety resulted from conducting the processing phase on a large industrial site, under controlled conditions.[454]

But it soon became apparent that it was much easier to get the tree-length system working in the bush than at the mill. With the tree-length system, the volume and size of the cull wood pieces increased.

The tree-length system required a new scaling method. Previous to 1976, the "stick scale" method was used. It required a scaler/tallyman and an assistant scaler on the other side of the skidway who hammered the end of the log to be measured to identify it, and then called out the diameter, amount of defect and any visible crooks. Bob McRae explains:

> In 1980, we installed a 70-foot weigh scale in front of the office. The weigh scale made life a lot more accurate. It is hard to imagine now how we could have operated without a weigh scale. Now all our incoming wood could be weighed and all payments are based on that accurate scale.[455]

Chapter 12: The Modern Mills of The McRae Lumber Company

McRae still employs a scaler. During the year, the scaler is most concerned with species distribution. He scales about 200 loads a year and, in doing so, he tries to get a varied distribution by township of origin. He scales the loads that come off the trucks and correlates his tally to the weigh-in amount. The MNR/AFA also sends scalers to work the slasher deck. These scalers try to work about half a day per month and scale for grade, defects and undersized stems.

All of this scaling is done to determine the stumpage fee McRae owes the AFA for the wood. Stumpage varies according to species and quality. After the AFA deducts its expenses, a net amount is paid to the Ontario government.

The scales are also used to weigh the McRae transport trucks making product deliveries. Measurements give a record of the weight of material shipped and give assurance that the loads carried do not exceed highway regulations.

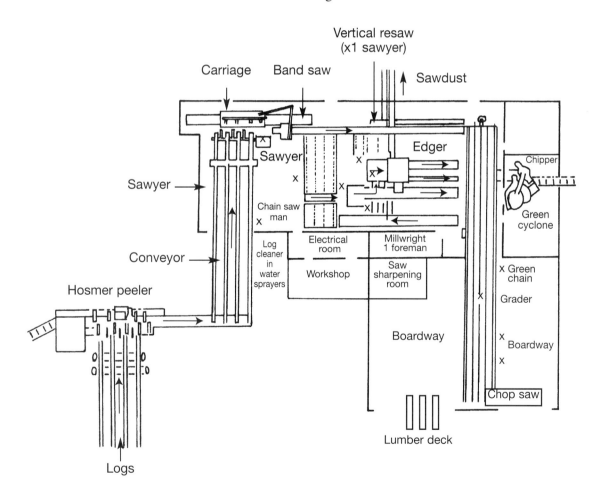

Figure 39 **High-Grade Band Mill Layout**

Source: After Jamie McRae

Duncan Fisher MacGregor

In December 1980, Duncan MacGregor was run over at the mill by a skidding transport truck. Dunc was 73. The event answered the question, "When was Dunc going to retire?"

Duncan MacGregor came out of Eganville, the second youngest child of Donald (Dan) and Annie (Keenan). Dan McGregor worked for the McLachlin Lumber Company of Arnprior. The McRaes lived next door.

Along with many others, the July 1911 Eganville fire wiped out Dan McGregor's investments. Later that fall he died, leaving his wife Annie with six children to raise. Education for the children became Annie's primary goal. Willie was sent to Queen's University in Kingston and Janet went off to Ontario Ladies College. Then World War I intervened. Willie joined the Canadian army, transferred to the British Black Watch regiment, was wounded several times, rose to the rank of major, and survived prison camp. After the war, Willie studied economics at Edinburgh, then law at Osgoode Hall in Toronto. Eventually he became a judge; his verdicts became famous as "Killaloe Law." Sometimes verdicts given by William MacGregor were not rendered strictly by the book or by acceptable precedence, but were appropriate to the case in a local sense.

Another brother, Ken, became an engineer

Photo 112 **Duncan MacGregor** **Source:** Mark Webber

Chapter 12: The Modern Mills of The McRae Lumber Company

and worked on the Kitimat power project in British Columbia.

Duncan MacGregor, while just as bright as the others in the family, had a different streak. While he read voraciously, he lacked the discipline to jump the required educational hurdles mandated by pedagogues. His psyche demanded space and he found that in the upper Madawaska Valley.

Dunc apparently inherited as much money as others in the family. In 1929, he had a store in Eganville. Perhaps his inheritance was invested in it, but the store was gone by the time he arrived in Whitney.

Dunc came to visit in Whitney on August 28, 1929 and ended up staying a lifetime. Joining the Whitney baseball team that day, he became an instant hero in the village by hitting a home run. To round out his first day in town, although bootless, he joined in with a crew fighting a forest fire at Madawaska Lake that took three weeks to put out. His ankles were badly burned and Dr. Gib Post treated him in the Red Cross Outpost hospital. The two men became lifelong friends, bound by their mutual love of classical literature and fishing.

Dunc started working for his brother-in-law, J.S.L., and spent the rest of his life employed by the McRae Lumber Company. In the bush, Dunc experienced all the phases of cutting and skidding logs. He even clerked for a while. When the mill at Lake of Two Rivers replaced the Airy mill destroyed by fire, Dunc went along. The Depression was then full blown; there wasn't much else to do.

At the new mill (Lake of Two Rivers) he began to specialize in mill yardwork, and he learned from an American scaler how to grade lumber.[456] Roy MacGregor writes:

> He had a remarkable eye for hardwood, was able to identify a cut board by eye quicker than most woodsmen could tell looking at the whole tree and leaf. The mill began producing on-line in early 1934, and Dunc took up his new duties grading the wood for shipping. He was paid twenty dollars, a nice raise from the fifteen dollars a month the regular mill workers were being paid.[457]

When Gib Post came down with tuberculosis, and he and his wife, Jay, who also served as his nurse, moved to Ottawa so that Gib could enter a sanatorium in Ottawa for nearly two years, it was the long weekly letters from Dunc that kept their spirits going. "They were the most wonderful letters I ever read," Jay would say more than a half-century later. "Duncan told stories so funny about the Whitney characters that Gib would lie there in tears while I read them aloud."

"Whitney was playing Barry's Bay down at the Bay," he wrote … "and Alfie Fox [the CNR station master of Whitney] was playing in the outfield. Alfie liked a few shots before the game and, as a result, he pulled off the greatest catch I ever saw. The Barry's Bay batter hit the ball, and Alfie made a one-handed grab at the ball and missed it. However, it landed on Alfie's head and bounced into the air and Alfie gloved it on the rebound. It was a wonderful catch and received great applause from the Whitney fans at the game."[458]

He returned to the mill at Lake of Two Rivers handling the high-priority birch shipments that were tagged for England and the manufacture of Mosquito fighters. The birch had to be of the highest grade, and he was now in charge of grading.[459]

An incident that happened when the mill was at Rock Lake underlined his expertise in grading wood. McRae sent a shipment of wood to a firm in Texas. The receiving company disputed the grade and Dunc was sent down to check it out. Dunc could tell by the saw cuts that the wood in question did not come from McRae. Further, he identified the McRae shipment in another pile of wood, and it was as billed.[460]

There was more to Dunc than his excellent grading skills. Because he was J.S.L.'s brother-in-law and was always on the job, Dunc, when he felt it necessary, would confront his brother-in-law.

> Dunc served a useful role for his brother-in-law. These were times of great growth in the lumber industry. J.S.L. was buying up more timber rights, expanding operations, and the work itself was changing, with the use of power saws in the bush and bulldozers to put in the roads and huge logging trucks to haul the timber out. "The Boss" was used to having his way and could be difficult to deal with, but Duncan, it seemed, had special permission to take issue with him if he felt something needed to be done. They had one spectacular argument at Hay Lake over, of all things, whether the cookery should have butter or margarine, but Dunc won the day, and the thanks of the other workers, when "The Boss" declared that they could have butter if they insisted.[461]

Andrew Siydock, another long-time McRae employee, frequently worked with Dunc. One very hot day they were counting, grading and loading wood into a boxcar at Wallace when the mill was at Hay Lake. Dunc worked 12 hours straight without complaint. A case of beer kept him going![462]

Yes, Dunc drank.

> We used to stop at Strawberries', you know. Just the other side of the west gate of the Park, towards Huntsville—he kept a tourist place, a half-assed one. He was a bootlegger. We never had a chance to buy any. You always had to work 'til six o'clock on Saturdays too. He sold that beer at cost, every damn bit of it. Beer was maybe $3 or $4 a case, and that's all he charged too. He told me that the breweries delivered it.[463]

Duncan continued to work as the mill was moved from Lake of Two Rivers to Hay Lake, to Mink, then to Rock Lake and finally to Whitney.

Gary Cannon remembered,

> Dunc drank quite a bit, but it didn't interfere too much with his running the yard. Once in a while on a Friday around four o'clock, Dunc would do a bit of staggering, but no, he was pretty good. By jeez. I've seen him, if we were shipping a two common pile of lumber and there was a one common or select or bet-

ter in the pile, why he'd carry that on his back 500 feet over to a pile where it should go. But he was that type of fellow. Maybe there was 50 cents more in that board. He was good at that, Dunc. Damn right. That's why they liked him. They know that he knew good work and they appreciated that, Donald and the boys too.[464]

In 1980, Donald McRae turned the business over to sons Bob and John, but Dunc refused retirement. He still came to work, so Bob and John McRae provided their great uncle Duncan with his personal trailer. Someone came from town to wash and change the sheets on his bed every week.

His last day in the mill yard was in December of 1980. He was 73 years old. The day was bitterly cold and with his parka hood pulled over his head he didn't hear the massive chip truck sliding backwards toward him. He was knocked down, pinned under the locked wheels of the skidding truck and dragged some 50 yards.

After a brief stop was made at St. Francis Memorial Hospital in Barry's Bay, an ambulance took him to Peterborough. From there he was moved to Toronto, where surgery was performed at Toronto Wellesley Hospital. His hip was pinned and his pelvis set. The family was told that he would likely not walk again. Huntsville Hospital provided over two months of therapy.

Shortly after the spring's snow had cleared from the sidewalks in Huntsville, he proved the prognosis incorrect, for he was able to walk from home to the Empire Hotel for a beer or three, then drop into the town library. At home he read and watched television.

Dunc's forced retirement lasted a lot longer than most expected. His charm worked its ways on nurses and doctors, family and friends, until his body finally gave out 15 years later.

Dunc had been part of the mill family virtually since its beginning. He was a brother-in-law, but also very much a friend to J.S.L., in addition to being a highly valued employee. Donald McRae was some eight years younger than Dunc—friendship and respect bonded them. With Bob and John, whether it was business or simply living, Dunc was always there for them, but without pressure. Roy MacGregor, Dunc's son and biographer, wrote these words on the flyleaf of his book, *A Life in the Bush:* "He loved you like sons, as you already know. Thanks for being so kind to him over the years." These feelings for Dunc were shared by all the family and the mill community. He may have been away from the mill for the last 15 years of his life, but his passing was keenly felt.

The Fire at the Low-Grade Mill

In the spring of 1983 a fire started in the low-grade mill.

Bob McRae describes the unforeseen changes.

> Then, in 1983, disaster struck. The original low-grade mill in Whitney burned on Easter weekend from a welding spark (we thought). John Mastine, using a bulldozer, created a firebreak between the chipper,

peeler and slasher, and the rest of the low-grade mill. This was successful, for the low-grade mill was destroyed but the chipper, peeler and the slasher were saved. As well, the high-grade band mill, geographically removed from the low-grade mill, was not affected. For the next year, we just chipped the lower grade wood. Markets were terrible for lumber, so we weren't in any hurry to rebuild. In 1985, we rebuilt the low-grade mill with two thin-kerf-technology, 42-inch head saws with one resaw and one resaw/edger. The new mill was in a steel and cement building, and it remains the same today.

The new mill was a scragg mill, so named because it cuts the large volume of low quality wood. The scragg mill still contains most of its original equipment from 1985: the twin circular saw and hydraulic overhead carriage, along with the original edger/resaw, resaw and multi-saw trimmers. The only addition has been the installation of a pre-cut trim saw to produce more pallet-stock. Since it was built, the scragg mill has continued to process the large volume of low quality wood arriving at the mill. The major products are construction timber, industrial pallet-stock for steel, blocking, tunnel lagging and pallet-stock.[465]

Another fire, in 1989, resulted in the loss of the old maintenance garage. Bob McRae remembers fighting the fire.

In 1989, we again had a major fire at the mill site. Our maintenance garage burned down, and it was the only building that wasn't sprinkled because we were planning on replacing the building. It started again with a spark during the workday, and despite having lots of water to fight the fire, we couldn't control it. All we could do was to stop the fire from spreading to the mills. We rebuilt the garage as fast as we could, but it still took 12 months. If it hadn't been for the hydrant system around the mills, we might have had a major disaster. Fire is an ever-present danger.[466]

The new garage was built on top of the hill near the band mill. It is a large garage capable of servicing multiple vehicles and is the cornerstone of McRae's maintenance abilities. It has six bay doors and inside there are two maintenance pits, a large welding shop, a tire room, our oil room and an office for John Mastine [fleet and maintenance manager]. This large garage provides space to work on multiple vehicles at a time and in a climate-controlled setting. It also is big enough to park all mobile equipment (loaders) at night during the winter. McRae equipment has always passed inspection by both the Ministry of Transportation and the Ministry of Labour.

John Mastine spends most of his days dealing with mechanical problems and commonly exhibits a gruff exterior. Beneath this exterior there is an impish streak which has occasionally surfaced. Jamie McRae recalled a story that Mastine told him:

Back in his early days as garage boss, there used to be lots of dynamite around for use in bush operations and the government was starting to crack down on who could use and possess the stuff. Well,

Chapter 12: The Modern Mills of The McRae Lumber Company

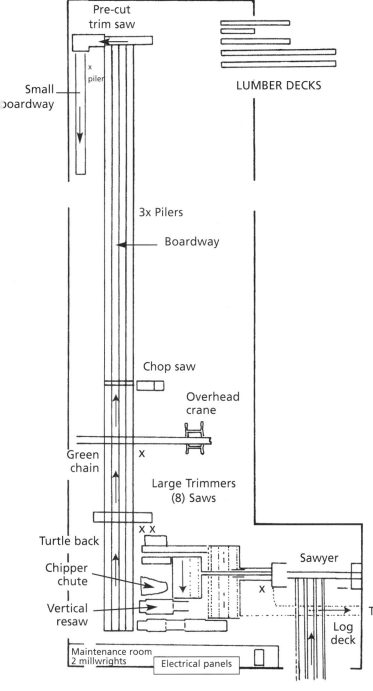

Figure 40 **Low-Grade (Scragg) Mill Layout**
Source: After Jamie McRae

John said that Gary [Cannon] gave him all the dynamite and told him to hide it somewhere, which he did, but Gary also gave him all the blasting caps. So I guess John decided that he would play a little prank on Johnny Yashinskie, one of the loader drivers.

Mastine had all of the loader operators trained that when they took their lunch break they had to turn off the master switch for the loader, which was located just beside the seat in the cab, and to turn it back on when they were ready to go after lunch. So what John did was take a blasting cap and bury it in the dirt beside the loader and connect it to the master switch so it would explode, make an awful bang and kick up a nice pile of dirt. Well, after lunch Johnny was walking back to his loader with Robert Cannon and Jerry Nicholas. Mastine told me he didn't know what to do as he didn't want to spoil the prank, but he knew that blasting caps were quite powerful and who knows what would happen with Robert and Jerry so close. Well, John said that the blasting cap exploded with a huge bang and dug a hole about a foot deep into the ground. Because the explosion was close,

> Robert and Jerry were so startled that they jumped up and started slipping around in the mud, eventually falling down into a big puddle all the while punching and grabbing at each other. John says that he just ran right out, grabbed what was left of the blasting cap and took off before anyone knew what had happened.[467]

When told that this story might appear in print, Mastine just chuckled. Boys will be boys.

Bush Operations

In 1975, the AFA was mandated to supply McRae Lumber with wood, while the MNR managed silvicultural activities such as tree marking, planting and thinning. The AFA contracted out the jobs connected to cutting and hauling because they did not own heavy machinery. In 1983, the Interim Forest Management Agreement transferred silvicultural work to the AFA. By 1986, while it remained the contractor with the AFA, McRae had subcontracted the work of cutting, skidding, hauling and road building to Stanley Kutchecoski. Stanley carried out this work for McRae both inside and outside the Park, where McRae had both private lands and timber leases from the Crown.

Bob McRae commented on changes that were made at both the mill and in bush operations.

> In 1986, we made major changes again, trying to cope with the wood coming from the bush. We installed a new slasher line at the mill to process tree-length hardwood. Gary [Cannon] had wanted to go tree-length for some time, and he and John had looked at [Grenville] Martin's lumber operation in Harcourt, where they did some tree-length. Gary foresaw a lot of advantages in the bush. In the bush, a tree-length operation required small landings and didn't leave debris on the landing. The footprint from logging was a lot less, and the utilization improved. However, what improved the bush operation complicated life at the mill, and complicated the trucking. In the bush and at the mill we had to upgrade our log loaders to handle the longer and bigger treelengths. We had to purchase tree-length trailers. The roads in the bush had to be straightened a bit and the grade on the hills lessened. But Gary was right that it made a simpler and lower impact operation in the bush. It didn't take long for the AFA to follow Gary's lead.
>
> At the mill—in 1987—we installed a sprinkler system for both mills powered by both diesel and electric fire pumps. This was an attempt to overcome the risk of fire. In the sawmill business, no matter how careful one is, fire remains an occupational hazard. Also, insurance was more reasonable and easier to obtain on a sprinkled property.
>
> Our existing 27-inch Brunette peeler and 75-inch chipper were very inadequate to handle the volume, size and roughness of the tree-length hardwood coming in from the bush. Once again, in 1988, we upgraded our log processing, by replacing the peeler with a drum debarker, and the 75-inch chipper with a 96-inch chipper.[468]

Safety

Although safety has always been an important issue within the logging and lumbering community, recently, worker safety has been given a much greater emphasis by the government. A concerted push by the Ministry of Labour forced many sawmills across Ontario to put a greater focus on safety and to make it a priority within the operation. In August of 2000, the McRae Lumber Company became recognized as a Safe Workplace Ontario sawmill with the Ontario Forestry Safe Workplace Association. This designation came after many physical and cultural changes were put into place throughout the site to improve employee safety.

The Second Slasher Line

In 1989–1990, a second slasher line was installed to increase the production of sawlogs, bolts and chipwood. Modelled on the earlier slasher, it incorporates an elevated log deck and a high visibility operator's cab, allowing for a high level of production with an optimizing of the quality of logs. Although the new slasher is the primary slasher, the older one is still used to keep pace with the steady inflow of trees from the bush operations, ensuring that the tree-lengths don't spoil sitting in the log yard.

In the summer of 2003, the drum debarker was replaced with a new Deal processor-peeler. This machine has an increased capacity and peels logs quicker, more fully and with less damage. These improvements were necessary to effectively manage the large volume of cull wood that still arrives at the mill from the bush operations.

Gary Cannon

When a company has been in business for as long as McRae Lumber has been, and has had the practice of keeping its good workers steadily employed, it is inevitable that unexpected deaths will occur sometime on the job. Gary Cannon's passing was one of these.

In June of 1996, the McRae Lumber Company family suddenly lost a long-time and valued member. Gary Cannon died of a heart attack in the bush. Gary came from a

Photo 113 **Slasher Line, Deal Processor and Chipper**　　　　　　**Source:** McRae Lumber Company

family with deep ties to McRae Lumber. His grandfather had come from Ireland and had a poor farm near Osseola some ten km northeast of Eganville. When his grandfather died at 42, the grandmother moved the family to Whitney, took in boarders and the older brothers went to work. His uncle Tom worked as a cook at Hay Lake about 1920, just before J.S.L. bought the Airy mill from Mickle, Dyment. Gary's father, Pat, was a helper in the kitchen. All four brothers worked in the kitchen: Joe, Tom, Pat and Bert. As well as working in the cookery, Gary's father drove a truck, and was an edger man.

Gary started working summers for McRae Lumber at 13 years of age, piling pickets. Mechanization was starting in earnest and soon Gary was working on all the McRae equipment: the two-man power saw, then the one-man saws. He was in the first group on the Blue Ox; then he worked on the International tractor and the Timberjack skidders, as well as the Prentice and Barco loaders. Gary knew machinery, and one of the first moves Donald

Photo 114 **Gary Cannon, Edmund and Philip Kuiack**
Three long-time McRae employees are shown here. In the background is the scragg mill and chipper.
Source: Steve Burns. Courtesy of General Store Publishing House.

Photo 115 **Jimmy Wojick**
In the background is a pile of white pine logs. Jim lives in Madawaska and has worked for McRae Lumber Company for over 30 years.
Source: Steve Burns. Courtesy of General Store Publishing House.

Chapter 12: The Modern Mills of The McRae Lumber Company

McRae made after taking over from his father in 1969 was to promote Gary to bush superintendent. He was then in charge of locating and building roads, organizing the jobbers and supervising all the rules and regulations that went with Workmen's Compensation and Ministry of Labour regulations.

Jamie McRae was looking forward to spending the summer of 1996 apprenticing with Gary Cannon.

Photo 116 Bob McRae and Gary Cannon
Source: Mark Webber

This was because Gary knew just about everything about the bush and how to get the timber out. I'd always known Gary was a great bush boss. I can't even recall how many days I'd spent driving around with him in the bush: summer days when school was cancelled, even days when Mom just needed a break. She'd drop us off at the mill, and within a short while we'd be picked up by big Gary in his huge red pickup truck.

I think waiting was the difficult part. My brother and I would always be sort of hiding in my dad's office. We'd sit around drawing pictures or trying on old hard hats. Then we'd hear that Gary was here and we'd make our way down the hall to see Gary standing behind the big office counter smiling that big beaming half-crooked smile. He'd say, "Well, ya ready? We can't stay round here; we've got lots to do." Then he'd let out a booming laugh, open the door and we'd be off.

The bush was a great place for a young boy; lots of big machines and trucks in a beautiful forest environment teeming with moose and deer—you really couldn't ask for anything more. You see, Gary knew the bush better than the back of his hand. He'd built just about every road on our side of Algonquin and knew each corner, hill and ridge intimately. He used to have a huge stack of aerial photos lying on the dash of his truck. Gary did everything by aerial photo. He'd have all the roads drawn on the photos with a red marker and he'd know what each photo was. He'd always show us the photo and say, "We're about here right now, but we're building the road over here," and he would point to a space full of trees on the photo.

When Gary died, it was a really big deal for this small town. I remember how shaken everyone seemed to be. Gary was a very well known, respected and admired man. The mill closed—I can't remember for how long, but I think it was for more than a couple of days. And I think it was partly out of respect and also because nobody knew how to function without Gary. I know my Dad [Bob] was really upset. I don't think I'd ever seen him so distressed—Gary was a great family friend. Gary used to take Dad and Uncle John into the bush as he did much later with my brother and me. They grew up with Gary and relied heavily on his knowledge and his expertise to keep the bush operations going smoothly.

The funeral was the first I'd ever been to and the Catholic church was so full—people were standing everywhere as every seat was full. The only part I remember about the service was that the choir sang "You Are My Sunshine," because that was what Gary was to so many people. He was always smiling, always happy, and always ready with a joke. He was a person that brightened up so many people's lives—he was that sunshine.[469]

Gary died Friday June 28, 1996. Jamie was to have started working with him the following Monday.

Products

The McRae Lumber Company mills about 25 per cent softwoods. Every winter the mill receives an allocation of red and white pine from the AFA. Most of this is sawlog and is cut at the band mill. Some spruce is milled, while one or two loads have come from Algonquin Park, most spruce comes from elsewhere.

Pine is usually cut as 4/4, 5/4, 6/4 and 8/4 lumber. What this refers to is that 4/4 lumber equals 1 inch boards, 6/4 is one inch and a half and 8/4 refers to two inch boards. The better quality pine boards are picked and sold as moulding for housing, while the rest is sold for a variety of building and furniture uses. All spruce is cut as dimensional timber for example 4"x 4"s, 6"x 6"s and used as construction timbers.

The band mill cuts hardwood into the general classes of lumber, industrial timber and pallet stock. Different grades of lumber are produced, sold to large kiln houses in the United States and, from there, distributed all over the world. The lumber is used to make

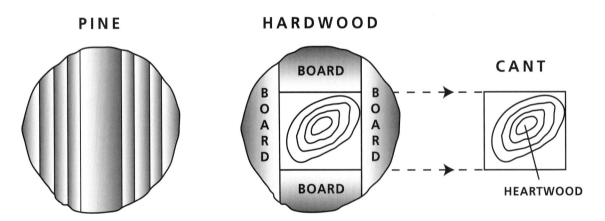

Figure 41 **Sawing Softwood and Hardwood Logs**
Source: McRae Lumber Company

everything from high quality furniture, moulding, doors and hardwood floors, to sports hardwood floors such as those found on basketball and squash courts. Lower quality wood goes to make truck-trailer flooring. Other products include dimensional lumber such as 3″x 3″s, 2″x 3″s, 3″x 4″s and 4″x 4″s, which are used to construct industrial pallets for steel companies.[470] Lagging or tunnel blocking is another product. Lagging is the term given to non-conductive material (such as wood) that is used in tunnels containing telephone or power cables under roads and highways. Pallets, the wooden platforms under stacked cases in a grocery store, are made of hardwood stock.

The scragg mill does not produce lumber but turns out all varieties of industrial timber, pallet stock and lagging.

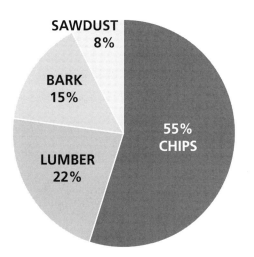

Figure 42 **Average Yield of Products from a Hardwood Log**
The products of the mills reflect the quality of the logs (tree-length stems) coming out of the forest. This is reflected in the average yield from a log, as a percentage
Source: McRae Lumber Company

Grading Hardwood

Hardwood grading is very complex. In North America, the rules are governed by the National Hardwood Lumber Association.

The chart below gives the grades that are applied, but actually doing so is a very complicated procedure.

Table 13 **Chart of Hardwood Grades**

CLEAR FACE CUTTINGS

   Highest Quality
      FAS (firsts and seconds)
      F1F (FAS 1 face)
      Selects
      #1 Common
      #2A Common
      #3A Common

   Sound Cuttings
      #2B Common
      #3B Common
      Sound wormy

   Lowest Quality

Source: McRae Lumber—after grading rules of National Hardwood Lumber Association.

All grades are determined from the poorer of the two faces (except for *FAS 1 face* and *Selects*, which are determined by both faces). *Firsts* (as the term suggests) are the clearest and most valuable boards; the ratings decline from there based on a very complicated set of rules. As you go down the list, the boards become less clear—with increasing defects.

Residues

With the transition to a managed forest and the great quantity of cull wood arriving at the mill, residues have become an increasingly important source of revenue. This importance can be seen in the average percentage of product (by weight) produced at the mill. Lumber makes up only about 22 per cent of all products, while chips are 55 per cent, sawdust 8 per cent and bark 15 per cent (Fig. 42).

In recent years, McRae trucks have shipped, on average, about 2,750 loads of residue per year for further processing. These residues consist roughly of 72 per cent wood chips, 10 per cent sawdust and 18 percent bark. The majority of wood chips go to the pulp mill at Portage du Fort. It produces exclusively kraft pulp from which is made a variety of products such as photo paper, packaging, and other paper products. Sawdust becomes particleboard, and bark is sold as garden mulch or hog fuel, which is burned to create power.

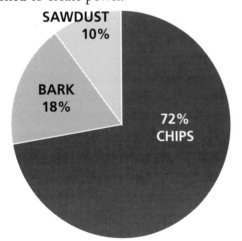

Figure 43 **Types of Residues Shipped by McRae Lumber**
Source: McRae Lumber Company

The Modern Sawyers

Sawyers have always been critical to the success of a lumber mill. John McRae commented:

> For almost a century, hardwood or mixed-wood mills used a system in which the initial breakdown was by carriage and circular saw, and then resawing after that, and edging. As we've gone to lower quality material, we've gone to less selection in the cutting process—in other words, trying to get the piece into a flat cant that then can be broken down through a gang saw. There is very little grade in today's hardwood trees, and so basically you put it into sizes that are marketable.
>
> The sawyers don't have any less skill than in the past. A lot of the timber is rougher today, and is more of a challenge just to get it down into useable pieces. Not every log is 16 feet long and basically round at both ends; there are a lot of crooked pieces and so forth. Probably it demands a higher concentration, if anything, but on different criteria.[471]

Prospects from the Bush

McRae has yet to see a significant increase in percentage of better stems coming from the bush. The growing cycle for hardwood is long and there has not been enough time to see a significant upgrade since 1974 when tree marking, and the selection and the uniform shelterwood systems were mandated for Algonquin Provincial Park by the Master Plan. The mill is still not receiving 100 per cent later-stage silvicultured wood because the AFA is still doing first-time tree marking and harvesting of some areas. Thus the AFA is still forwarding to the mill large quantities of cull wood.

After 30 years of silvicultural treatment Bob McRae was still prompted to say: "It is important to note that even after 30 years of managed forests, on any day we still have more than enough pulpwood to plug up the chipper on a regular basis."[472]

Bush Operations in 2005

Fifty years ago, McRae Lumber had scores of men in the bush with teams of horses—all living through the winter in remote bush camps nestled deep in the forest. Since those days, there has been a steady decline in the overall presence of the company directly in the bush. Now skidding and cutting have been contracted out to local jobbers, and within the past

Photo 117 Robert Cannon
Robert Cannon is slashing in this photo.
Source: Jamie McRae

Photo 118 Feller-Buncher
This machine is more efficient and is safer to use than a chainsaw.
Source: Somi Oh

three years the company has stopped using its own trucks to haul timber to the mill. As of 2003, McRae is no longer the contractor to cut, skid and haul allocated wood to the mill. Bob Robinson Logging was hired as the contractor by the AFA. Gravelling operations are still run by McRae Lumber Company; its operators are subcontracted out by McRae to Robinson for building roads, along with a sand truck and grader for winter road maintenance. The relationship between the AFA, McRae Lumber and Robinson is quite simple. The AFA holds the only timber licence to cut Algonquin wood. They contract with Robinson to cut and haul the amount allocated to McRaes.

The machinery in the bush continues to change, with the goal of increasing production efficiency, maximizing operator safety, and leaving a smaller footprint in the forest. One of the recent changes has been the introduction of the feller-buncher. This machine is capable of cutting as much timber as two to four chainsaw cutters—with much greater safety. Robinson currently uses one feller-buncher in its operations and uses, on average, three skidders to keep up with it. The particular feller-buncher used by Robinson is a small Tigercat model, which uses a circular saw as the cutting device on the grapple arm.

The Bancroft-Minden Forest: Another Source of McRae Wood

The Bancroft-Minden Forest is one of five units that make up the forest management units of central and eastern Ontario. An independent cooperative company manages each of these units. The Bancroft-Minden Forest Company is a cooperative of ten sawmills, one pulp and paper mill, one particle-based mill and two associations of 25 independent loggers. The fifth unit, Algonquin Park, is managed by the Algonquin Forestry Authority (Map 23).

The McRae Lumber Company holds timber licences in the townships of Airy, Sabine and McClure, and is therefore a partner in the Bancroft-Minden Forest Company. McRae Lumber is very familiar with this area as J.S.L. McRae cut timber here on leases he took over from Dennis Canadian when the latter company closed up operations. Lands privately owned in the area by the company are not included in this arrangement (Map 2).

The government mandated this type of arrangement across the province when it passed the *Crown Forest Sustainability Act* in 1994. By this act, Crown forest management units of the Ministry of Natural Resources were turned into privately run and financed businesses—such as the Bancroft-Minden Forest Company. This legislation provided regulations for forest planning, public involvement, information management, operations, licensing and the management of trust funds for reforestation. By this means, the government simply turned over to private enterprise the expense side of forest management. Government revenue was maintained through the retention of Crown dues or stumpage fees, and ultimate control was maintained through the MNR's power to issue the licences to operate and through its power to supervise company operations. As was found when the AFA started up in Algonquin Park, no matter how frugally the business was managed, the expenses to the members increased over previous arrangements.

The Bancroft-Minden Forest is required to have a Crown Sustainable Forest Licence (SFL). It is renewable every 5 years up to a period of 20 years, so long as the company obeys all terms and conditions and obtains approval of its Forest Management Plan (FMP). A required Forest Management Plan, prepared at company expense, sets out management and silvicultural operations. The MNR assists in formulating the plan and approves it. A Local Citizens Committee provides input to the FMP and sits in an advisory capacity to the MNR. The MNR ensures that the plan is followed by undertaking its own compliance audit, while an independent audit is also conducted before renewal of the SFL

Chapter 12: The Modern Mills of The McRae Lumber Company

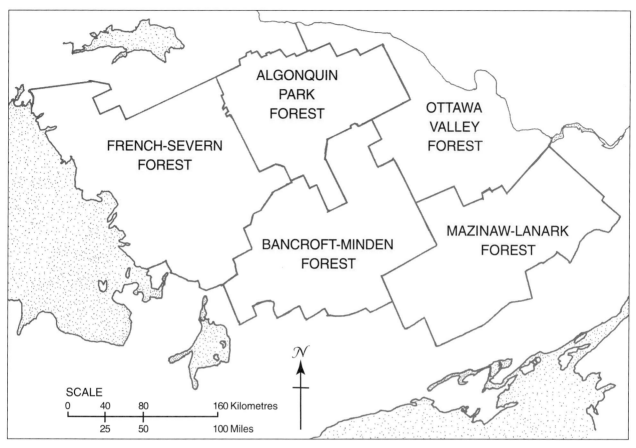

Map 23 **Forest Management Units of Central and Eastern Ontario**
Source: Ontario Investment Service, Province of Ontario and Ministry of Natural Resources, 2001.

every five years. The Sustainable Forest Licence permits the harvesting of forest resources within the forest management unit, but requires renewal and maintenance activities necessary for the long-term health of the forest. These activities include the planting of seedlings and the tending of young plantations.

Tree marking, followed by harvesting, is carried out using the selection and uniform shelterwood systems as described previously.

Changing Markets and Perceptions

Introducing new products into new markets has been a challenge. John McRae described the new ways the company has dealt with this challenge.

> I've a fairly good concept of the land base and what it can yield—the products that will flow from it—and I try to see how that fits into the marketplace. I worked, for instance, on selling blocking materials that

go to the steel mills for their shipping. It's kept me busy, and the company as a whole busy, just handling the transition from high-quality to low-quality production.

The biggest change has been that we've gone from producing high-quality, high-value products to producing low-value products and more of them. We've made this transition because the forest-management system has dictated that's what comes out of the forest. The province, the Crown, has insisted on this forest-management system. The clearest manifestation of this policy is the actual physical marking of the trees that come out of the woods by either the MNR or the AFA, which is a Crown agency. That system basically started in '68–'69 on an experimental basis and steadily grew. By '74–'75, it had become total marking. Initially, the company resisted this policy change. But I could see that it was a necessary transition. That doesn't mean that it's not painful though, in terms of economics. For sustained cutting you need to go this way. Although the company had been high-grading for decades, the options were getting more limited.

The Rock Lake mill was in existence for 22 years. That was the company mill with the highest longevity; it lasted longest because of the advent of mechanization in the woods, the fact that you're bringing wood over long distances to the mill site by truck. The rivers around here aren't big enough to move timber over long distances, and the railroad followed one route only. We consider this mill site [Whitney] to be permanent; it's in the centre of our limits. We will continue on a rotation basis to sustain this operation off those acres which we now have a right to cut.

We ship mainly to southern Ontario, especially the Golden Horseshoe area, our biggest market by far. We ship very little to the furniture companies per se. We ship a fair amount into distribution yards, which in turn would probably, in small lots, distribute it to people using it for furniture. We ship to hardwood distribution yards in Toronto, like Oliver Lumber or Ontario Hardwoods. They inventory wood for distribution. They break down our shipments and sell it off in smaller lots—the dimension lumber that's used for manufacturing furniture parts by furniture companies. We also ship to Thompson-Hyland in Burks' Falls—part of a larger furniture group, known as the Strathearn Group, owned by Dylex Corporation. They make a lot of furniture parts within the group. We ship a fair amount of wood into people making upholstered furniture in Toronto. The rough wood or dressed wood generally goes into frame shops, which make the frames and ship them to upholsterers, often another arm of the same company. Again, Dylex is very much involved in both frame and upholstery shops. We do some business directly with Algoma Steel Corporation in Sault Ste. Marie. In the Hamilton area, we sell mainly to two or three large distribution yards there, because the steel companies' buying habits don't suit our capabilities. Those distribution yards are, in turn, tightly tied to Dofasco and Stelco.

At maximum, there would probably be 25 per cent going to the furniture business, directly and indirectly. Probably 10 to 15 per cent goes into the flooring business, mainly into truck flooring—truck decks and containers. We sell to a manufacturer in North Bay, Laidlaw Industries. Probably 50 per cent goes into the steel and packaging business. Another 10 to 20 per cent is softwood; we deal with a variety of people, to distribution yards in the Toronto markets and lumber wholesalers who resell the yards. It's a volume, low-margin business generally.

Most of our tractor-trailer loads are running at the maximum limits. We probably ship eight to ten loads of wood chips a day, two or three loads of sawdust a day, a load of bark and three or four loads of lumber for a total of 15 or 16 loads a day. The wood chips go to three end-users in the kraft or pulp and paper business, the major one being Consolidated Bathurst at Portage du Fort, in Quebec. After that would be CIP at Gatineau, Quebec, and then some into Temoix in Temiskaming, Quebec. The bark goes to Alltreat Farms, who take it and make garden mulch for sale in garden centres. The majority of the sawdust goes to Domtar in Huntsville. They're making a particleboard out of it.[473]

John commented on the change in markets.

We've gone from being solely a lumber producer to today, when residues are about 30 per cent of our dollar sales, but in terms of physical volume, our lumber production probably represents 25 per cent. Depending on the species, woodchips would be upwards of 50 to 60 per cent. I guess bark would be about five to ten per cent, and the remainder being sawdust, again depending upon the species. In my father's generation and my grandfather's, 70 per cent of their volume moving was in lumber.

Today, our most critical markets are our residue markets. If we have marketing problems, they tend to influence our production almost immediately, for it's very difficult to put a great amount of our residues in inventory for any length of time because the physical problem of moving them when the market does exist is almost too great to handle. They pile up so quickly.

Because we're much more into commodity items, or low-quality items, the markets are very, very competitive. It's not like selling all high-quality products, where just by virtue of the fact that you are one of the only people selling, you have control in the market place. We don't have that luxury anymore. Most of the things that we sell today can be supplied readily by other people. So I guess that's one of the reasons why today we have to be reasonably efficient relative to our competition and provide a reasonable degree of service and reliability to our customers. We try to differentiate ourselves from the competition on that basis.[474]

On Financing: Prudence

The thrifty Scot still lies deep in the McRae genes. The attitude is reflected in Donald McRae's statement: "The only time to do business with a bank is when they owe you money!"

Bob and John are just as cautious.

> Probably one of our distinguishing characteristics, as far as financial management, is that we have an aversion to overloading ourselves with debt, because we see our industry as being highly cyclical in terms of earnings. It doesn't fit very well to have a great deal of debt. Or at least in my brother's opinion and mine—it just makes it too difficult to manage. It locks you into too many decisions on a financial basis, rather than on a purely operational or marketing basis. We've attempted to avoid any substantial degree of long-term debt. We feel that it's afforded us the flexibility we've needed to be reasonably successful in our operations. It hasn't tied our hands. As well, it's given us the satisfaction of being able to do what we want when we want. Perhaps it hasn't allowed us to grow when we could. But that's not one of our principal objectives: growth for the sake of growth.
>
> The pattern of the industry tends to follow interest rates. As interest rates have risen, well, house building is counter-cyclical to that, and consequently expenditures on furniture and flooring, and things like that are down at the same time. Most recently, in the early eighties, when interest rates peaked at 22 per cent, well, there wasn't a lot of house building going on. That tended to depress prices, as well as volumes, substantially.
>
> Yes, we had to borrow to build the new mill in '79, but not on a long-term basis. As you plan these things, you plan on how you are going to retire the debt if possible. And we've been fortunate in that ... we planned our major expenditures so that you come on the market as the market improves. That's always the objective. After we opened the mill here, we had a good year before the market slowed down substantially.[475]

Looking back from the year 2005, Bob and John can see 25 years of successful management. McRae Lumber Company continues to enjoy a solid reputation for the timely delivery of well-manufactured lumber according to customer specifications, as well as for meeting or exceeding industry standards as set out by manufacturing associations such as the Canadian Lumbermen's Association (CLA) and the National Hardwood Lumber Association (NHLA-USA).

Relations with the Labour Force and Community: Reward Good Work with Steady Employment

The McRae Lumber Company has had good relations with its workforce since it began operation in 1911. At that time, J.S.L. worked side-by-side with his hired hands. Phil Roche commented,

> Everybody had to work; there was no unemployment insurance then; if you wanted to eat, you worked—there was a very stable group, a pretty large group that lived here. McRae tried to keep them all year-round, even though there were times when there wasn't much work for them.[476]

In common with many businesses, the lumber industry is subject to downturns in the economy that affect sales. For many companies, the onset of a soft market is the signal for layoffs in their labour force. McRae has resisted doing this, and the loyalty of their employees is a direct result. John McRae gave one reason for doing so.

> We did not layoff during the recession of the late 1980s. We try to maintain a very steady course if we possibly can. I think more than anything it translates into lower earnings, in that we choose to probably maintain the labour force at a constant level.[477]

Discharge of Community Responsibilities

Both Bob and John McRae have personally served their community.

Bob McRae has served as a Director of the Friends of Algonquin Park since its inception in 1983.

> The McRae Lumber Company helped greatly in creating the new Algonquin Logging Museum. McRae's provided the notcher to use on the logs for the camboose shanty and loaned the lifter used in erecting the walls of the building. McRae also contributed the material, and Bernie Stubbs Sr., after he retired, built the furnishings that went into the camboose shanty. McRae also refurbished the locomotive and the donkey engines, and provided expertise in setting up the blacksmith shop. The traditions of the McRae family from the square-timber days through the modern era live on in this museum.[478]

As previously noted, the McRae Lumber Company, in conjunction with the AFA, refurbished the old railway bridge that crosses the Madawaska River at the south end of Whitefish Lake. This bridge, and another across the Madawaska River upstream at the old airfield, makes possible a family bicycle trail from Mew to Rock Lake along the roadbed of J.R. Booth's Ottawa, Arnprior and Parry Sound Railway.

John McRae serves as a member of the Local Citizens Committee of the AFA/Algonquin Provincial Park forest. This citizens group gives input to the AFA and MNR relative to the forest management of Algonquin.

Within the Whitney community, the McRae Lumber Company is held in high respect, as it has been since coming to the community in 1922. Usually, when help in a particular area is needed, it is just carried out quietly. The J.S.L. McRae Scholarship Fund began in 1970 with awards of $250; today they are $500 each. The intent is to help students who are pursuing post-secondary education (college or university). Originally they were for students who had a parent or grandparent employed at the mill, but they have since been expanded to include local young people. They can be awarded for every year of attendance at a post–secondary school, although a student must apply every year. Four scholarships were awarded in the first year (1970) while 16 were awarded for the school year 2004–2005.[479]

Adapting to Changing Times

A large part of McRae's success has been due to its ability to adjust to changing times. John made the following comments in 1986.

> Our philosophy has been to try at least to be cooperative, and to a degree we try to be progressive too. We've moved in that direction very quickly, whereas a lot of our competitors have resisted the movement. They have been very reluctant to accept the change in the management system. They've given very little cooperation and have been embroiled in a continuous confrontation with the Crown. And they try to continue with their old methods. Even though the forest is marked, they tend to take the high-grade logs out and leave the rest. Our philosophy has been to try and take everything that has been marked for removal. That's one of the things that differentiated us from our competitors. We felt that cooperation will, in the long run, be to our benefit, because the sooner we are able to clean up the acres that we're going to operate *in perpetuity*, the sooner we're going to bring them up to high production.

> We've operated in good faith. The Crown isn't in a position where it wants to enter into management agreements with companies like ourselves—smaller operators. But I think we're confident that as long as we operate as responsible forest users, the likelihood of our losing our rights to operate on the acres we traditionally have, our right to is likely not threatened. We've seen that from what has gone on, the way the Crown has treated other people. I couldn't see any reason why the Crown would change its approach.[480]

John made the following comment 20 years ago and it holds true for today.

> In the short space of time that I've been on the scene, some of them [other mills] have gone out of business for one reason or another. We, as managers, realize that in this business you have to operate on a long-term basis. Of course, in the short-run, you have to perform also. But if you don't have an eye to the future, your future is limited. We've also been able to make the transition because the company is relatively well-financed, and hasn't been at the mercy of lenders. Also, we have the good fortune to be relatively close to our resource base. We haven't had to go out and bring timber great distances to our mill. The longest distance that we bring timber would be about 30 miles, whereas lots of mills are bringing wood in a couple of hundred miles. Granted, it's higher quality material. A lot of them have chosen to maintain quality standards and reach out much farther. Our philosophy is that we have to learn to exist on what

we have around us. The critical thing is learning to come to terms with the resource base that we have. Probably it has been easier because Bob and I are younger too. We're probably a little more adaptable. It's also been a good thing that we have had exposure to what goes on in the woods and what the woods are like in this area. It's given us a good appreciation for what is out there, and an understanding that under any situation you're not going to change that. If you want to know what you're going to be doing 50 years hence in the lumber business, you can go and look at it growing on the ground today.[481]

McRae Maxims for Continuing Success

Over the past four generations, the McRaes have done very well timbering and milling, and there is yet another generation being groomed to take over. How have the McRaes managed to have this exceptional run?

This book chronicles the history of McRae Lumber Company—the times and the environment in which they have operated. The book has dealt with things—crosscut saws and teams of horses, ginpoles and skyline jammers, band saws and alligators, as well as perceptions and politics—but mainly it has been about the lives of exceptional people both within and outside of the family and company. They have all earned their way through their daily work in a demanding industry where failure to perform quickly brings harsh consequences. What is truly exceptional in this story is that the McRae family has had outstanding success over four generations. The case could be made for saying that six generations are involved in this continuity.

Obviously there has been ability and this has been passed on through the generations. This required marriage partnerships. Some glimpses of the women in the families have been given—enough, it is hoped, that the book shows that they have been exceptional and equally as important as the men. Cissie Campbell, wife of J.D. was described as a delightful person; Janet MacGregor, wife of J.S.L. showed obvious spirit and spunk, and was said by her nephew, Roy MacGregor, to be the "most wonderful woman that we will ever know"; and Helen McRorie, wife of Donald, sustained him through his trials of managing the company through the socio-political crisis centred on timber leases in Algonquin. All of these women were supports to their husbands, and they all played a huge role raising and stimulating the education of their children. The McRae men have chosen well—or was it the other way around!

The McRaes have consistently showed the capacity to adapt to given situations and to anticipate and plan for the future.

- J.S.L. devised the cribbing of hardwood logs and in doing so made possible the harvesting of logs far beyond the hauling range of horse and sleigh to the Airy mill;

- The McRae Lumber Company has been quick to use new machinery such as trucks to haul sleighs and the skyline power jammer to build log dumps;

- The company installed a chipper in the early 1960s and developed a market for this product;

- Donald McRae anticipated the increase in cull wood that would come from the bush as a result of the Master Plan of 1974 and built a low-grade scragg mill in 1973 in south Whitney to cope with this;

- McRae Lumber developed markets for the construction and pallet lumber produced by the low-grade scragg mill as well as for residue products such as chips, sawdust and bark;

- McRae Lumber started adapting their bush and mill operations in the early 1960s to what the forest was capable of then producing;

- McRae Lumber, early on, realized that the sooner the bush received silvicultural treatment the sooner it would be that it could be brought up to high quality production;

- McRae Lumber, in 1968, was the first company in the Algonquin area to have a timber lease marked and cut in what became known as the selection system in hardwoods and to carry this through to the point of adapting their mill to the lower quality of wood coming from the forest;

- McRae Lumber has avoided any substantial long-term debt thus maintaining flexibility in financing;

- During economic downturns, McRae Lumber has chosen to accept lower earnings rather than lay off employees. This is a development of a policy started by J.S.L., who retained key employees year round at a time when seasonal layoffs were typical of the industry;

- McRae Lumber is very active in the community through the Friends of Algonquin Park, in connection with the AFA and MNR, with the Local Citizens Committee, and a scholarship program for young people graduating from high school and wanting to attend a community college or university.

Quite possibly the most important factor in the McRae success through the generations has been the way the young men have been brought into the business—in effect they have served apprenticeships. In the case of Bob and John, they were taken into the bush camps as youngsters to see what was going on. Stopping for lunch at the cookery was a big deal. All these trips were very happy experiences. In their early teens they worked during their holidays from school at all aspects of bushwork as well as in the mill. They proved that they could do the work, to the other workers, to their father and grandfather and most importantly to themselves. Both Bob and John are university educated and both have held jobs other than in lumbering. They joined the company full-time because they wanted to. Jamie, Bob's son, has recently gone through this process and has joined the company full-time because he enjoys it.

West and south of Whitney lies the upper watershed of the Madawaska River which has yielded its timber to sustain the McRae mills. With continuing good forestry and prudent management, the McRae Lumber Company looks to maintaining its place as a producer of fine wood products and its respected place in the community long into the future.

McRae Lumber Company Limited in 2006

P.O. Box 160
Hay Creek Road
Whitney, Ontario Canada
K0J 2M0

Main Office   (613) 637 5352
Main fax   (613) 637 5395

Web Site: www.mcraelumber.ca

Mill, Office & Loading: Monday-Thursday 7 a.m.–5 p.m. – Friday 7 a.m.–3 p.m.

Directory

John McRae	Sales & Land Manager	johnmcrae@mcraelumber.ca
Robert McRae	Production Manager	robertmcrae@mcraelumber.ca
Jamie McRae	Production, Safety	jamesmcrae@mcraelumber.ca
Mervin Dunn	Office Manager, A/P	mervindunn@mcraelumber.ca
John Mastine	Fleet Manager	(613) 637 2977
Sandra August	Payroll, A/R, IT	august@mcraelumber.ca
Dan Martineau	U.S. Export Sales	(705) 946 1937

End Notes

1. Roderick MacKay, *A Chronology of Algonquin Park History, Algonquin Park,* Technical Bulletin No. 8, Whitney [ON]: Friends of Algonquin Park, 2002, 9.
2. The names given to the first owners of the mill at Airy vary. Figure 2 shows the name as Dyment, Mickle, 1920. Yet bills of lading from the same time period show it as Mickle, Dyment. Locally the mill has always been referred to as Mickle, Dyment or even colloquially as Nickel, Diamond. Mickle, Dyment has been used throughout the book.
3. High-grading is the term given to the cutting of only choice trees. Some control was present, usually in the form of a diameter limit.

Section I: **SETTING THE SCENE**

4. Square timber is the term that was given to the long logs, usually of white or red pine or oak, that were hewn square with a broad-axe for export to Britain in the 19th century where they were sawn into planks or used as "dimension" or construction timber.

Chapter 1: **Peopling the Ottawa, Bonnechere and Madawaska River Valleys**

5. Timber rafts:
The basic unit of a timber raft on the Ottawa River was the crib, a rectangle of red pine cross pieces or traverses 26 feet wide, and in length as long as a matched pair, again of red pine. Tightly wedged ironwood or oak pins in augured holes fastened the corners. Then, parallel to the side timbers, sticks of timber were wedged under the traverses. Two or three additional traverses were added. On this floating raft were added top grade large sticks on the sides and one or two in the middle. These prize timbers were not pinned, but "calumet" pins were driven into the traverses on either side of the timbers. Rowlocks were placed on the outside loading sticks. Iron hole pins to take oars were placed at each corner. Cribs were joined by a cap piece, again usually of red pine, to bolt stakes or pickets, with a three-foot space between the cribs. The cribs were joined together on the sides by binding chains. The "band" would be three to five cribs wide, and three to seven cribs long. A number of bands would make up a raft of one to two hundred cribs.
Charlotte Whitton, *A Hundred Years A- Fellin',* Ottawa: Runge Press, 1974, 125-126.
An average Ottawa raft of 100 cribs contained 2,000 to 2,400 timbers—80,000 to 120,000 cubic feet of sound white pine.
Whitton, 127.
On the St. Lawrence River, rafts were bound together with "wythes." They were ropes made out of twisted birch saplings. The Calvins used pine to float large volumes of oak and other hardwoods down the St. Lawrence from Garden Island, which was located off shore from Kingston, from 1836 to 1914.
Marion Calvin Boyd, *The Story of Garden Island,* Reprinted, Kingston [ON]: Brown & Martin, 1983.
J.S.L. McRae took the "start of the crib," and modified it to move hardwood logs.

6. Sandra J. Gillis, *The Timber Trade in the Ottawa Valley, 1806-54,* Report Series, Volume 153, Ottawa: Department of Indian and Northern Development, n.d., 70.
7. John W. Hughson and Courtney Bond, *Hurling Down the Pine,* Third Edition, Chelsea [QC]: The Historical Society of the Gatineau, 1987, 41-43.
8. Roderick MacKay, *Spirits of the Little River Bonnechere: A History of Exploration, Logging, and Settlement:1820 to 1920,* Pembroke [ON]: Friends of Bonnechere Parks, 1996.
9. R.L. Jones, *A History of Agriculture in Ontario,* Toronto: University of Toronto Press, 1946, 116-117.

10. The Ottawa-Opeongo Road began at Farrell's Landing on the Ottawa River at Lac des Chats, two miles west of the mouth of the Bonnechere River. At Renfrew, the road proceeded inland on a course about halfway between the Bonnechere and Madawaska Rivers, until it reached the Madawaska River at Barry's Bay on Lake Kamaniskeg.
11. Herb Berger, personal communication with author, February 2003.
12. Bob Lyons, *Whitney: Island and Sabine Township*, Bancroft [ON]: Bancroft Times, 1986.
13. Niall MacKay, *Over The Hills To Georgian Bay: A Pictorial History of the Ottawa, Arnprior and Parry Sound Railway*, Erin [ON]: Boston Mills, 1981.

Chapter 2: **Early Mills of the Whitney Area**

14. Perhaps the greatest influence on the development of the upper Madawaska was the construction of the Ottawa, Arnprior and Parry Sound Railway. J.R. Booth started by building the Canada Atlantic Railway (1881-1883) from Ottawa, through Coteau Landing on the St. Lawrence River, to the Canada-United States border near East Alburg, where it joined with the Central Vermont Railway. In 1888, Booth incorporated two railway companies: the Ottawa, Arnprior and Renfrew Company and the Ottawa–Parry Sound Railway Company. Two disputes with the Canadian Pacific Railway had to be settled. The first involved a level crossing to the west of Arnprior. The second involved Haggarty Pass, a narrow pass through the Opeongo Mountains to the west of Renfrew, some two miles west of Wilno. Both situations were settled in favour of Booth.

    In the west, the Parry Sound line was acquired and amalgamated with the Parry Sound Colonization Railway. Depot Harbour on Georgian Bay became the western terminus. In 1899, the OA&PS was absorbed into the Canada Atlantic. This entire line was sold to the Grand Trunk Railway in 1905; it in turn became part of the Canadian National Railway in 1923. For more on this, see Clyde C. Kennedy, *The Upper Ottawa Valley—A Glimpse of History*. Pembroke [ON]: Renfrew County Council, 1970.
15. *Canada Lumberman*, September 1895.
16. *Canada Lumberman*, September 1895.
17. Ian Radforth, *A History of the McRae Lumber and the Community of Whitney, Ontario*, unpublished manuscript, 1990.
18. Brian D. Westhouse, *Whitney: St. Anthony's Mill Town on Booth's Railway*, Whitney[ON]: The Friends of Algonquin Park, 1995, 19.
19. *Canada Lumberman*, November 1896, 18.
20. *Canada Lumberman*, November 1896, 18.
21. Richard S. Lambert and Paul Pross, *Renewing Nature's Wealth*, Toronto: Ontario Department of Lands and Forests, 1967, 421.
22. Ontario Ministry of Natural Resources, *Annual Report on Forest Management, 1999–2000*, Information Series, Toronto: Queen's Printer for Ontario, 2001.
23. D.W. Wyatt, *A History of the Origins and Development of Algonquin Park: A Background Paper Presented to Algonquin Park Task Force, Ontario Department of Lands & Forests*, Toronto: Queen's Printer for Ontario, 1971.
24. Roderick MacKay, *Chronology*, 13.
25. Ottelyn Addison, *Early Days in Algonquin Park*, reprinted, Whitney [ON]: The Friends of Algonquin Park, 1985, 41.
26. Addison, 45.
27. Audrey Saunders, *Algonquin Story*, Second Edition, Whitney [ON]: Friends of Algonquin Park, 1998, 105.
28. Roderick MacKay, *Chronology*, 12.
29. Westhouse, 21.
30. Westhouse, 22.
31. Westhouse, 24.
32. *Canada Lumberman*, August 15, 1913. Reprinted in Westhouse, 23.
33. *Canada Lumberman*, August 15, 1913. Reprinted in Westhouse, 23–24.
34. Bob McRae, personal communication with author, 2004.

35 *Canada Lumberman,* August 15, 1913. Reprinted in Westhouse, 24.
36 The Standard Chemical Company, already located in South River, enlarged that plant and built a lumber railroad deep into northwestern Algonquin Park and harvested wood from there until after World War II.
37 Lyons, 11.
38 Lyons, 11.
39 *Canada Lumberman,* March 1903.
40 *Canada Lumberman,* March 1907.
41 *Canada Lumberman,* February 1903.
42 *Canada Lumberman,* August 1926.

Chapter 3: The McRae Family Background.

43 Roderick MacKay, *Spirits,* 48.
44 Roderick MacKay, *Spirits,* 90.
45 *Reflections of a Century, Stories and Photos, 1902–2002,* Eganville: The Eganville Leader, 2002, 5.
46 *Reflections of a Century,* 64-65.
47 Roderick MacKay, *Spirits,* 90.
48 Roy MacGregor, *A Life in the Bush - lessons from my father,* Toronto: Penguin, 1999.
49 Duncan MacGregor (with Donald McRae), interview by Ian Radforth, tape recording, June 1986.
50 Donald McRae (with Duncan MacGregor), interview by Ian Radforth, tape recording, June 1986.
51 Janet (MacGregor) McRae, interview by Mark Webber, tape recording, June 1976.
52 McRae Lumber Company
53 McRae Lumber Company
54 McRae Lumber Company
55 McRae Lumber Company
56 Albert Perry, interview by Ian Radforth, tape recording, June 1986.
57 Perry, interview by Radforth.

Section II: An Overview of the Operations of the McRae Lumber Company 1922 to 1980

58 The Crown Lands Department administered lands until 1908, when it was folded into the Department of Lands, Forests and Mines. Reorganization created the Department of Lands and Forests in 1920; it was renamed the Ministry of Natural Resources in 1972.
59 *Annual Report on Forest Management, 1999–2000,* 6.
60 See Appendix IV: "Timber License 122 for 1937–1938 to J.S.L. McRae."
The timber lease detailed the location of the cut, spelled out the ground rent (a basic charge for the lease) and fire protection charges. The timber cut is measured in the bush by government agents in a procedure known as scaling. Charges, payable by species, were established by rates termed the "Minimum Bonus." These were the lowest prices acceptable by the government. A company seeking the timber could raise this rate and thereby make its bid for the lease more attractive and perhaps outbid a competitor for the lease. Crown dues were prices fixed by the legislature as the upset price at $ per M (thousand) ft. BM (Board Measure 12″ x 12″ x 1″).

Chapter 4: The Airy Mill Operations: 1922-1933

61 Gerald Killan, *Protected Places: A History of Ontario's Provincial Park System,* Toronto: Ontario Ministry of Natural Resources, 1993, 40-41.
62 Joe Mason, *My Sixteenth Winter: Logging on the French River,* Cobalt [ON]: Highway Bookshop, 1974.
63 Duncan MacGregor (with McRae), interview by Radforth.
64 Donald MacKay, *The Lumberjacks,* Toronto: McGraw-Hill Ryerson, 1978.
65 Thomas Nightingale, *The Diary of Thomas Nightingale: Farmer and Miner,* N.P.: Malcolm Wallbridge, 1967. Excerpt printed in *Glimpses of*

Algonquin, compiled by G.D. Garland, Whitney [ON]: The Friends of Algonquin Park, 1989, 69-72.

66 Gary Cannon, interview by Ian Radforth, tape recording, June 1986.

67 M.A. Adamson et al., *U of T Logging Report 57: McRae Lumber Company, 1926,* Unpublished paper.

68 Lambert and Pross, 400.

69 Duncan MacGregor (with McRae), interview by Radforth.

70 Donald McRae (with MacGregor), interview by Radforth.

71 T. Wayne Crossen, *A Study of Lumbering in North Central Ontario 1885-1930 with Special Reference To the Austin-Nicholson Company,* Cochrane [ON]: Ministry of Natural Resources, Northern Regional Office, 1976. 193-194.

72 Crossen, 194.

73 A.J. Herridge, J.A. Hawtin and J.H. Jamieson, "McRae Lumber Company, Hay Lake, Sabine Township," *U of T Logging Report 77,* 1948.

74 Dan Strickland, *Algonquin Park Logging Museum: Logging History in Algonquin Provincial Park,* Reprinted, Whitney [ON]: The Friends of Algonquin Park, 1996.

75 Bob McRae, personal communication with author, 2000.

76 Bob McRae, personal communication with author, 2004.

77 Donald McRae, interview by Bob McRae, tape recording, 2002.

78 Strickland, *Algonquin Logging Museum.*

79 Donald McRae, interview by Bob McRae, tape recording, 2002.

80 Marjorie McRae McGregor (with Alex McGregor), interview by Ian Radforth, tape recording, June 1986.

81 Killan, 60.

82 Killan, 66

83 Duncan MacGregor (with McRae), interview by Radforth.

84 "1926: Briefs," *The Eganville Leader,* reprinted in *Reflections of a Century,* 148.

85 "1931: Briefs," *The Eganville Leader,* reprinted in *Reflections of a Century,* 179.

86 Radforth, *A History of the McRae Lumber Company.*

87 Bob McRae, personal communication with author, February 2006.

88 Jamie McRae, personal communication with author, February 2006.

89 Bob McRae, personal communication with author, February 2006.

90 Duncan MacGregor (with McRae), interview by Radforth.

91 Duncan MacGregor (with McRae), interview by Radforth.

92 "Smallpox Outbreaks in the Lumbering Districts of Ontario," *Statistics of Disease in Ontario,* 20 (1901): 38, quoted in *The Mississagi Country: A Study in Logging History* by Graham A. MacDonald, Toronto: Ministry of Natural Resources, 1974, 40–43.

93 Crossen, 44.

94 Crossen, 181.

95 MacDonald, *The Mississagi Country,* 37–38.

96 Janet MacGregor McRae, interview by Webber.

Chapter 5: The Lake of Two Rivers Operations: 1933-1942

97 Donald McRae (with MacGregor), interview by Radforth.

98 Donald McRae (with MacGregor), interview by Radforth.

99 G.A. Mountain, "An Inventory of Existing Historic Sites in Algonquin Park," N.p.: APM, 1976.

100 Minnesing Lodge was built in 1913. It was affiliated with the Highland Inn as a CNR wilderness outpost. Guests were driven there by horse-drawn wagon from Highland Inn on Cache Lake. In 1923, Dr. Henry Sharman took over the location, renaming it Camp Minnesing and running it as a religious retreat. The site was abandoned for a time, then, in 1937, it was taken over by Manley

Sessions, who operated it as a lodge. The road was then in rough shape. Guests and supplies were at times brought in by water from Joe Lake Station to Doe Lake, then across the portage to Burnt Island. A jeep was used on the portage. Minnesing Lodge was bought out by the government in 1956 and the buildings removed.

101 *The Eganville Leader*: "50 Years Ago…", January 31/90.
102 Duncan MacGregor and Donald McRae, interview by Radforth.
103 Duncan MacGregor (with McRae), interview by Radforth.
104 George Garland, personal communication with author, March 2004.
105 Donald McRae (with MacGregor), interview by Radforth.
1066 Duncan MacGregor (with Frank Shalla), interview by Ian Radforth, tape recording, June 1986.
107 Donald McRae (with MacGregor), interview by Radforth.
108 Donald MacKay, 113.
109 Donald McRae (with MacGregor), interview by Radforth.
110 Quoted in *Algonquin: The Park and Its People*, Toronto: McClelland & Stewart, 1993, 40.
111 Donald McRae (with MacGregor), interview by Radforth.
112 Shalla (with MacGregor), interview by Radforth.
113 Radforth, *A History of the McRae Lumber Company*.
114 Radforth, *A History of the McRae Lumber Company*.
115 Janet McRae, interview by Webber.
116 Cannon, interview by Radforth.
117 Bernie Stubbs Sr. (with Bernie Stubbs Jr.), interview by Bob McRae, 2003.
118 Janet McRae, interview by Webber.
119 Lorne Boldt, interview by Ian Radforth, tape recording, June 1986.
120 Bob McRae, personal communication with author, 2003.
121 Donald McRae, interview by Ian Radforth, tape recording, June 1986.
122 Donald McRae, interview by Radforth.
123 Bob McRae, personal communication with author, 2000.
124 Donald McRae, interview by Radforth.
125 Donald McRae, interview by Radforth.
126 Shalla, interview by Radforth.
127 Ted Kuiack, interview by Ian Radforth, tape recording, June 1986.
128 Donald McRae (with MacGregor), interview by Radforth.
129 Donald McRae, interview by Radforth.
130 Duncan MacGregor with Frank Shalla, interview by Ian Radforth, tape recording, June 1986.
131 P.J. Whelar, letter to J.S.L. McRae, November 23, 1936.
132 Duncan MacGregor (with Shalla), interview by Radforth.
133 Donald McRae, interview by Radforth.
134 Duncan MacGregor (with McRae), interview by Radforth.
135 Shalla (with MacGregor), interview by Radforth.
136 Duncan MacGregor (with McRae), interview by Radforth.
137 Henry Taylor, interview by Ron Pittaway, tape recording, Bancroft ON, December 1979.
138 Taylor, interview by Pittaway.

Chapter 6: The Hay Lake Mill Operations 1942-1952

139 Pete Kuiack, interview by Ian Radforth, tape recording, June 1986.
140 Ted Kuiack, interview by Radforth.
141 Alex Cenzura, (with Mrs. Cenzura), interview by Ian Radforth, tape recording, June 1986.
142 Donald McRae, interview by Radforth.
143 Don George, telephone conersation with author, May 2004.
144 Cannon, interview by Radforth.
145 Bob McRae, personal communication with author, May 2004.

Chapter 7: **The Mink Lake Operations: 1952-1957**

146 Cenzura (with Mrs. Cenzura), interview by Radforth.
147 Ted Kuiack, interview by Radforth.
148 Donald McRae (with MacGregor), interview by Radforth.
149 Cannon, interview by Radforth.
150 Dan Strickland recounts the history of Cockburn's invention:
Cockburn's design was an instant success and he was soon turning out some 200 boats a year from his shop in what is now downtown Pembroke. The boats ranged from less than 20 feet (6 m) to as much as 50 feet (15m). Built of heavy pine timber to withstand the battering they got on the rivers, large pointers weighed more than half a ton but drew only a few inches of water and could be pivoted by just one tug of an oar. The upswept bow and stern enabled men to perch over the water and work on hung-up logs below them. Dan Strickland, *Algonquin Park Logging Museum,* Whitney [ON]: The Friends of Algonquin Park, 1993.
151 Cannon, interview by Radforth.
152 Cannon, interview by Radforth.
153 Cannon, interview by Radforth.
154 Duncan MacGregor with Frank Kuiack, interview by Ian Radforth, tape recording, June 1986.
155 Duncan MacGregor (with Kuiack), interview by Radforth.
156 Mrs. Tim Devon of Sault Ste. Marie, quoted in Donald MacKay, 201.
157 Frank Kuiack (with Margaret Kuiack), interview by Ian Radforth, tape recording, June 1986.
158 R.J. Taylor, quoted in Donald MacKay, 202.
159 Cannon, interview by Radforth.
160 L. Cahill, D.W. MacGregor and S.W. Lukinuk, "McRae Lumber Company, Hay Lake, Sabine Township," *U of T Logging Report 121,* unpublished report, 1949.
161 Donald McRae, interview by Radforth.
162 Cannon, interview by Radforth.
163 Cannon, interview by Radforth.
164 Donald McRae, interview by Radforth.
165 Eddie Levean (with Jenny Levean), interview by Ian Radforth, tape recording, June 1986.
166 Duncan MacGregor (with Shalla), interview by Radforth.
167 Donald McRae, interview by Radforth.
168 J.E. Bier and A Crelock, "McRae Lumber, Company, Cranberry Lake, Canisbay Township," *U of T Logging Report 75,* unpublished report, 1930–1931.
169 Ted Kuiack, interview by Radforth.

Chapter 8: **The Rock Lake Mill Operations: 1957-1980**

170 Ted Kuiack, interview by Radforth.
171 Brent Connelly, e-mail to Roy MacGregor, February 15, 2003. Used by permission of Brent Connelly.
172 Cenzura, interview by Radforth.
173 Bob McRae, personal communication with author, 2003.
174 Fred Parks, interview by author, tape recording, 2003.
175 Insurance report describing fire protection measures, Dale & Company, September 27, 1968.
176 Bob McRae, personal communication with author, 2003.
177 Donald McRae, interview by Radforth.
178 Boldt, interview by Radforth.
179 Boldt, interview by Radforth.
180 Bob McRae, personal communication with author, 2003.
181 Jamie McRae, personal communication with author, 2006.
182 Bob McRae, personal communication with author, 2003.
183 Bob McRae, personal communication with author, 2003.
184 Alex Cenzura (with Mrs. Cenzura), interview by Radforth.
185 Cannon, interview by Radforth.

186. Bob McRae, personal communication with author, 2003.
187. Radforth, *A History of the McRae Lumber Company*, 185.
188. Radforth, *A History of the McRae Lumber Company*, 185.
189. Parks, interview by author.
190. Cannon, interview by Radforth.
191. Cannon, interview by Radforth.
192. Cannon, interview by Radforth.
193. Radforth, *Bushworkers and Bosses*, 186.
194. Radforth, *Bushworkers and Bosses*, 186.
195. Radforth, *Bushworkers and Bosses*, 187.
196. Cannon, interview by Radforth.
197. Gary Cannon, quoted in *A History of the McRae Lumber Company*, Ian Radforth.
198. Radforth, *A History of the McRae Lumber Company*, 210.
199. Radforth, *A History of the McRae Lumber Company*, 210.
200. Bob McRae, personal communication with author, 2003.
201. Bob McRae, personal communication with author, 2003.
202. Gary Cannon, personal communication with Ian Radforth.
203. Cannon, interview by Radforth.
204. Felix Voldock, quoted in Stanfield and Lundell, 40.
205. Ian Radforth, *Bushworkers and Bosses: Logging in Northern Ontario,* Toronto: University of Toronto Press, 1987, 185.
206. Cannon, interview by Radforth.
207. Felix Voldock, quoted in Stanfield and Lundell, 40.
208. Cannon, interview by Radforth.
209. Cannon, interview by Radforth.
210. The Joe Lavally mentioned here is the son of the "Hay Lake" Joe Lavally who guided Don McRae's ex- POW friends.
Bob McRae, personal communication with author.
211. Bob McRae, interview by Ian Radforth, tape recording, June 1986.
212. John McRae, interview by Ian Radforth, tape recording, June 1986.
213. John McRae, interview by Radforth.
214. Bob McRae, interview by Radforth.
215. E. Ray Townsend, personal communication with author, October 2005.
216. Brent Connelly, personal communication with author, September 2005.
217. Bill Brown, personal communication with author, 2002.

Section III: **Interaction with the Province of Ontario**

Chapter 9: **Reconciling Timber Harvesting with Recreational Use in Algonquin Provincial Park**

218. Lambert and Pross, 394.
219. Lambert and Pross, 398.
220. Lambert and Pross, 401.
221. Lambert and Pross, 408.
222. Lambert and Pross, 401.
223. Lambert and Pross, 408.
224. Lambert and Pross, 410.
225. Ed Charest, personal communication with author, 1998. This route basically follows the valley of Forbes Creek for much of its length as did the "warpath" trail which started from Charest's Depot on Black Bay on the Barron River. It operated from the latter half of the 1800s until the Canadian Northern Railway (later the Canadian National Railway) was constructed in 1915. This put Charest out of business.
226. Dan Strickland, *Big Pines Trail*, Whitney [ON]: The Friends of Algonquin Park, 2001.
227. Don George, *View from the Finish Line: Memoirs of a Dirt Forester,* unpublished manuscript, 2001.
228. Bob McRae, personal communication with author, 2003.
229. Don George, telephone conversation with author, December 2003.

230 Don George, telephone conversation with author, December 2003.
231 Roderick MacKay, *Chronology,* 200.
232 Harvey Anderson, personal communication with author, October 2004.
233 Ontario Forest Research Institute, *Regenerating Yellow Birch,* video recording, 1995.
234 Forestry Branch, Research Section, *Forest Research Reserve–Swan Lake: 1950-1968,* March 1968.
235 Don George.
236 F.H Eyre and W.M. Zillgitt, *Partial Cuttings on Northern Hardwoods, The Lake States: Twenty Year Experimental Results,* Technical Bulletin 1076, Washington: U.S. Department of Agriculture, September 1954.
237 Don George.
238 *Regenerating Yellow Birch.*
239 H.W. Anderson, personal communication with author, January 2006.
240 Many Algonquin Lakes have raised water levels. Most of these levels were established between 1890 and 1920 and drowned former land areas. Some of the wooden lumber dams were replaced by concrete structures as late as the 1960s. The resultant change in biotic conditions led to the producing of methylmercury that is now found in Algonquin fish as well as fish found in similarly affected areas. See Robert E. Hecky, *Northern Perspectives* 15.3, October 1987.
241 Killan, 60.
242 Roderick MacKay, *Chronology*, 16 .
243 Killan, 121-122.
244 OA, RG-1. IA-2, box 37, file 27-1501-5, memo from Leslie Frost to Clare Mapledorm, March 29, 1956.
245 Aldo Leopold, *A Sand Lake County Almanac. With Essays on Conservation from Round River,* New York: Ballantine Books, 1970, 262, quoted in Killan, 131.
246 Lambert and Pross, 481.
247 Killan, 133.
248 Killan, 128.
249 Killan, 129.
250 Killan, 131.
251 Quoted in Killan, 155.
252 Quoted in Killan, 155.
253 Quoted in Killan, 158.
254 *Canadian Geographical Journal,* 72. 1965.
255 Rachel Carson, *Silent Spring,* Boston: Houghton Mifflin, 1962.
256 Donella H. Meadows et al., *The Predicament of Mankind.* A Potomac Associates Book. New York: Universe Books, 1972.
257 Abbott Conway was the manager and vice-president of the Anglo-Canadian Leather Company Ltd of Huntsville, Ontario. In its day it was said to be the largest tannery in the British Empire. The water used at the tannery was said to be the best in the world for tanning leather. The source of this water was the Big East River, which flowed from the western highlands of Algonquin Park, via Lake Vernon to Huntsville. Tannin, used to cure the leather, was extracted from hemlock bark. This bark came from the forests of the area, much of it from Algonquin Park. Mickle, Dyment's Airy mill was a supplier. J.S.L. McRae continued to do so when he bought that mill in 1922 and later when the mill moved to Lake of Two Rivers. Canada Packers bought the tannery in 1953 and closed out the operation in 1960.
258 Killan, 172.
259 Quoted in Killan, 173.
260 Killan, 176.
261 Quoted in Killan, 177–178.
262 Killan, 180.
263 Killan, 185.
264 *Algonquin Provincial Park Advisory Committee Report. Government Policy,* 14.
265 Quoted in Killan, 193.
266 Killan, 200.
267 Roderick MacKay, *Chronology,* 21.
268 Ontario Ministry of Natural Resources, *Algonquin Provincial Park Master Plan,* Toronto: Queen's Printer for Ontario, October 1974.
269 Ontario Ministry of Natural Resources, *Algonquin Provincial Park, Master Plan Review,* Toronto:

Queen's Printer for Ontario, 1989–1990, 1.
270 Roderick MacKay, *Chronology,* 23.

Chapter 10: The Algonquin Forestry Authority: Making a New Management System Work

271 E. Ray Townsend, *Algonquin Forestry Authority: A Twenty Year History, 1975–1995*, Huntsville [ON]: private printing, n.d., 10.
272 Consulting Team for the Implementation of the Algonquin Forestry Authority, "Report," March 31, 1974, 9.
273 Consulting Team for the Implementation of the Algonquin Forestry Authority, "Report," March 31, 1974.
274 Bernard Reynolds, personal communication with author, September 2002.
275 Bernard Reynolds, personal communication with author, September 2002.
276 Bernard Reynolds, personal communication with author, September 2002.
277 I.D. (Joe) Bird, "Algonquin Forestry conceived from the compromise of logger and recreations," *Pulp and Paper Canada,* January 1979.
278 Townsend, 16.
279 Carl Corbett, personal communication, June 2005.
280 Townsend, 17.
281 Townsend, 17.
282 Townsend, 17.
283 Townsend, 18.
284 John Simpson, personal communication with author, January 2006.
285 Don George, personal communication with author, 1983.
286 Joe Bird, quoted in Townsend, 29.
287 Carl Corbett, personal communication, September 10, 2004.
288 Quoted in Townsend, 16.
289 John Simpson, e-mail to author, January 24, 2006.
290 Townsend, 15.
291 Brent Connelly, personal communication with author, 2003.
292 Bob McRae, personal communication with author, 2003.
293 Carl Corbett, personal communication with author, June 2005.
294 Ontario Ministry of Natural Resources, "Executive Summary," *Annual Report on Forest Management 1999/2000,* Toronto: Queen's Printer for Ontario, September 2001, 1.
295 Carl Corbett, personal communication with author, June 2005.
296 KBM Forestry Consultants, *Algonquin Forestry Authority Audit,* March 2003.
297 Roderick MacKay, *Chronology,* 23.
298 Townsend, 18. An "eight-holer" is a latrine that could accommodate eight people at one time. Efficiency and time saving was always a consideration in operations —modesty was not!
299 Carl Corbett, personal communication with author, June 2005.
300 John McRae, interview by Radforth.
301 Townsend, 42.
302 Carl Corbett, personal communication with author, June 2005. "The cutover map currently shows an estimated 30-35 % of the recreation/utilization zone has not been cut in AFA history. It is estimated that two-thirds of this area is in the central and eastern parts of the Park and the balance is on the west side."
303 Townsend, 42.
304 Townsend, 43–44.
305 Bob McRae, personal communication with author, 2003.
306 Cannon, interview by Radforth.
307 Townsend, 44.
308 Harvey Anderson, personal communication with author, November 2002.
309 Harvey Anderson, personal communication with author, November 2002.
310 Tree-markers are required to take a provincial

course and pass an examination.
311 Townsend, 39.
312 Townsend, 39.
313 Jamie McRae, personal communication with author, 2005.
314 Townsend, 41.
315 Townsend, 41.
316 Strickland, *Trees of Algonquin Provincial Park,* 38.
317 Townsend, 41.
318 Strickland, *Trees of Algonquin Provincial Park,* 38.
319 Townsend, 41.
320 Keith Fletcher, personal communication with author, July 8, 2005.
321 *The Raven* 38.9 (August 14, 1997).
322 Dan Strickland, *Big Pines Trail: Ecology and History of White Pines in Algonquin,* Whitney [ON]: The Friends of Algonquin Park, n.d.
323 Townsend, 51.
324 Townsend, 51.
325 Townsend, 34.
326 A. Herridge, personal communication with author, February 19, 2001.

Chapter 11: **Whitney, Its People and Algoquin Park During and After World War II**

327 Location of the Mickle, Dyment bush camp and calculation of the distances involved by Jamie McRae, July 2004.
328 *Eganville Leader,* January 29, 1926.
329 *Reflections of a Century.*
330 George Garland, *Names of Algonquin: Stories Behind the Lake and Place Names of Algonquin Provincial Park,* Algonquin Park Technical Bulletin No. 10, Whitney [ON]: The Friends of Algonquin Park, 1991. Billings Lake, a widening of the north branch of the southward flowing York River was the destination of the two men, not the scene of their deaths. This small lake, however, was the only one close to the scene that could reasonably be renamed to memorialize a local hero (Map 10).
331 Garland, 52.
332 Roy MacGregor, 94.
333 Janet McRae Webber with Cathy McRae Freeman, interview by author, tape recording, November 2003.
334 Helen McRae (with John McRae), interview by Ian Radforth, tape recording, June 1986.
335 Janet McRae Webber (with Cathy McRae Freeman), interview by author.
336 *The Raven* 23.4 (July 15, 1982).
337 Bernard Shaw, *Lake Opeongo: Untold Stories of Algonquin Park's Largest Lake,* Burnstown [ON]: General Store Publishing House, 1998, 30.
338 *The Raven* 23.4 (July 15, 1982).
339 Shaw, 30.
340 Janet McRae, interview by Webber.
341 Shaw, 32.
342 Shaw, 38.
343 Shaw, 34.
344 Shaw, 35.
345 Ernie Martelle, personal communication with author, August 2003.
346 Mrs. Cenzura (with Alex Cenzura), interview by Ian Radforth, tape recording, June 1986.
347 Janet McRae, interview by Webber.
348 Mrs. Cenzura (with Alex Ccnzura), interview by Radforth.
349 Bob McRae, personal communication with author, 2004.
350 Mrs. Henry Taylor, interview by Ron Pittaway, tape recording, November 1979.
351 Phil Roche, interview by Ian Radforth, tape recording, June 1986.
352 Bernie Stubbs Sr. (with Bernie Stubbs Jr.), personal communication with Bob McRae, July 2003.
353 Roderick MacKay, *Chronology,* 7.
354 Bernie Stubbs Sr. (with Bernie Stubbs Jr.), personal communication with Bob McRae, July 2003.
355 Lyons, 7.
356 Janet McRae, interview by Webber.
357 Janet McRae, interview by Webber.
358 Janet McRae, quoted in Radforth.

359 Janet McRae, interview by Webber.
360 Janet McRae Webber (with Cathy McRae Freeman), interview by author.
361 Roderick MacKay, *Chronology.*
362 Boldt, interview by Radforth.
363 Mrs. Cenzura (with Alex Cenzura), interview by Radforth.
364 H. Eleanor (Mooney) Wright, *Trailblazers of Algonquin Park,* Eganville [ON]: HEW Enterprises, 2003.
365 Wright, 71–72.
366 Wright, 80.
367 Lambert and Pross, 197.
368 Arthur A.J. Herridge, "Frank MacDougall," *Aski* 4.6 (Summer 1975): 3.
369 Janet McRae, interview by Webber.
370 Janet McRae, interview by Webber.
371 Killan, 62.
372 Killan, 66.
373 Killan, 63.
374 Federation of Ontario Naturalists, *Sanctuaries and the Preservation of Wildlife in Ontario.* Toronto: FON, 1934.
375 Killan, 63.
376 Wright, 72–75.
377 Killan, 62.
378 Frank MacDougall, "Annual Report for Algonquin Park for year ending 31 March 1937."
379 *Globe & Mail* November 3, 1938.
380 Killan, 67.
381 Killan, 67.
382 Killan, 70.
383 Shalla, interview by Radforth.
384 Alex Cenzura (with Mrs. Cenzura), interview by Radforth.
385 Marjorie McGregor (with Alex McGregor), interview by Ian Radforth, tape recording, June 1986.
386 Duncan MacGregor (with Frank Shalla), interview by Ian Radforth, tape recording, June 1986.
387 Duncan MacGregor (with Frank Shalla), interview by Radforth.
388 Janet McRae, interview by Webber.
389 Duncan MacGregor (with Frank Shalla), interview by Radforth.
390 Cannon, interview by Radforth.
391 Cannon, interview by Radforth.
392 Roy MacGregor, 195.
393 Marjorie McGregor (with Alex McGregor), interview by Radforth.
394 *Year Book of Canada*, Ottawa: Queen's Printer for Canada, 1942.
395 Janet McRae Webber (with Cathy McRae Freeman), interview by author.
396 The Township of Airy was incorporated in 1961.
397 Rachel McRae, personal communication with author, 2005.
398 Roy MacGregor, 189.
399 Donald McRae, interview by Bob McRae. Bob McRae made six tapes of his father's wartime experience.
400 Donald McRae, interview by Bob McRae.
401 Alan Burgess, *The Longest Tunnel,* Hammondsworth [UK]: Penguin, 1990, 219.
402 Burgess, 20–28.
403 Donald McRae, interview by Bob McRae.
404 Donald McRae, interview by Bob McRae.
405 Donald McRae, interview by Bob McRae.
406 Scott Young, "A Prisoner is Lonely," *MacLean's Magazine* 1944. The story of Ft. Lt. Don Morrison DFC, DFM. Don McRae was one of Morrison's roommates for a time.
407 Donald McRae, interview by Bob McRae.
408 Donald McRae, interview by Bob McRae.
409 Eric Williams, *The Wooden Horse,* New York: Viking Penguin, 1957.
410 Donald McRae, interview by Bob McRae.
411 Donald McRae, interview by Bob McRae.
412 Donald McRae, interview by Bob McRae.
413 Donald McRae, interview by Bob McRae.
414 M.R.D. Foot and J.M. Langley, *MI9: Escape and Evasion, 1939-1945,* London: Book Club Associates, 1979, 115–116.
415 Donald McRae, interview by Bob McRae.
416 Donald McRae, interview by Bob McRae.

417 Donald McRae, interview by Bob McRae.
418 Donald McRae, interview by Bob McRae.
419 Donald McRae, interview by Bob McRae.
420 Donald McRae, interview by Bob McRae.
421 Donald McRae, interview by Bob McRae.
422 Donald McRae, interview by Bob McRae.
423 Burgess, 219.
424 Helen McRae, interview by Radforth.
425 Douglas Lloyd, *The Statten Camps: Some Memories from the First Forty Years, 1921-1962*, Private printing, 1996–2000. Lloyd wrote,

Many of the Park Rangers had enlisted and their departure left the Park seriously understaffed. To ease this crisis the Park authorities developed a program known as the Algonquin Park Auxiliary Rangers, or A.P.A.R. for short.

Dr. George Garland, a prominent geophysicist and former Chairman of the Friends of Algonquin Park wrote,

The canoe trips of the Taylor Statten Camps had made significant contributions to maintenance of the park interior from the earliest days of the Camps. Trips of older campers from both Ahmek and Wapomeo marked and cleared portages, explored new routes and improved campsites on many lakes. These activities were formalized in 1942 when the Park Administration requested all camps to assist in maintenance of the interior as a wartime measure. (Despite what is stated in Algonquin Story, by Audrey Saunders, page 148, the date is 1942, with preliminaries in 1941. I was at camp in 1942 but did not take part in the A.P.A.R. program until the following year; I was thus in the 2nd large batch of the participants, not the first. RDL) Each camp in the Park was assigned a definite area, and a former ranger, Mr. Maurice Kirkland, was employed for the summer as the liaison officer. He was a high school teacher from Oshawa and worked well with the campers; he was also a cottager on Cache Lake and thus could easily arrange to travel to all the camps.

The Taylor Statten Camps were assigned the area between Canoe Lake and McCraney Lake, and between McCraney and Islet Lake, although it was not possible to devote much attention to the latter area. The main area became known as the Ahmek District, although Wapomeo campers were not excluded. In 1942, the area was a region little travelled by canoe trippers, probably because of its small lakes and its poorly marked, demanding portages.

At a meeting arranged by John Hall in July of 1942, head of canoe trips at Ahmek, and attended by Maurice Kirkland, it was agreed to encourage the Mountaineer canoe trips of that month to go into the Ahmek District. Groups camped on Drummer, Namakootchie and Panther lakes and worked on the intervening portages. Communications was maintained between groups, and on occasions, fresh supplies from camp were forwarded to them. The trails were initially very difficult to locate, as they had not been marked for years. After similar but smaller campaigns in 1943 and 1944, trails were well marked through to McCraney Lake and good campsites were prepared on several lakes. The Park's Canoe Routes map still shows these trails and sites as "low maintenance."

Maurice Kirkland had been equipped with some tools and supplies to assist the camps. Among these supplies was a stock of badges reading "Ontario Forestry Branch—Fire Warden." The early versions were all numbered; later on only the year was shown. John Hall proposed that these badges be given to campers who had contributed to the work in the Ahmek District, and who had passed a test on woodcraft.

426 Lambert and Pross, 553–557.
427 Killan, 77.
428 Killan, 76–77.
429 Roderick MacKay, *Chronology*, 19.
430 Roy Wilston, quoted in *Telling the Tales of the Old*

Timers, by Joan Finnegan, Burnstown [ON]: General Store Publishing House, 1998, 77.

431 Roy Wilston, quoted in *Telling the Tales of the Old Timers*, 73-74.
432 Lyons, 21.
433 Roche, interview by Radforth.
434 Donald McRae (with Helen McRae), interview by Radforth.
435 Lyons, Bob. p. 21
436 Roderick MacKay, *Chronology*, 19.
437 Personal communication from township office, 2003.
438 Helen McRae (with Donald McRae), interview by Ian Radforth, June 1986.
439 Cathy McRae Freeman interview with author, tape recording, November 2003.
440 Helen McRae (with Donald McRae), interview by author, tape recording, 2003.
441 Bob McRae, personal communication with author, 2005.
442 Cannon, interview by Radforth.
443 Roy MacGregor, 307.
444 Cathy McRae Freeman, interview by author.
445 Roche, interview by Radforth.
446 Marjorie McRae McGregor (with Alex McGregor), interview by Radforth.
447 *A History of the Public Library in Whitney*, Whitney [ON]: Private printing, n.d.

Section IV: **The Recent Years of the McRae Lumber Company**

Chapter 12: **The Modern Mills of the McRae Lumber Company**

448 McRae Lumber Company
449 Bob McRae, personal communication with author, 2003.
450 John Mastine, personal communication with Jamie McRae, 2006.
451 Bob McRae, personal communication with author, 2003.
452 Bob McRae, personal communication with author, 2003.
453 Jamie McRae, personal communication with author, 2005.
454 Bob McRae, personal communication with author, 2003.
455 Bob McRae, personal communication with author, 2003.
456 Roy MacGregor, 98.
457 Roy MacGregor, 103–104.
458 Roy MacGregor, 95–96.
459 Roy MacGregor, 104.
460 Bob McRae, personal communication with author, 2003.
461 Roy MacGregor, 190.
462 Andrew Siydock, interview by author, tape recording, August 2003.
463 Duncan MacGregor (with Frank Shalla), interview by Radforth.
464 Cannon, interview by Radforth.
465 Bob McRae, personal communication with author, 2003.
466 Bob McRae, personal communication with author, 2003.
467 Jamie McRae, personal communication with author, December 2005.
468 Bob McRae, personal communication with author, 2003.
469 Jamie McRae, personal communication with author, December 2005.

470 See endnote 88.
471 John McRae, interview by Radforth.
472 Bob McRae, interview by Radforth.
473 John McRae, interview by Radforth.
474 John McRae, interview by Radforth.
475 John McRae, interview by Radforth.
476 Roche, interview by Radforth.
477 John McRae, interview by Radforth.
478 Ron Tozer, personal communication with author, 2005.
479 Bob McRae, personal communication with author, December 2005.
480 John McRae, interview by Radforth.
481 John McRae, interview by Radforth.

Appendices

I: Taped interviews

McRae Collection of Tapes

Boldt, Lorne, interview by Ian Radforth, tape recording, June 1986.
Cannon, Gary, interview by Ian Radforth, tape recording, June 1986.
Cenzura, Alex (with Mrs.Cenzura), interview by Ian Radforth, tape recording, June 1986
Hutchison, Frank, interview by Ian Radforth, tape recording, June 1986.
Kuiack, Pete, interview by Ian Radforth, tape recording, June 1986.
Kuiack, Ted, interview by Ian Radforth, tape recording, June 1986.
Frank Kuiack (with Margaret Kuiack), interview by Ian Radforth, tape recording, June 1986.
Eddie Levean (with Jenny Levean), interview by Ian Radforth, tape recording, June 1986.
MacGregor, Duncan with Frank Kuiack, interview by Ian Radforth, tape recording, June 1986.
MacGregor, Duncan with Donald McRae, interview with Ian Radforth, tape recording, June 1986
MacGregor, Duncan with Frank Shalla, interview with Ian Radforth, tape recording, June 1986.
McGregor, Marjorie (McRae) with Alex McGregor, interview by Ian Radforth, June 1986.
McRae, Donald, interview by Bob McRae, tape recording, 2003.
McRae, Donald, interview by Ian Radforth, tape recording, June 1986.
McRae, Helen with Donald McRae, interview by Ian Radforth, tape recording, June 1986.
McRae, Janet (MacGregor), interview by Mark Webber, tape recording, June 1976.
McRae, John D, interview by Ian Radforth, tape recording, June 1986.
McRae, Bob (Robert), interview by Ian Radforth, tape recording, June 1986.
Perry, Albert, interview by Ian Radforth tape recording, June 1986.
Roche, Phil, interview by Ian Radforth, tape recording, June 1986.
Shalla, Frank, interview by Ian Radforth, tape recording, June 1986.

Lloyd Collection of Tapes

Freeman, Cathy (McRac), Janet (McRae) Webber and Mark Webber, interview by the author, tape recording, November 2003.
Kmith, Fred, interview by the author, tape recording, August 2003.
Parks, Fred, interview by the author, tape recording, August 2003.
Siydock, Andrew, interview by the author, tape recording, August 2003.

Algonquin Park Visitor Centre Collection of Tapes

MacKay, Rory, interview by George Furlong, tape recording, South River, June 4, 1976
Taylor, Henry, interview by. Ronald Pittaway, tape recording, Bancroft, November 27, 1979.
Taylor, Mrs. Henry, interview by. Ronald Pittaway, tape recording, Bancroft, November 27, 1979.

Appendix II: **Mills of the McRae Lumber Company**

	Airy	Lake of Two Rivers	Hay Lake	Mink Lake	Rock Lake	Whitney
	1922–1933	1933–1942	1942–1952	1952–1957	1957–1980	1973–Present Scragg mill–1973 Band mill–1979
Mill Operation	May to end of Sept.	May to end of Sept.	May to end of Sept.	May to end of Sept.	Year round	Year round
Hot Pond	No	Yes	Yes	Yes	Yes	No
Power	3 boilers	2 Watrous boilers; 2 Dutch ovens	2 Watrous boilers; 2 Dutch ovens	Diesel	2 Watrous boilers; 2 Dutch ovens	Diesel; Electricity
Fuel	Sawdust & Residue	Sawdust & Residue	Sawdust & Residue	Diesel oil	Diesel oil; Sawdust & Residue	Electricity
Saws	Double line Double cut band saw; Resaw; Edger	Single line Band head rig; Resaw; Edger	Single line Double cut band saw; 2 edgers	Single line Double cut band saw; 2 edgers	Single line Double cut band saw; 2 edgers	
Workers	200–300 in bush; 50 in mill	200 in bush; 60 in mill	200 in bush; 60 in mill	150 in bush; 60 in mill	100 in bush–decreasing to 30; 60 in mill	77 in mill
Drying of Lumber	Weather seasoning	Weather seasoning	Weather seasoning	Weather seasoning	Weather seasoning	None
Transportation	CNR	CNR	CNR	CNR	CNR; Truck	Truck
Limits	Nightingale, Canisbay, Eyre, Airy, Sabine	Canisbay	Sproule, Bower	Bruton, Sabine, Clyde	Canisbay, Clyde	Algonquin Park via AFA, Lawrence, McClure
Cut	14'-16' logs with crosscut saw	14'-16' logs with crosscut saw	14'-16' logs with crosscut saw	14'-16' logs with power saw	14'-16' logs with power saw Tree length stems	Tree length stems
Skidding & Hauling	Horse	Horse and Truck	Horse and Truck	Horse and Truck	Timberjack and Truck	Jobber
Product	Dimensional lumber Lath	Dimensional lumber Lath	Dimensional lumber	Dimensional lumber	Dimensional lumber	Dimensional lumber; Chips Residue
	2-3 million bd. ft.	6-7 million bd. ft.	5-7 million bd. ft.	5-7 million bd. ft.	7-10 million bd. ft.	12 million bd. ft.

Appendix III: **Order-in-Council: Reduction in Stumpage Fees**

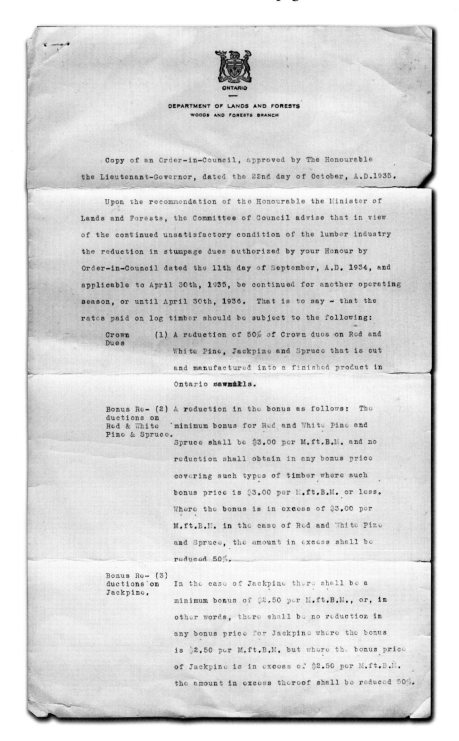

ONTARIO

DEPARTMENT OF LANDS AND FORESTS
WOODS AND FORESTS BRANCH

Copy of an Order-in-Council, approved by The Honourable the Lieutenant-Governor, dated the 22nd day of October, A.D. 1935.

Upon the recommendation of the Honourable the Minister of Lands and Forests, the Committee of Council advise that in view of the continued unsatisfactory condition of the lumber industry the reduction in stumpage dues authorized by your Honour by Order-in-Council dated the 11th day of September, A.D. 1934, and applicable to April 30th, 1935, be continued for another operating season, or until April 30th, 1936. That is to say - that the rates paid on log timber should be subject to the following:

Crown Dues (1) A reduction of 50% of Crown dues on Red and White Pine, Jackpine and Spruce that is cut and manufactured into a finished product in Ontario sawmills.

Bonus Reductions on Red & White Pine & Spruce. (2) A reduction in the bonus as follows: The minimum bonus for Red and White Pine and Spruce shall be $3.00 per M.ft.B.M. and no reduction shall obtain in any bonus price covering such types of timber where such bonus price is $3.00 per M.ft.B.M. or less. Where the bonus is in excess of $3.00 per M.ft.B.M. in the case of Red and White Pine and Spruce, the amount in excess shall be reduced 50%.

Bonus Reductions on Jackpine. (3) In the case of Jackpine there shall be a minimum bonus of $2.50 per M.ft.B.M., or, in other words, there shall be no reduction in any bonus price for Jackpine where the bonus is $2.50 per M.ft.B.M. but where the bonus price of Jackpine is in excess of $2.50 per M.ft.B.M. the amount in excess thereof shall be reduced 50%.

Appendix III: **Order-in-Council: Reduction in Stumpage Fees** (*continued*)

-2-

Hardwood and other types.	(4) A reduction of 50% of the Crown dues on hardwood and other types of timber not mentioned in (1), (2) and (3), and such reductions in the bonus thereof in each case, as may in the opinion of the Minister of Lands and Forests be deemed justified in the public interests.

The Committee further advise that such reductions be granted on the clear and distinct understanding that:

(a) The reductions are of only a temporary measure and applicable to the bush operations carried on during the present operating season of 1935-36, which season expires on the 30th of April, 1936.

(b) The reductions, subject to the rights granted thereunder, shall in no way operate as an impairment of the original contract entered into between the Licensee operator or purchaser and the Crown.

(c) The timber cut in accordance with or under or by virtue of such reductions shall be manufactured in Ontario sawmills into the finished product, such as lumber, ties, lath or such other product as is generally deemed to come within the scope of sawmill operations.

(d) Fair wage rates shall be paid to workmen, reasonable prices charged for van or other supplies and the Regulations of the Department of Health approved by the Lieutenant-Governor in Council dated 17th of April, 1934, made under the Public Health Act, shall be adequately enforced.

(e) The Minister may demand that each Licensee or operator be called upon to make a survey of his possibilities and to submit a statement showing the types and quantities of timber to be cut, the area or areas on which proposed cutting shall take place,

Appendix III: **Order-in-Council: Reduction in Stumpage Fees** (continued)

-3-

and the location of the mill or mills where the manufacturing shall be conducted and to sign an agreement obligating himself to employ such number of men, to purchase such equipment and supplies, and to cut such quantities of timber as shall be agreed upon between them and the Minister, and to meet such other terms and conditions as form part of any reductions in, or modifications to, the contract price.

(f) Monthly or periodic submissions as may be determined by the Minister shall be made by each licensee or operator showing the number of men employed, the kinds and quantities of timber cut and such other information as in the opinion of the Minister may be deemed desirable.

(g) Licensees or operators under any reduction arrangement shall be required to pay all necessary ground rent and fire protection charges at $5.00 and $6.40 per square mile respectively.

(h) The Minister may in such cases where licenses are held in suspense, due to the non-payment of charges, and where he is of the opinion that sufficient security in the way of collateral is lodged in the Department, permit operations for the ensuing season of 1935-36.

(i) The preceding reductions shall in no way apply to timber sales made subsequent to September 11th, 1934.

(j) Where any grievances arise or dispute occurs in respect of these provisions, the Minister shall be the arbiter and his decision shall be final and conclusive.

Certified, C. H. Bulmer (sgd.)
Clerk, Executive Council.

Appendix IV: **McLaughlin Township Timber Licence, 1937**

Form-O-2
1,000—June, 1936

SALE OF NOVEMBER 9TH, 1937

Timber License No. _122_ for 19_37_-19_38_ (see No. ____ of 19__-19__)

SUBJECT: To all Terms and Conditions of Sale dated October 21st, 1937, a copy of which is hereto attached and forms part of this license.

ONTARIO

TIMBER LICENSE

Know all men by these presents:

THAT by authority of the Crown Timber Act, and Regulations made thereunder, and for and in consideration of the payments made and to be made to the Crown, subject to the conditions hereinafter mentioned.

I hereby give leave and license unto _S. L. McRae_

Whitney, Ont.

his, their Agents and Workmen, from the _first_ day of _December_ 1937 or its to the thirty-first day of March, 19_38_, and no longer, to cut the kind or kinds of timber specified and on the lands described on the back of this page so far as the said lands so described are ungranted public lands or patented lands on which the said timber remains the property of the Crown; with the right, as against all persons other than the Crown, its Agents, Officers and Servants and those acting under or by license, permission or other authority from the Crown, to hold and occupy said lands for the purpose of, and so far only as may be necessary for the cutting and removal or the preservation of said timber; and with the further right of conveying away the said timber through any ungranted lands of the Crown, and of exercising such other rights as are provided by the Regulations under the Public Lands Act.

The Licensee has the right to all timber unlawfully cut on said lands covered by and during the term of this License, with full power to seize, recover and remove the same.

This License is subject to the following conditions, namely:

To the respective manufacturing conditions set forth respectively in Schedules A, B and C of the Crown Timber Act.

To securing from the Crown by License or Occupation, Lease or otherwise, right to construct and use mills, logging camps, office structures, depots, or any other buildings that may be necessary in connection with operations.

To the withdrawal from said lands, of lands for which patent may be granted on the ground that, prior to the date or issue of this License, the conditions of settlement in respect to said lands so withdrawn have been complied with, and that three years have elapsed from the date of location or sale of said lands.

That nothing herein contained shall prevent any person, who first obtains therefor the permission of the Minister, taking, without compensation for the timber so taken, from any land included in this License, any kind of timber which it is intended shall, by or on behalf of the Province of Ontario, be used in the making of roads or bridges, or in any "Public work" as defined by "The Public Works Act."

That any person may, under the authority of the Crown, at any time make and use roads upon, and travel over, the ground hereby licensed.

That no person lawfully upon or using any portion of the lands included in this License shall be hindered in the lawful occupation or use of the land so occupied or used.

The Licensee shall not cut any timber until he gives notice to, and gets the approval of the District Forester.

Before any timber cut under this License passes out of the agency within which it has been cut, all timber dues and bonus, if any, for timber cut hereunder shall be paid to the Treasurer of the Province of Ontario or security given satisfactory to the Minister; and all timber dues and bonus, if any, for timber cut hereunder shall be paid to said Treasurer not later than the 30th day of September following the season in which it was cut.

When required by the Minister or by any officer or employee authorized by the Minister, the Licensee, (his, their or its) officers, agents, servants or employees, shall submit all timber, cut under the provisions of this License, and the manufactured product thereof, to be counted and measured, and shall pay the dues thereon when so required by the Minister or any officer or employee thereunto authorized.

In addition to all other penalties, upon failure by the Licensee, (his, their or its) officers, agents, servants or employees to comply with any of the conditions of this License, the Minister may declare all or any portion of timber cut under this License, or the manufactured product thereof, to be forfeited to the Crown and the same shall thereupon be forfeited accordingly.

All disputes with reference to Government road allowances shall be decided by the Minister, who may define the portion of any such road allowance included in a License, and his decision shall be final and binding.

This License is subject further to all the provisions of the Acts cited at the foot hereof and Regulations made thereunder.

In this License "Minister" shall mean the Minister of Lands and Forests of the Province of Ontario.

GIVEN, in duplicate, under my hand, at Toronto, this _first_ day of _December_ in the year of our Lord One Thousand Nine Hundred and _thirty-seven_

PAYABLE under this License:

Ground Rent .... $ 95.00
Fire Protection .... $ 121.60
                         $ 216.60

Also Timber Dues and Bonus.

Deputy Minister of Lands and Forests.

SEE—
- The Public Lands Act, R.S.O. 1927, Ch. 35.
- The Crown Timber Act, R.S.O. 1927, Ch. 38.
- The Mining Act, R.S.O. 1927, Ch. 45.
- The Veterans' Land Grant Act, Ont. Stat. 1901, Ch. 6.
- The Timber Cutting Regulation Act, Ont. Stat. 1928, Chap. 15.
- The Provincial Forests Act, Ont. Stat. 1929, Chap. 14.
- The Forest Fires Prevention Act, Ont. Stat. 1930, Chap. 60.
- The Forestry Act, R.S.O. 1927, Chap. 41.
- The Mills Licensing Act, R.S.O. 1927, Chap. 39.
- The Pulpwood Conservation Act, Ont. Stat. 1929, Chap. 13.
- The Provincial Parks Act, R.S.O. 1927, Chap. 82.
- The Woodmen's Employment Investigation Act, Ont. Stat. 1934, Chap. 66.
- The Forest Resources Regulation Act, Ont. Stat. 1936, Chap. 22.

Appendix IV: **McLaughlin Township Timber Licence, 1937** *(continued)*

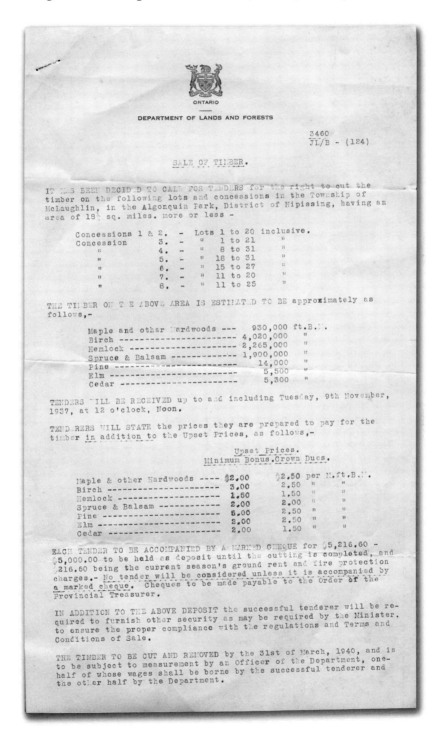

ONTARIO
DEPARTMENT OF LANDS AND FORESTS

3460
JL/B - (124)

SALE OF TIMBER.

IT HAS BEEN DECIDED TO CALL FOR TENDERS for the right to cut the timber on the following lots and concessions in the Township of McLaughlin, in the Algonquin Park, District of Nipissing, having an area of 18½ sq. miles. more or less -

```
Concessions 1 & 2. - Lots 1 to 20 inclusive.
Concession    3.  -   "   1 to 21     "
    "         4.  -   "   8 to 31     "
    "         5.  -   "  18 to 31     "
    "         6.  -   "  15 to 27     "
    "         7.  -   "  11 to 20     "
    "         8.  -   "  11 to 25     "
```

THE TIMBER ON THE ABOVE AREA IS ESTIMATED TO BE approximately as follows,-

```
Maple and other Hardwoods ---   930,000 ft.B.M.
Birch ------------------------ 4,020,000   "
Hemlock ---------------------- 2,265,000   "
Spruce & Balsam -------------- 1,900,000   "
Pine -------------------------    14,000   "
Elm --------------------------     5,500   "
Cedar ------------------------     5,300   "
```

TENDERS WILL BE RECEIVED up to and including Tuesday, 9th November, 1937, at 12 o'clock, Noon.

TENDERERS WILL STATE the prices they are prepared to pay for the timber in addition to the Upset Prices, as follows,-

```
                              Upset Prices.
                         Minimum Bonus. Crown Dues.

Maple & other Hardwoods ----  $2.00      $2.50 per M.ft.B.M.
Birch ----------------------   3.00       2.50    "    "
Hemlock --------------------   1.50       1.50    "    "
Spruce & Balsam ------------   2.00       2.00    "    "
Pine -----------------------   5.00       2.50    "    "
Elm ------------------------   2.00       2.50    "    "
Cedar ----------------------   2.00       1.50    "    "
```

EACH TENDER TO BE ACCOMPANIED BY A MARKED CHEQUE for $5,216.60 - $5,000.00 to be held as deposit until the cutting is completed, and $216.60 being the current season's ground rent and fire protection charges.- No tender will be considered unless it is accompanied by a marked cheque. Cheques to be made payable to the Order of the Provincial Treasurer.

IN ADDITION TO THE ABOVE DEPOSIT the successful tenderer will be required to furnish other security as may be required by the Minister, to ensure the proper compliance with the regulations and Terms and Conditions of Sale.

THE TIMBER TO BE CUT AND REMOVED by the 31st of March, 1940, and is to be subject to measurement by an Officer of the Department, one-half of whose wages shall be borne by the successful tenderer and the other half by the Department.

Appendix IV: **McLaughlin Township Timber Licence, 1937** (continued)

- 2 -

Pt. McLaughlin Twp.

NO TIMBER TO BE REMOVED FROM THE AREA until after it has been scaled by a licensed Government Scaler.

SUBJECT TO THE EXCEPTIONS HEREINAFTER PROVIDED no Birch or Maple less than 15" in diameter 18" from the ground, nor any Basswood, Ash or Elm, less than 10" in diameter 18" from the ground, shall be cut.

NO HEWN TIES SHALL BE MADE upon the limit.

ALL CUTTING OF TIMBER BY A LICENSEE shall be subject to the supervision and control of a person appointed by the Minister from time to time for that purpose, but in the event of a disagreement between the person so appointed and the licensee, the matter may be referred to the Minister whose decision shall be final.- Such cutting shall take place only and when directed by the Minister and subject to such diameter limits, sequence of cutting, removal and disposition of slash, provision for re-seeding, and such further and other restrictions, as may be determined by the Minister from time to time.

FOR THE PURPOSE OF WATERSHED PROTECTION, beautification of park, fire protection, game preserves or game shelters, or for any other purpose that from time to time the Minister may deem advisable, the Minister, out of the areas included in any timber license, may withdraw certain timber from cutting and direct that such timber shall be left standing, and the licensee shall not be entitled to any compensation for such timber so withdrawn unless directed by the Lieutenant-Governor in Council.

WHEREVER PARCELS OF LAND within the Limit are covered by Licenses, Leases or Patents, the rights of the holders of said Licenses, leases or Patents shall be respected and the successful tenderer shall not cut any timber thereon.

TO ENSURE EFFECTIVE CO-OPERATION IN BUSH METHODS, no cutting shall be allowed until licensee advises the District Forester, and obtains the necessary permission.

NO CUTTING WILL BE ALLOWED within 150 feet of South shore of Island Lake except in vicinity of Camp Minnesing, where reservations adjacent to the Camp will be 300' from shore line for a distance on the lake of the lease, plus 150' on each side of the lease, and further, no cutting within 150' of shore line of Little Otterslide and Otterslide Lakes.

TO CARRY ON SUMMER OPERATIONS between the 15th of May and the 15th of September, special permission must first be obtained from the Minister of Lands and Forests.

IN THE EVENT OF ANY DISPUTE ARISING as to the measurement, the Minister of Lands and Forests shall be the sole arbiter,- No application for a rescale will be entertained unless the whole season's output is available for measurement.

AN ANNUAL GROUND RENT of $5.00 per sq. mile, or fraction of a mile, and an Annual Fire Protection Charge of $6.40 per sq. mile, or fraction of a mile, are to be paid.

THE TIMBER IS TO BE SOLD SUBJECT TO "The Manufacturing Condition", that is to say, it is to be manufactured in the Dominion of Canada.

Appendix IV: **McLaughlin Township Timber Licence, 1937** *(continued)*

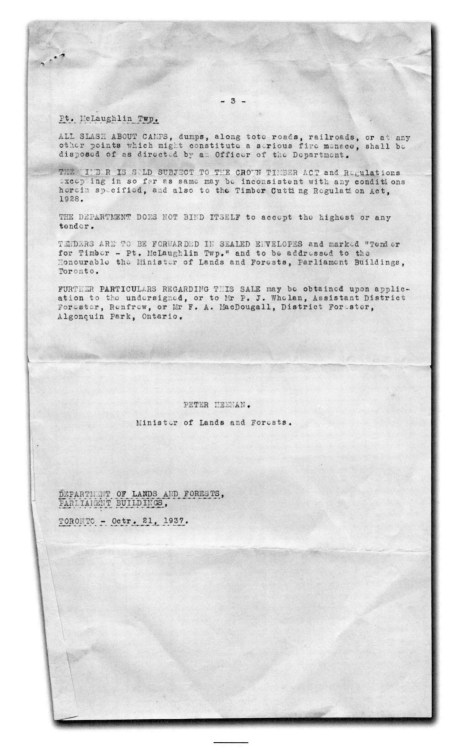

- 3 -

Pt. McLaughlin Twp.

ALL SLASH ABOUT CAMPS, dumps, along tote roads, railroads, or at any other points which might constitute a serious fire menace, shall be disposed of as directed by an Officer of the Department.

THE TIMBER IS SOLD SUBJECT TO THE CROWN TIMBER ACT and Regulations excepting in so far as same may be inconsistent with any conditions herein specified, and also to the Timber Cutting Regulation Act, 1928.

THE DEPARTMENT DOES NOT BIND ITSELF to accept the highest or any tender.

TENDERS ARE TO BE FORWARDED IN SEALED ENVELOPES and marked "Tender for Timber - Pt. McLaughlin Twp." and to be addressed to the Honourable the Minister of Lands and Forests, Parliament Buildings, Toronto.

FURTHER PARTICULARS REGARDING THIS SALE may be obtained upon application to the undersigned, or to Mr P. J. Wholan, Assistant District Forester, Renfrew, or Mr F. A. MacDougall, District Forester, Algonquin Park, Ontario.

PETER HEENAN,

Minister of Lands and Forests.

DEPARTMENT OF LANDS AND FORESTS,
PARLIAMENT BUILDINGS,

TORONTO - Octr. 21, 1937.

Appendix IV: **McLaughlin Township Timber Licence, 1937** *(continued)*

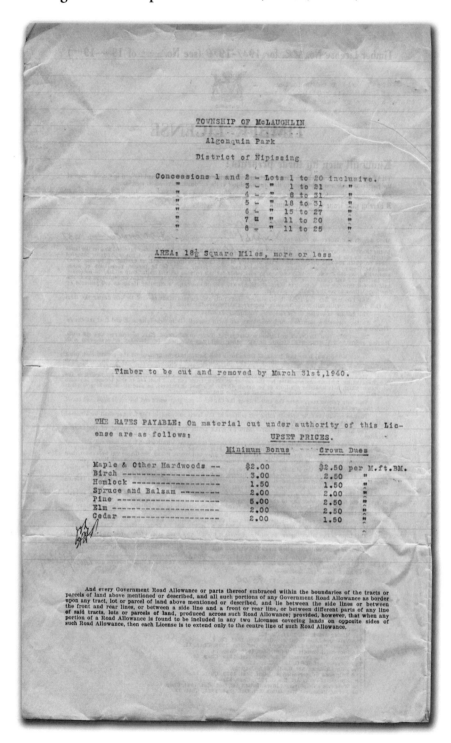

Appendix IV: **McLaughlin Township Timber Licence, 1937** *(continued)*

No. 122 of Season 19 37-38
No. _____ of Season 19 ____

LICENSE

Name J. S. L. McRae
Whitney, Ont.
Locality McLaughlin Twp.
- part -
District of Nipissing (Alg. Park
Area 18½ square miles

Ground Rent, $ 95.00

Fire Protection, $ 121.60

Date of License Dec. 1st 19 37

Appendix V: **McRae Lumber Company Employment Report, 1943**

DEPARTMENT OF LABOUR - NATIONAL SELECTIVE SERVICE

Manpower Statistics Unit
Ottawa.

(Required Under the National Selective Service Civilian Regulations)

REPORT ON EMPLOYMENT (DLR 25)

Name of Firm.... John A. L. McRae ....

Head Office Address.... Whitney, Ontario ....

This report covers:

Plant Name............ John A. L. McRae ....

Plant Address........... Whitney, Ontario ....

NOTE:

    Please be certain that three copies of this return are mailed to the above address before July 7, 1943. The fourth copy is for your files. A separate report must be made for each plant and should be completed even though the plant is not engaged in WAR work. Reports should cover all employees, including plant office staff. A separate report should be returned for the head office unless the head office is located at one of the plants and covered in the report for that plant. If some of your employees are engaged in both WAR and CIVILIAN work, divide this number of employees between WAR and CIVILIAN roughly on the basis of time spent on WAR and on CIVILIAN work. Refer to the accompanying letter for definition of WAR work.

A. Employment as of June 30, 1943 or nearest previous pay date:

	MALE	FEMALE	TOTAL
1. Employment on WAR Work	150		150
2. Employment on CIVILIAN Work			
3. Total Employment, WAR and CIVILIAN, (1 plus 2)	150		150

B. What percent of your current production do you estimate is for WAR purposes?...... 100 %
State the basis of this estimate, e.g. Sales Value, Material Used, Working Time of Staff, etc ....

C. Estimated employment necessary for your production as of November 30, 1943. (State estimated total employment needed at November 30, 1943, not net increase from June 30 to November 30, 1943.)

	MALE	FEMALE	TOTAL
1. Employment on WAR Work	250		250
2. Employment on CIVILIAN Work			
3. Total Employment, WAR and CIVILIAN, (1 plus 2)	250		250

D. Does your WAR Production as now planned call for a change in the size of your WAR employment after November 30, 1943? Check (x) below.

    1. Increase ( )   2. Decrease ( )   3. No Known change (x)

E. If you checked "increase" or "decrease" under D, state the approximate month during which the increase or decrease should be completed and the estimated WAR employment needed at that date.

    Month................Year 194... Employment on WAR Work............

(over)

Appendix V: **McRae Lumber Company Employment Report, 1943** (continued)

ALGONQUIN HARVEST

Appendix VI: **Inspector's Report: Provincial Department of Health, August 1945**

100— MARCH, 1943 (B 3154

PROVINCIAL DEPARTMENT OF HEALTH — **DIVISION OF INDUSTRIAL HYGIENE**

ONTARIO

Code #1

INSPECTOR'S REPORT

Date of visit: 30 Aug 45

- Owner: J. S. L. McRae Lumber Co.
- Jobber: —
- Township: Sabine
- Camp No.: 1 only
- Foreman: Mr. J. S. L. McRae
- Operation: Sawmill
- Men: 60
- M. & S. Contracts: Yes
- Dr.: R. L. McKinnon
- Buildings: 6
- Cookery: Yes
- Bunkhouse: 3
- Office: 1
- Scalers: —
- Stables: Yes
- Bath: No
- Toilets: Yes
- Laundry: Yes
- Walls: Lumber
- Floor: Lumber
- Roof: Roofing
- Capacity of Bunkhouses: 4 Beds in room 10 x 10 x 9
- Capacity of Cookery: Adequate
- Cook's quarters: Good
- Lighting: Electric
- Ventilation: Good
- Screens: Yes
- Cleanliness: Good
- Bunks: Lumber D. Height
- Bedding: Ticks & Blankets
- Washroom: Good
- Drainage: Good
- Food: Good Storage
- Garbage: Fed to pigs & burned
- Water supply: A
- Manure: None
- Toilets: Satisfactory
- Medical supplies: Good
- Date Camp Closes: 10 Oct. 45
- General: An electric refrigerator 10 x 10 x 7 is now being installed at kitchen.

Under the Regulations of the Ontario Department of Health you are required to make the following corrections:

1. Double height bunks situated in upper & lower hallways are to be removed.

2. Sleeping rooms are only large enough to sleep three men. Four men must not be crowded into rooms this size.

Appendix VII: **Pembroke Administrative District: Objectives and Staffing, 1967**

The wood industries dependent on the Petawawa Management Unit wanted an increase in the production of red and white pine. To achieve this, the Department of Lands and Forests initiated a program in the mid-1960s. Its three main objectives were:

1. **Restocking of Current Cutting Areas**: In order to maintain their present production of the forest soils and to start toward improved production it will be necessary to secure regeneration of the cutover stands within the time allotted.

2. **Restocking of Barren and Scattered Areas:** Areas which fall into this category are to be brought back into production through suitable means of regeneration, either naturally or artificially.

3. **Stand Conversion:** A white pine conversion program of considerable magnitude is to be undertaken within the twenty-year period covered by the management plan. The following table shows the present and proposed distribution of the working groups in the unit.

Species	Present Area (acres)	% of Total Area	Proposed Area (acres)	% of Total Area
White Pine	95,702	29	194,574	58
Red Pine	26,872	8	26,872	8
Jack Pine	9,120	3	9,120	3
Spruce	7,592	2	10,576	3
Balsam	2,984	1	—	—
Hemlock	2,664	1	—	—
Other Conifers	1,074	1	1,074	—
Maple	23,664	7	23,664	7
Yellow Birch	4,514	1	7,178	2
Poplar	124,088	37	38,830	12
White Birch	11,350	3	—	—
Other Hardwoods	11,708	4	9,444	3
Protection Forest	11,508	3	11,580	4
Total Productive Forest	**332,840**	**100**	**332,912**	**100**

Source: Internal mimeographed report: Pembroke Management Unit, 1967.

From the latter 1950s through 1967, silviculture methods and treatment included: thinning to improve spacing where pine was regenerating quite well in patches; the use of scarification and planting of white pine, which was successful until it was heavily browsed by deer; the use of frilling or girdling of a hardwood overstory (poplar) to release white pine. Where a strip shelterwood harvest was used, the strips were scarified to promote natural seeding, in areas classified as barren, direct planting was employed. White Pine blister rust was present in the area and a program of Ribes control was started in 1957 and extended through 1964.

The severe browsing of the seedling trees by the large deer herd of the area led the Timber Branch to declare a deer hunt in the area. This provoked a storm of protest and the hunt was cancelled.

Staffing of the Pembroke Administrative District, 1967

District Forster								
2 District Foresters								
District Accountant	Fish & Wildlife Supervisor	Forest Protections Supervisor	Lands Supervisor	Stenographic Services Supervisor	Park Supervisor	Construction Supervisor	Timber Supervisor
Asst. District Accountant	Asst. Senior Conservation Officer	Clerk	Asst. Lands Supervisor	3 Asst. to Supervisor		12 Caretakers	4 Foresters
2 Clerks Officer	Biologist	Communic. Stenographs.					Scaling Supervisor
	Hatchery Manager	Pilot					Scaling Clerk
	2 Hatchery Assts.	Engineer					
		Mech. Supt.					
		3 Mechanics					
		Radio Operator					

Chief Ranger Divisions (Deputy Headquarters Staff)

Pembroke	Stonecliffe	Whitney
Chief Ranger	Chief Ranger	Chief Ranger
2 Conservation Officers	Conservation Officer	2 Conservation Officers
Caretaker	Clerk	Biologist
Carpenter	Radio Operator	Carpenter
Stockman		Clerk
Clerk		Typist
2 Sign Shop Workers		Radio Operator

Chief Ranger Divisions (Field Staff)

Pembroke			Stonecliffe	Whitney			
Deputy Chief Ranger			Deputy Chief Ranger	Deputy Chief Ranger			
Pembroke	Achray	Round Lake	Stonecliffe	Brent	Parks	Whitney E.	Whitney W.
2 Rangers	10 Rangers	6 Rangers	5 Rangers	3 Rangers	3 Rangers	13 Rangers	Museum

Appendix VIII: **KBM Forestry Consultants, "Algonquin Forestry Authority Audit Report: 1997–2002"**

Summary Table of Recommendations, Suggestions and Best Practice: 37–38.

The text in italics is the reply of the Algonquin Park Superintendent, John Winters, Wed. September 15, 2004.

Principle 1: Commitment

Suggestion (1): The Ministry of Natural Resources should clearly describe its operating relationship in the Memorandum of Understanding with the Algonquin Forestry Authority.

Recommendation (1): The Ministry of Natural Resources must undertake a review and assessment of the capacity and resources of Algonquin Park to fulfill its forest management mandate, as it applies to the collection and management of resource values information and compliance monitoring. Deficiencies found must be addressed.

Algonquin Park, through the office of the Superintendent, has replied to one criticism. By receiving additional funding from the MNR, a full-time position of Senior Compliance Technician has been filled, along with a winter position of Forest Compliance Inspector. Hopefully, these additions will provide the information needed in the areas of resource values and compliance monitoring.

Principle 2: Public Participation

Recommendation (2): The Local Citizens Committee must meet more frequently between planning periods in order to more effectively monitor plan implementation and to provide ongoing advice to the Park Superintendent.

Suggestion (2): The Local Citizens Committee, in conjunction with Algonquin Park and the Algonquin Forestry Authority, should develop a strategy for improving its communications with the public on forest management issues in the Park.

Suggestion (3): Algonquin Park and the Algonquin Forestry Authority should make the Annual Work Schedule summary available annually for inspection at the same locations as the Forest Management Plan and/or on the organizations' web sites.

Principle 3: Forest Management Planning

Suggestion (4): The Sustainable Forest Management Modelling for the 2005–10 Forest Management Plan should include:

a) Land base reconciled with Tables FMP-1, 2, and 3.

   The discrepancies in base numbers between these tables have been resolved.

b) Natural succession for all current and potential forest unit/silvicultural combinations.

   Use "INF" upper limit for harvest flow policies to capture any potential volume increases to maximize socio-economic benefits. Ensure that strategic model projections are accurately documented in the plan (e.g. Tables FMP-3 and FMP-12).

c) Refine operability ages and product proportions to reflect projected gains through more intensive renewal treatments.

d) Investigate use of volume targets to maximize timber production within bounds of forest sustainability and other management objectives.

e) Eliminate broad use of maximum harvest area constraints to allow SFMM to calculate optimal forest unit harvest areas and timing for harvest activities.

The balancing of timber volumes cut and the marketing thereof is very tricky because the market place is in constant adjustment to factors external to Algonquin Park, such as the impact of softwood tariffs with the United States and the number of housing starts in either or both countries.

Recommendation (3): Algonquin Park must ensure that the planning team meeting minutes and Local Citizen Committee meeting minutes prepared during the development of the 2005-2010 FMP include documentation that the Local Citizens Committee and the planning team examined the Socio-economic impact Model (SEIM).

Done: Housekeeping.

Recommendation (4). The Algonquin Forestry Authority must develop and implement a strategy to return areas previously strip cut on the forest to a natural forest condition to emulate natural disturbance and landscape patterns, as required by the *Crown Forest Sustainability Act.*

This is in progress but it will take at least another fifty years to grow out the strips.

Recommendation (5): Algonquin Park must ensure that:

a) All available, mapable cultural heritage information pertinent to the Park is captured on a Natural Resources Value Inventory System (NRVIS) values layer and made available for forest management planning.

b) All newly discovered cultural heritage values are documented in a values layer, that Areas of Concern planning is conducted, and that the Forest Management Plan is amended where necessary.

c) All areas with a likelihood of habitation (persons having lived on-site) are thoroughly investigated by or under the direction of a qualified cultural heritage specialist before operations are permitted, to define Areas of Concern boundaries.

d) *Many cultural sites were identified in the 1960s and have been flagged as no cut areas. Some discovered sites have since been difficult to find in the field as recorded data has proven to be quite variable in accuracy.*

Recommendation (6): Algonquin Park, in cooperation with the Algonquin Forestry Authority, must ensure that tree markers and other forest workers are trained by cultural heritage experts to identify cultural heritage artefacts/structures within the Park.

The intent is to avoid disturbance in all such sites, discovered, known or as yet unknown. To make the latter possible, skidder operators have been given training to recognize "new" potential sites.

Suggestion (5): The Algonquin Forestry Authority should modify its Standard Operating Procedure with respect to Area of Concern and the line marking such that the direction is to stop operations (as was done in the Egan Farm example) when operators encounter unidentified cultural heritage values.

This is standard procedure.

Suggestion (6): Algonquin Park should prepare a Cultural Resources Management Plan as was called for in the Algonquin Provincial Park Management Plan.

This initials a huge and expensive effort that is not currently available.

Recommendation (7): Algonquin Park must:

a) Conduct pre-harvest aerial surveys to confirm and categorize moose aquatic feeding areas. Until potential moose aquatic feeding sites are confirmed as used by moose and deer, they must be re-labelled on Natural Resource Values Information map layers as "potential moose aquatic feeding areas."

b) Conduct heron and osprey nesting site surveys, in conjunction with MAFA surveys, for areas that are proposed for forestry operations.

To cover the situation, protection is being put on all such "potential" areas.

Principle 4: Plan Implementation

Suggestion (7): The Algonquin Forestry Authority should reinforce with its staff the procedure for the correct identification and reporting of bypass harvest areas.

A continuing process.

Suggestion (8): The Algonquin Forestry Authority should minimize the amount of woody debris left at landings following harvest operations.

A continuing process.

Suggestion (9): Algonquin Park and the Algonquin Forestry Authority should pursue opportunities for using prescribed burns, where the costs and benefits are reasonable.

The preference here is to scarify and expose sandy soil to the sun. The sun's rays on a warm day are sufficient to open the jack pine cones.

Recommendation (8): In preparation for the 2005–2010 Forest Management Plan, Algonquin Park must lead a review of the existing Park roads strategy with the Algonquin Forestry Authority and the Local Citizens Committee.

The location of roads is a difficult problem that receives near continuous attention by all constituents of Algonquin Park.

Suggestion (10): The Algonquin Forestry Authority should consider minimizing the use of gravel in the construction and maintenance of tertiary roads.

It is a standard practice to closely monitor both the need for and the quantity of gravel needed for each section of road.

Best Practice (1): The Algonquin Forestry Authority is to be commended on the practice of phasing out culverts at stream crossings in favour of portable bridges.

The general principle applied here relative to culverts is that fewer is better.

Suggestion (11): The Algonquin Forestry Authority should review its culvert installation practices to ensure that potential impacts on water quality are minimized and that best management practices are routinely considered during all installations and decommissionings.

Recommendation (9): The Algonquin Forestry Authority must ensure compliance with all conditions for Category 9 and 14 aggregate pits.

The rehabilitation of disused portions of pits is being carried out.

Suggestion (12): In future, the Algonquin Forestry Authority should indicate on gravel pit site plans that the restriction on gravel pit size is for the active portion of the gravel pit. The clarification should also be made in the Algonquin Provincial Park Plan when it is next revised.

This will be done.

Principle 5: Systems Support

Best Practice (2): The audit team commends the Algonquin Forestry Authority for its recognition of the value of training in meeting the objectives of sustainable forest management.

Principle 6: Monitoring

Recommendation (10): The Algonquin Forestry and the Algonquin Park compliance inspectors must provide accurate location information on Forest Operation Inspection Reports for "Not in Compliance" instances when follow-up action is required.

This is being done.

Suggestion (13): Algonquin Park should complete the verification section of tree marking inspection forms fully, indicating whether verification was conducted pre-harvest or post-harvest, on-the-ground or as a paper review.

Park staffing additions now make this possible.

Recommendation (11): Algonquin Park must prepare and implement a five-year Compliance Plan (2000–05) as an Annual Compliance Operations Plan (2003–04) for the Algonquin Park Forest in a timely manner.

Clarification needed.

Suggestion (14): The Algonquin Forestry Authority and Algonquin Park should ensure that any audit material that is date-sensitive bears evidence of date of submission or receipt.

This procedure will be followed.

Suggestion (15): The Annual Report conclusions and summary should be more strongly associated with overall plan objectives and how a particular year's operations are contributing to the achievement of those objectives

Principle 7: Achievement of Management Objectives and Forest Sustainability
None.

Principle 8: Contractual Obligations

Recommendation (12) The Algonquin Forestry Authority must re-examine Recommendations 5, 6 and 7 from the 1997 Independent Forest Audit and either ensure full compliance with the original Action Plan or develop new actions as part of this audit, whichever is most appropriate.

These items were leftovers from the previous audit, but not yet addressed in full or in a new plan of action.

Conclusion

It is the finding of the audit team that, with the exceptions noted, management of the Algonquin Park Forest conformed to program direction and legislation during the 1997–2002 period, and that the Algonquin Park Forest is being managed effectively and in a manner consistent with accepted criteria of forest sustainability. In addition, the audit team finds that the Algonquin Forest Authority has complied with the provisions of the Algonquin Park Forestry Agreement and recommends that the term of the Agreement be extended for a further five years.

This audit gave the AFA a good report. The Park and the MNR need to make improvements in compliance areas relative to Cultural Heritage and this process is underway. The audit is tabled in the Legislature and is therefore "out for all to see."

Appendix IX: **Map of Townships In and Adjacent to Algonquin Provincial Park**

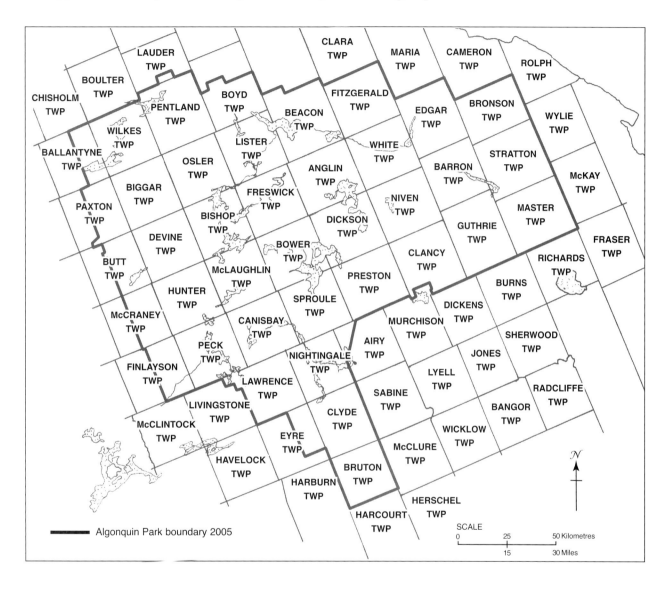

Appendix X: **Map of Boundary Changes to Algonquin Provincial Park: 1893-1993**

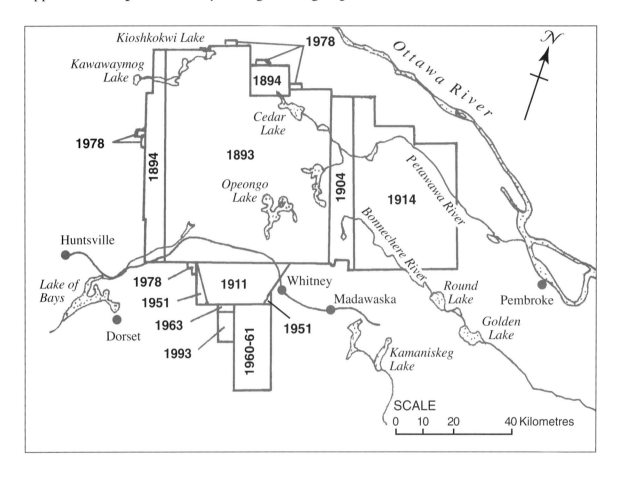

GLOSSARY OF TERMS

The cutting of the eastern North American forest was an operation that began at the beginning of the 1800s. While the English language dominated, French was common too. Irish, Swedish, Finnish, Polish, German and other languages were also significant in certain areas. The industry created its own terminology, and the language was colourful.

With the development of mechanization, an additional layer of language has developed.

Alligator	A steam-powered, flat-bottomed warping tug used in the assembly and towing of booms (rafts) of logs; amphibious so it could be warped (moved) from lake to lake.
Backbreaker	A big log, usually more than 14 inches in diameter.
Bag	A quantity of logs, surrounded by booms, readied for towing.
Band mill	A sawmill that uses band saws to cut planks from logs.
Barber chair	A stump with a high projection, like the back of a chair, on one side.
Bark mark	An ownership mark, made with a stamping hammer, placed back about two feet from the end of a log.
Bang juice	Dynamite.
Bank	A percentage of the daily count of logs cut and skidded purposely withheld by log makers and teamsters. If the weather turned bad, the crew could draw on the bank for their daily count without loss of pay.
Bear trap	A device used to secure wrapping chains on a sleigh-load of logs.
Beaver	To hack a tree down in an amateurish fashion.
Bench	The wooden bridge connecting two sleigh runners or bobs.
Bitch chains	Chains that go up and over a sleigh load of logs.
Bitch hooks	Hooks that hold bitch chains (see Fig. 20).
Burling	A two-man footwork contest on floating logs. The loser ended up in the water.
Block & tackle	Two pulleys through which a rope or chain is passed—used to increase lifting power.
Blocking materials	Industrial hardwood used, for example, in the steel industry to stop or prevent the rolling of coils of hot steel.
Blue bottle	A container of horse drugs.
Boardway	The area used for sorting wood after it has been trimmed by the saws.
Bobs	A pair of sleigh runners joined by a wooden bridge, termed a bench.
Bobsled	A short sled, usually used as a pair in-line, to transport logs.
Boom-rat	A worker that walks the logs in the water and rafts moving logs into booms.
Bow saw	A buck saw with a bow-shaped frame.

Boy's axe	A three-quarter axe with a 2 1/2-pound head and a 28-inch handle.
Bridge	The narrow section of a tree remaining before it falls.
Broad axe	A large axe, the blade of which is bevelled on one side only, used in squaring logs.
Boomage	The 6″ allowance on log lengths permitted by provincial regulations to compensate for damage and losses on river runs.
Buck	To saw, or to carry anything such as water or wood.
Bucking	To cut a tree stem, usually to a 16-foot length.
Buckbeaver	The boss of the main haul road.
Buckeroo	One who rides and moves (burls) logs in the water and rafts them into booms.
Bucksaw	A hand saw for bucking or cutting wood.
Bug	A lantern made by inserting a candle in a hole in the side of a can, or a lantern of any type.
Bull	The camp boss or foreman.
Bull buckeroo	The head man of a crew of fallers and buckeroos.
Bull cook	A cook's helper or choreman; he had to carry all the wood and water.
Bull edger	The mill worker who ran the edger saw or an edger with more saws and horsepower.
Bunkie	A bed partner.
Burl	A knot-like growth common on the trunk of a tree.
Bush superintendent	A walker or timber cruiser. An employee who would assess the quantity and quality of timber on a prospective licence. He would also lay out the bush roads, locate the bush camps and supervise operations.
Bushtail	Horse.
Butt	Trunk of a tree from which the branches have been trimmed.
Butt-shake	Separation of growth rings between late fall and spring wood at the base of the tree.
Cackler	Camp clerk.
Cadge	To move a burden, e.g., move supplies along a bush road to a bush camp.
Cadge crib	A device used to move a boom of logs across a lake. It consisted of a floored raft of logs usually 30-feet (or more) square. At the centre was a winch, commonly turned by a horse that walked in a circle, which wound up a long rope attached to an anchor. The anchor was rowed out from the cadge crib in the direction the men wanted to move the raft. The winding up of the rope then slowly moved the raft and the boom of logs that was attached to it.
Carriage	A platform that moved down rails, taking the logs into the saws.
Camboose	Served during the early days as a stove. Four hardwood logs and a few stones formed a square containing the fire, while smoke escaped through a hole in the roof. Baking was done in kettles buried in the hot ashes.

Glossary of Terms

Canadian-style trimmer	A set of saws that cut rough lumber to length as directed: e.g., from 16 feet to 4 feet.
Canned daylight	Coal oil torch.
Cant	The interior section of a log after the outside boards have been cut off.
Cannon	A log that dropped off the end of a skidding pile and stuck up in the air.
Cant hook	A tool consisting of a hook and a dog on a wooden handle, used to turn or roll logs.
Cat-skinner	Caterpillar tractor operator.
Cheat-stick	A wooden log scaler's rule.
Check	Separation of wood between spring and fall growth rings.
Chickadee	A worker who cleans off the road apples left behind by the horses.
Chicot	A standing dead tree.
Chipper	A machine with a rotating disc with blades, used to make wood chips out of residue wood.
Chokerman	The man who fastened the choker, a loop of cable, around a log, then attached it to the cable from the winch on the skidder.
Circular mill	A sawmill using a circular saw to manufacture lumber.
Clapboard	A thin board split from a log, used for boarding or roofing.
Clevis pin	The U-shaped iron bracket on the tongue of a sled used to attach it to a truck.
Cold skidding	Moving logs out of the woods long after they have been cut.
Come-along	Winching device.
Conductor	The man on the tanker who handled the barrel and pulled the plugs to put water on the road.
Cookee	The dishwasher and assistant cook.
Corner binds	Chains at the four ends of a sleigh-load of logs.
Cow's mouth	A notch chopped in a tree to fell it.
CPR strawberries	Prunes.
Crook	A curve in a log, also termed sweep.
Crown	The top of a tree.
Crown fire	A forest fire burning the tops of trees.
Cruiser	A timber cruiser; an estimator of the cut in an area.
Cruiser's axe	A double-bitted axe of three-quarter size.
Crazy wheel	A braking device that eased sleighs down steep hills.

Cross-haul	A method of piling logs on a skidway. A chain decking line was fastened to a top log on the skidway with a dog or sharpened hook, and they passed around the log to be moved; it was then passed over the pile, put through a pulley and then attached to the harness of a horse. When the horse moved forward, the log moved up a pair of skid logs. The roller kept the log straight. When the log passed the dog, this hook pulled out.
Cull trees	Defective trees that will not yield lumber.
Deacon's seat	A large seat by the fireplace in a bunkhouse, regarded as a place of honour.
Deadhead	A log partially sunk in the water.
Decking	The placing of logs in a skidway by a chain decking line.
Decking line	A chain used to roll logs up skids and pile them.
Dieback	The dying off of branches starting at the top of the trees.
Dog	Hook at end of a decking chain.
Donkey engine	Steam engine used to pull log sleighs up hills; later, a portable gas engine used to power Sky Jammers.
Donkey puncher	One who operates a donkey engine.
Down log	A log that has fallen of its own accord.
Double barrel	A bunk bed for two men.
Drum debarker	A rotating drum used to remove bark from logs.
Drumlinoid	A hill shaped like a hard-boiled egg cut on the long axis. It is composed of mixed sand, silt and clay that have been moulded in the last stages of a glacier's advance. The hills tend to be orientated in the direction of ice flow with the large end pointing into the approaching ice.
Dump or Banking Ground	The area along a riverbank where the logs were piled in preparation for the river drive.
Edgers	The saws that trimmed the boards to give them parallel edges after going through the head saw.
Epsom salts	Genuine Jimmy—the camp doctor.
Fawnfoot	The knob at the end of an axe handle.
Faller	A lumberjack who cuts down trees.
Frilling	Cutting or girdling around a tree to kill the tree.
Frog	To loiter, loaf.
Flat cant	A two- or four-sided piece of wood.
Flume or chute	A wooden chute through which logs were funnelled by water.
Flying axe handles	Diarrhoea.
Gangsaw	Designed in England in the late 1800s, a device consisting of a frame into which as many as 40 saws were fitted.

Glossary of Terms

Get a saddle	A warning not to ride the saw—in other words, to get to work.
Glut	A wooden wedge for splitting logs.
Grab	The hook attached to the doubletrees used for skidding.
Grandpa	The general superintendent.
Graybacks	Body vermin.
Grease burner	A locomotive engineer.
Green sorter	The table where a lumber inspector grades freshly cut timber.
Green wood	Freshly cut wood that contains a high percentage of moisture.
Groundhog	A logger who works on the ground.
Grubhoe	A mattock used by workmen while road building.
Grunt	A helper.
Gut wagon	A wagon or sleigh used to take food to men dining in the bush.
Gypo outfit	A small, independent logging company.
Hair pounder	A teamster.
Hang	The angle at which an axe head is placed on the handle.
Haw	Command from a teamster to turn the horses left.
Hayburners	Camp horses.
Head saw	The first saw in a mill to cut a log.
Helve	An axe handle.
Hewer	A man skilled with a broad axe.
Hog	A revolving drum fitted with knives that reduce waste to useable fuel for the mill.
Hot pond	A heated area or vat of water, used to store and thaw logs during the winter before the logs were cut in the mill.
Hot skidding	Skidding logs out immediately after being cut.
Housewife	A kit of sewing supplies.
Jack ladder	A conveyor device used to lift logs into the mill.
Jackpot	Three or more trees felled on top of each other, usually by accident, and difficult to untangle.
Jam crew	The crew in change of a log drive.
Jammer	Invented by Blind River lumberman Peter W. Wallace, a log-lifting set-up used to load sleighs—an improvement on the gin-pole.
Jerk	To dry thin strips of meat in the sun.
Jerky	Sun-dried meat.
Jobber's sun	Coal oil torches used to get work started in the dark.

Jumper	The front bob of a sleigh used to skid oversized logs.
Kerf	The space made by a saw in cutting a log.
Kickback	The action of a falling tree, the butt of which jumps backward over the stump.
Kitchen mechanic	The camp cook—also called a hash burner, a mulligan-mixer and a stomach-robber.
Live log	A log that will float.
Loco	Crazy.
Lodged tree	A felled tree held up by a standing tree.
Logging berries	Prunes.
Log boom	A floating collection of hundreds of loose logs surrounded by large logs chained together. Such booms were pulled by a cadge crib, alligator or steam tug down lakes to a mill.
Log wrench	A cant hook.
Long butt	The rotted part of a log cut off as waste.
Long clear	Salt pork.
Lop	To trim off branches.
Log-hook	A tool for carrying logs, consisting of a pair of hooks hung in the middle of a handle.
Macaroni	Sawdust.
Maul	A club-shaped wooden tool for driving wedges.
Mill run	Lumber marketed without being graded.
Morning glories	Pancakes.
Muzzle-loading bunk	Old type of sleeping bunks, whose only access was from the front end. Two men usually occupied the bed, which was furnished with a straw mattress and two blankets.
Over-run	The expected gain in footage hoped for by sawmill operators when their log production was sawn into boards. This was caused by deficiencies in the Doyle Rule when it was used to scale smaller diameter wood.
Pay-cheater	A timekeeper.
Peavey	A tool for handling logs—similar to a cant hook, except that it has a pike at the lower end of the handle.
Peckerwood mill	A small portable sawmill.
Pig's foot	A hook attached to the bull ropes which, when jammed into each end of a log, enable the bull rope operators to transfer the log to the waiting top loaders.
Pike	A metal point or spike usually attached to the end of the pole.
Piling	The process of air-drying the cut lumber in stacks for 60 to 90 days.
Piling skid	Logs laid against a log pile, up which other logs are rolled.

Glossary of Terms

Pinch	Pressure in the kerf, which prevents sawing.
Pitch-off	In areas too rough for horses, logs were pitched down hills. Stones that became embedded into such logs made trouble for sawmillers.
Pole axe	A single-bit axe with a hammer-head.
Pup	A hook at end of a decking chain—also termed a dog.
Ramrod	Camp boss.
Rear sweep	An operation conducted to clean up the river after the main drive had passed by.
Resawing	Use of a downstream resaw to produce lumber from a cant.
Ring rot	A severe defect in logs.
Road apples	Round balls of horse manure.
Road monkey	Also chickadees—men used to clean and maintain the main haul road.
Rollers	Cant-hook men at the skidways.
Sandhill man	See *Sandpiper*.
Sawbill	Sawing instructions for the sawyer, edgerman and trimmerman.
Sandpiper	On steep grades, this man sanded the iced ruts.
Scoring	The act of notching corner-rounded waney or square timber prior to hewing the sides flat.
Scragg mill	A primary breakdown machine with two saws for cutting off the slabs from a log, used to make a two- or four-sided cant.
Set	The cutting thickness of a saw, determined by the amount of bend given the teeth.
Shake	A separation in the growth rings in pine and hemlock.
Sharp shod	Sharp-caulked shoes for horse working on iced roads.
Siltation	Deposits of silt in a streambed, washed out from a disturbed upper slope.
Sidewinder	A tree that is knocked down by a tree being felled.
Silver tip	An old man or a blonde Swede.
Skid	Logs on which logs are piled off the ground; part of a platform (skidway) on which logs are piled.
Skyline power jammer	Two A-frames about 65 feet high and about 200 feet apart, joined by two wire cables on pulleys. A spreader bar is attached to the lower cable by means of two crotch lines. A pulley is attached to the spreader bar directly above it; the upper cable runs through this pulley. Tightening this cable lifts the load. From each end of the spreader bar hang two lengths of wire cable. These cables are put around the logs and the ends of each are fastened together by means of a hooking system which holds firmly as long as there is tension applied to it. The cables go slack when the logs are deposited on the pile. Power is provided by an 8-cylinder engine.
Slash	The debris and branches from trimmed logs.

Slasher	The operator of a mobile or fixed-line slasher who cuts treelengths into logs.
Smoke-eater	Forest fire fighter.
Snubber	A rope attached to a tree that was tied to sleighs loaded with logs to ease them down slopes.
Sorting jack	A series of long pockets in a river's mouth into which logs were sorted by species or ownership.
Sprinkler system	A fire protection system, usually a spray system from a ceiling, that triggers automatically with a rise of temperature above a set level.
Spud	A tool for removing bark from logs.
Squint-eye	One who sharpens saws.
Stamping hammer	A tool used to impress the company mark on the ends of logs.
Stick	A (big) log.
Stomach robber	A poor cook.
Strawberries	Beans.
Strip	An area blazed by the field boss for a crew to cut.
String	The term applied to the average girth of a quantity of waney or square timber.
Stump detective	One who estimates the amount of waste in stumps and tree tops.
Sway bar	Bars that helped keep the large sleighs lined up.
Swedge or swage	The width of the cutting edge of the saw.
Sweep (1)	An operation to clean up stray logs after the main drive had passed.
Sweep (2)	A defect in a log caused by the log curving — also termed crook.
Swage set	A tool that causes the teeth of a crosscut saw to flare out so that the saw made a cut wider than the blade. This kept the saw from binding.
Swamper	One who makes roads.
Swamper's joy	Canned jam that was mostly turnips with the odd red strawberry.
Swedish saw	See *Bow saw*.
Tan bark	Peeled hemlock bark, cut into 4-foot strips, used in the tanning of leather.
Timber cruiser	An employee of a timber company or the government who would assess the type, quality and quantity of timber in a licence or tract of bush.
Toothpick	A small tree.
Top loader	An expert cant-hook man who arranged the loads on sleighs.
Tote road	A road used to bring in supplies, separate from the haul road. "Take the tote road" was an expression used to dismiss an employee.
Trimmer	Saw that cut the boards into uniform lengths.

Glossary of Terms

Trace (fly) team	An extra team that was used to get logs to the main line.
Travois	A skid road.
Turkey	A pack or bag (frequently a potato sack) in which a logger carried his belongings.
Turtleback	A type of chain that moves lumber in a mill; it is sloped so that boards slide over top of it.
Undercut	The notch cut in a tree to fell it in the desired direction.
Walker	See *Bush Superintendent*.
Walking bull	A foreman who directs several camps.
Waney	Partially squared timber with four corners left rounded. Skidding waney to the mill resulted in less wasted wood in the bush.
Walking the corner	The development of a cutting plan for a season, which could include the location of camps, tote road, haul road and dumps.
Wanigan	A portable kitchen.
Warp	To move a boat by means of a rope or cable and winch. This method was employed for moving the steam-warping tug or alligator. Also a term describing crooked lumber.
Winch	A device to wind up a rope or cable, frequently with gears, which increased the pulling power.
Wolf tree	A large, often poorly formed tree that robs smaller ones of light.

Sources:

Anderson H.W. and J.A. Rice, *A Tree-marking Guide for the Tolerant Hardwoods Working Group in Ontario,* Toronto: Ministry of Natural Resources, Forest Resources Branch, 1993.

Graham MacDonald, *The Mississagi Country: A Study in Logging History,* Toronto: Ministry of Natural Resources, Parks Division, 1974.

Donald MacKay, *The Lumberjack*s. Toronto: McGraw-Hill Ryerson. 1978.

SELECTED BIBLIOGRAPHY

Books

Addison, Ottelyn. *Early Days in Algonquin Park*. Whitney [ON]: The Friends of Algonquin Park, 1985.

Burgess, Alan. *The Longest Tunnel*. Hammondsworth [UK]: Penguin, 1990.

Carson, Rachel. *Silent Spring,* Boston: Houghton Mifflin, 1962.

Crossen, Wayne. *A Study in North Central Ontario 1885-1930, with Special Reference to the Austin-Nicholson Company.* Toronto: Ontario Ministry of Culture & Recreation/Ministry of Natural Resources, 1976.

Eganville Leader. *Reflections of a Century, Stories and Photos, 1902–2002.* Eganville [ON]: Eganville Leader, 2002.

Eyre, F.H. and W.M. Zillgitt. *Partial Cuttings on Northern Hardwoods, The Lake States: Twenty Year Experimental Result.* Technical Bulletin 1076. Washington: U.S. Department of Agriculture, September 1954.

Finnegan, Joan. *Tallying the Tales of the Old-Timers.* Burnstown [ON]: General Store Publishing House, 1968.

Foot, R.D. and J.M. Langley. *MI 9: The British secret service that fostered escape and evasion, 1939–1945.* London: Book Club Associates, 1979.

Gillis, Sandra. *The Timber Trade in the Ottawa Valley, 1805-54.* Manuscript Report Series 153. Ottawa: Department of Indian Affairs and Northern Development, n.d.

Haggart, George, Ivah (Van Meter) Munro, Tom Saunders, Harry Stubbs, Russell Van Meter and the pupils of Grade VII and VIII, Whitney Public School, Mrs. Mary Hollett, Principal. *History of Whitney.* Whitney [ON]: mimeograph, n.d.

Hughson, John W. and Courtney C.J. Bond. *Hurling Down the Pine.* Third Edition. Chelsea [QC]: The Historical Society of the Gatineau, 1987.

Jones, R.L. *A History of Agriculture in Ontario.* Toronto: University of Toronto Press, 1946.

Kennedy, Clyde C. *The Upper Ottawa Valley - A Glimpse of History.* Pembroke [ON]: Renfrew County Council, 1970.

Killan, Gerald. *Protected Places: A History of Ontario's Provincial Parks System.* Toronto: Ontario Ministry of Natural Resources, 1993.

Lambert, Richard S. and Paul Pross. *Renewing Nature's Wealth.* Toronto: Ontario Department of Lands and Forests, 1967.

Leopold, Aldo. *A Sand Lake County Almanac, With Essays on Conservation from Round River.* New York: Ballantine Books. 1970.

Lloyd, R. Douglas. *The Statten Camps: Some Memories from the first Forty Years, 1921-1962,* Private printing, 1996.

Lyons, Bob. *Whitney: Island in the Shield and Sabine Township.* Bancroft [ON]: Bancroft Times, 1986.

MacDonald, Graham A. *The Mississagi Country: A Study in Logging History.* Toronto: Ontario Ministry of Natural Resources, Park Division, 1974.

MacKay, Donald. *The Lumberjacks.* Toronto: McGraw-Hill Ryerson, 1978.

MacKay, Niall. *Over the Hills to Georgian Bay: A Pictorial History of the Ottawa, Arnprior and Parry Sound Railway*: Erin [ON]: Boston Mills, 1981.

MacKay, Roderick. *A Chronology of Algonquin Park History.* Algonquin Park Technical Bulletin Series. No. 8. Whitney [ON]: The Friends of Algonquin Park, 2002.

———. *Spirits of the Little Bonnechere: A History of Exploration, Logging and Settlement, 1820 to 1920.* Pembroke, [ON]: Friends of the Bonnechere, 1996.

MacKay, Roderick and William Reynolds. *Algonquin.* Erin [ON]: Boston Mills, 1993.

Mason, Joe. *My Sixteenth Winter: Logging on the French River.* Cobalt [ON]: Highway Bookshop, 1974.

Meadows, Donella, et al. *The Limits to Growth: A Report to The Club of Rome's Project on the Predicament of Mankind.* A Potomac Associates Book. New York: Universe Books, 1972.

Mountain, G.A. *An Inventory of Existing Historic Sites in Algonquin Park.* Whitney [ON]: Algonquin Park Museum Archives, n.d.

Radforth, Ian. *A History of the McRae Lumber Company and the Community of Whitney, Ontario.* Unpublished draft, 1990.

———. *Bushworkers and Bosses; Logging in Northern Ontario, 1900-1980,* Toronto: University of Toronto Press. 1987.

Runtz, Michael. *Explorers Guide to Algonquin Park.* Toronto: Stoddart, 1993.

Saunders, Audrey. *Algonquin Story.* Second Edition. Whitney [ON]: The Friends of Algonquin Park, 1998.

Saunders, Tom et al. *History of Whitney.* Unpublished report, n.d.

Shaw, Bernard. *Lake Opeongo, Untold Stories of Algonquin Park's Largest Lake.* Burnstown [ON]: General Store Publishing House, 1998.

Stanfield, Donald and Elizabeth Lundell. *Algonquin, The Park and Its People.* Toronto: McClelland and Stewart, 1993.

Strickland, Dan. *Logging History in Algonquin Provincial Park—Algonquin Park Logging Museum,* Whitney [ON]: The Friends of Algonquin Park, 1993.

Strickland Dan and Russ Rutter. *The Best of* The Raven. *The Algonquin Provincial Park Newsletter.* Whitney [ON]: The Friends of Algonquin Park, 1996.

Townsend, E. Ray. *Algonquin Forestry Authority: A Twenty Year History, 1975-1995.* Huntsville [ON]: Private printing. n.d.

Westhouse, Brian D. *Whitney: St. Anthony's Mill Town on Booth's Railway.* Whitney [ON]: The Friends of Algonquin Park, 1995.

Williams, Eric. *The Wooden Horse.* New York: Viking Penguin, 1949.

Whitton, Charlotte. *A Hundred Years A-Fellin'.* Ottawa [ON]: Runge Press, 1974.

Wickstead, Bernard. *Joe Lavally and the Paleface.* 1948. Reprinted. Whitney [ON]: The Friends of Algonquin Park, 1993.

Wright, H. Eleanor (Mooney). *Trailblazers of Algonquin Park.* Eganville [ON]: HEW Enterprises, 2003.

Periodicals and Articles

Canada Lumberman. July 25, 1895.
Canada Lumberman. December 1895.
Canada Lumberman. November 1896.
Canada Lumberman. March 1903.
Canada Lumberman. August 1913.
Chington, Michael. "Algonquin Park: Canoeist Paradise or Tree Farm." *Canadian Geographic* 113:6 (1993).
Lee-Whiting, Brenda. "Opeongo Road - An Early Colonization Scheme." *Canadian Geographical Journal* 74.3 (March 1967).
Young, Scott. "A Prisoner is Lonely." *Maclean's Magazine,* 1944.

Government Reports

Ontario Ministry of Natural Resources. *Algonquin Provincial Park, Master Plan Review. 1989-90.* Toronto: Queen's Printer for Ontario, 1991.

Ontario Ministry of Natural Resources. *Annual Report on Forest Management, 1999–2000.* Forest Information Series. Toronto: Queen's Printer for Ontario, 2001.

Ontario Ministry of Natural Resources. *Annual Report on Forest Management, 2000–2001.* Forest Information Series. Toronto: Queen's Printer for Ontario, 2002.

Ontario Ministry of Natural Resources. *Forest Research Reserve - Swan Lake.* Toronto: Queen's Printer for Ontario, 1968.

Ontario Parks. *Algonquin Provincial Park Management Plan.* Toronto: Queen's Printer for Ontario, 1998.

Wyatt, D.W. "A History of the Origins and Development of Algonquin Park." A Background Paper Presented to the Algonquin Park Task Force. Ontario Department of Lands and Forests. 1971.

Other Reports

University of Toronto, Faculty of Forestry. *Logging Reports.* Unpublished papers.

1926:	Adamson, M.A., W.E. McCraw, D. McLaren, D.M. Parker.
1930–1931:	Bier, J.E. Choate and A. Crelock.
1948:	Herridge, A.J., J.A. Hawtin and J.H. Jamieson.
1949:	Cahill, L., D.W. MacGregor and S.W. Lukinuk.
1950:	Brackenbury, S.W. Cleaveley and F.T. Collict.

INDEX

A Silvicultural Guide for the Tolerant Hardwoods Working Group in Ontario, 190

A Tree Marking Guide for the Tolerant Hardwood Working Group in Ontario, 193

Aballah, George B., 190

Addison, Peter, 183

Airy, xxii, 2, 21-24, 31, 32, 34, 36, 37 43, 46, 47, 63, 65, 70, 72-77, 79, 83-85, 89, 92, 94, 96, 99, 104, 107 117, 125, 131, 133, 139, 150, 222, 223, 226, 228, 232-236, 238, 256-258, 268, 271, 278, 284, 291

Airy, Township of, 254

A.L. Dennis Salt & Lumber Company, 18-21

Algonquin Forestry Authority (AFA), xx, xxiv, xvi, 34,120, 143, 154,155, 157, 158, 207, 209-215, 263, 268, 269, 276, 280, 283, 284, 286, 289, 292

Algoma Steel Corporation, 286

Algonquin National Park Act, xxiii

Algonquin Park Advisory Committee, 183

Algonquin Park Forestry Authority Act, 189

Algonquin Provincial Park Master Plan, xx, xxiv, 34, 35, 153, 154, 157, 160, 161, 167, 175, 177, 179, 180, 182, 184-187, 189, 191, 193, 194, 200, 214, 215, 263, 282

Algonquin Provincial Park: Management Plan, 186, 187

Algonquin Park Museum Archives (see Visitor Centre)

Algonquin Provincial Park, xx, xxii, xxiii, xxiv, 4, 15-17, 25, 34, 42, 68, 154, 159, 160, 161, 163, 165, 168, 72, 73, 80, 86, 90, 107, 142, 143, 169, 174, 176-179, 181-189, 193-200, 203, 207, 208, 211-217, 219, 221, 223, 235, 237, 239, 241, 243, 245, 247, 249-251, 253-255, 257, 259, 261, 263, 268, 280, 284, 289, 292

Algonquin Park Independent Forest Audit, 199, 323-326

Algonquin Park Station, xvii, xxii, 17, 72

Algonquin Park Task Force (APATAF), 182

Algonquin Park Visitor Centre (see Visitor Centre)

Algonquin Wildlands League, 178, 179, 182-184

Amable Creek, 20

Anderson, Harvey, 169, 170, 172, 173, 205, 215

Areas of Undertaking, 197

Arnprior, xiv, xxiii

Avery, Joseph, 221

Aylen Station, 18

Baltic Sea, 3

Bancroft, xxiii, 1, 10 125, 174, 216, 223, 255, 259, 264, 284

Bark Lake, 18, 27

Barnet, Alex, 5, 6

Barras, Dave, 212

Barry's Bay, 5, 6, 11, 27, 30, 31, 36, 85, 90, 101, 102, 120, 131, 132, 216-218, 221, 226, 238, 255, 259, 261, 271, 273

Bartlett, George W., 17, 18, 173

Basin Depot, 25, 212

Bélanger, Grégoire, 5

Belaria, 241, 247

Bennett, R.B., 228

Bice, Ralph, 214

Billings, John, 216, 217

Bird, I.D. "Joe", 190-192, 195, 215

Bissonette, Auguste, 94

Bodsworth, Fred, 177, 178

Boldt, Lorne, 93, 119, 132, 137, 138, 228

Bonnechere River, 1, 3-7, 9, 24-26, 175, 256, 294

Booth, J.R., xxii, 1, 6, 17, 68, 71, 72, 123, 154, 212, 220, 232, 289

Bower Township, 107

Bowers, Ned, 130

Boyle, Basil, 125

Bracebridge, 172

Brackenburg, Keith, 249

Bristol, 96, 242

Brown, W.J. "Bill", 191, 193, 194, 197, 213, 214

Brudenell Township, 5

INDEX

Brunelle, Rene, 182
Bruton Township, xxiv, 107, 117, 133, 174, 203
Burton, Don, 169, 215
Buckman, Jed, 36
Buchman, Isaac, 94, 103, 104
Buckman, John, 104
Burnstown, 27
Burnt Island Lake, xxiv, 86, 88
Bushell, Roger, 242

Cache Lake, 16-18, 71-73, 94, 173, 229-231, 234, 251, 252, 254
Cameron & Company, 18
Camp Minnesing. 71
Campbell & McNab, 25
Campbell, Isabella "Cissie", 25, 291
Campbell, Peter, 24, 118, 120, 131, 134, 183, 229, 237, 242, 247
Campbell, Robert, 2, 24-26
Calman, Ned, 130
Calvert, Bill, 215
Canada Atlantic Railway, 18, 295
Canada Veneers Ltd., 193
Canada Splint, 193
Canadian International Paper (CIP), 287
Cain, Walter C., 36, 232
Canisbay Lake, 71
Canisbay Township, 40-42, 47, 48, 50, 51, 71, 72, 81, 83, 86, 173
Cannon, Bert, 130, 278
Cannon, Gary, 203, 213, 235, 255, 273, 275, 276, 277
Cannon, Joe, 130, 278
Cannon, Pat, 130, 278
Cannon, Robert, 275
Cannon, Tom, 278
Canoe Lake, xxi, 16, 17, 71, 217
Carleton Place, 91, 229
Carson, Harry J., 190
Carson Lake, 193

Carson, Rachel, 177
CBC, 177
Central Ontario Railway, 2, 10, 102
Cenzura, Alex, xviii, 104, 118, 132, 135, 141, 229, 233
Cenzura, Mrs. Alex, 222, 223
Chippior, Martin, 98
Class Environmental Assessment for Timber Management in Ontario, 197
Close, Charlie, 130
Clyde Township, 174, 203
Clancy (Little Hogan) Lake, 4, 5, 46, 63, 107, 174, 199, 203
Coghlan, Thomas, 6
Colonel By, 3
Colonial preferences, 3
Colonization roads, 1, 3
Connelly, Brent, 192, 213
Conroy Marsh, 25
Conservation Council of Ontario, 175
Consolidated Bathurst, 193, 287
Consolidated Revenue Fund, 197
Conway, Abbott, 177, 178, 184, 214
Conway, Sean, 214
Corbett, Carl, 195, 198, 204, 211, 213
Coulas, Paul, 130
Cranberry (Canisbay) Lake, xvi, 40-42, 47, 48, 50, 51, 71, 72, 81, 83,
Crossbill Lake, 86, 88, 92
Crown Timber Act, 189
Crown Forest Sustainability Act, 189, 197
Cudall, Howard, 244

Davis, William, 182, 197
Day, Harry "Wings", 241
Dennis Canadian Lumber Company, xxiii, 2, 18 20, 21, 30, 32, 36, 79, 217, 221, 226, 255, 284
Dennison Farm, 92
Department of Health, xvi, 82, 106
Doering, John "Bud", 211

Dofasco, 286
Domtar, 287
Dorset, 17, 163, 250
Duego, Dave, 172, 173
Dunn's, 6
Dylex Corporation, 286
Dymond, J.R., 214, 231
Dyment, Symon, 22

Eady, Stewart, 125
Ebbs, Adele, 214
Egan, John, xxiii, 4-6, 17, 24, 71, 199
Egan Estate, 1, 5, 6, 17, 71
Eganville, xxiii, 1, 2, 5, 25-27, 30, 43, 73, 221, 226-228, 238, 255, 256, 270, 271, 278
Emergency airfield (Lake of Two Rivers), xii, 34, 93
Environmental Assessment Act, 1997, 1998
Etmanski, Dominic, 131
Eyre Township, 107, 139, 142, 188

Fairfield Farm, 5
Falls, Bruce, 177
Fanshawe, Peter "Hornblower", 242
Farm Bay, 6, 38, 46
Farrell's Landing, 1, 5
Federation of Anglers and Hunters, 178
Federation of Ontario Naturalists (FON), 174, 175
Ferguson, J.A.S. "Fergie", 245, 247, 249
Finlayson, William, 173, 229, 230
Fletcher, Keith, 210
Floody, Wally, 243, 249
Forest Bay, 38
Forest Commission, 229
Forest Management Plan, 193, 194
Forest Management Understanding, 195
Forest Resources Inventory, 162, 193
Foster, Dr., 30, 91
Found Lake, 88, 251
Fraser, Alexander, 6, 13
Free Homestead Act, 5

Freeman, Cathy (McRae), 256
Friends of Algonquin Park, 213, 289
Frost, Leslie, 17, 182, 215
Fuller, Leonard, 99

Galeairy (Long) Lake, 1, 6, 11, 13, 32, 38, 41, 46, 61, 63-65, 68, 69, 74, 226, 264, 265
Galt, 43, 62
Garland, George, 213, 214, 216, 217
George, Don, xiii, 108, 120, 140, 166-168, 171, 191, 193-195, 205, 215
Giles, Warren, 215
Gilles Lumber Company, 232
Gilmour Lumber Company, 16
Glengarry County, 2, 3, 24, 25
Godwin, Gordon, 190
Golcar, Anthony, 131
Golden Horseshoe, 286
Golden Lake Lumber Company, 24-27, 73, 74
Goodman Stanforth, 193
Gordon, Al, 169, 215
Gotowski, 242
Grand Trunk Railway, 18, 30, 74
Gravenhurst, xxiii
Gray, Walter, 178
Great Depression, 21, 34, 71, 83-85, 99, 102, 125, 130, 221, 226-230, 232, 236, 271
Greenwood, Ben, 174

Haggart, "Sandy", 9, 220
Harper, Dave, 168
Hagarty Township, 6
Hamilton, 286
Hampton, Howard, 198
Hardy, Patrick, 178
Harris, Michael, 197
Harsh, George, 242
Hay Creek, 38, 63, 70, 224
Hay Creek Road, 265
Hay Lake, xvii, 20, 32, 38, 68, 70, 102-105, 107,

INDEX

109-113, 115-121, 132, 139, 140, 142, 216, 233, 249, 250, 272, 278
Head Lake, 86, 88, 101
Heintzman, George, 125
Henderson, Gavin, 177, 184, 215
Herridge, Arthur, 214, 215
Highland Inn, xxii, 16-18, 71, 173, 251
Hill, Arthur, 13
Hogan Lake, 210
Holmberg, George, 125, 220
Howe, C.D., 96
Hueston, William, 176, 179, 180, 214
Huntsville Memorial Hospital, 273
Hyde, Arlene, 211

Ide, E.B., 231
Interim Forest Management Under-standing (IFMU), 195-197
Irish Navvy, 3

Johnson, L.B., 177

Kaminiskeg Lake, 30, 31
KBM Forestry Consultants, 199
Killarney Lodge, 230
Keen, Rob, 213
Keenan, James, 179
Keenan, William, 24, 26, 27
Kelly, A.C., 6
Kelly, Frank, 130
Kennedy, John F., 177
Kent Brothers, 193
Kenyon Township, 2, 3
Kerio, Vince, 197
Killan, Gerald, 179
Killarney Lodge, 88, 230
Killarney Provincial Park, 183
Kirkwood Forest, 229
Knight's sawmill, 25
Kolanowski, 242

Kuiack, Frank, 130
Kuiack, Pete, 103
Kuiack, Ted, 97, 104, 119, 128, 131, 132, 134
Kutchecoski, Stanley, 276

Lacombe, John, 125
Laidlaw Industries, 287
Lake Louisa, 13, 38-41, 44, 46, 151
Lake Superior Provincial Park, 183
Lake of Two Rivers, xv, 6, 13, 14, 16, 18, 92-95, 97-102, 105, 107, 110, 130, 139, 152, 154, 182, 213, 228, 230, 234, 238, 271, 272
Lake Opeongo, xxiv, 6, 13, 14, 16, 18, 92, 168, 219, 220, 231
LaRochelle, Greg, 233
Lake St. Peter, 229
Lake Sasajewan, 86
Lake Travers, 200
LaValley, Ethel, 214
Lavally, Joe, 249
Lavieille-Dickson Wilderness Area, 188
Lawrence Township, 3, 36, 46, 61, 66, 239
Leopold, Aldo, *174*
Levean, Eddie, 130
Lindeiner, Kommandant von, 247
Littlejohn, Bruce, 177
Local Citizens Committee, 209, 289
Lock, Arthur, 190
Logger's Day, 213
Long Lake (see Galeairy Lake)
Louisa Creek, 39-41, 46, 53, 55, 63, 69
Luckasavitch Lake, 10
Lumb, E.T. & J., 99, 138, 226, 269, 274, 278
Lynch, Ned, 130

MacDougall, Frank, 159, 161, 163, 168, 173, 184, 214, 229, 251, 252
MacGregor, Duncan "Dunc" Fisher, xiv, xvii, 27, 36, 43, 55, 72, 79, 80, 89, 98, 99, 100, 128, 131, 139, 140, 154, 234, 235, 263, 270, 271
MacGregor, Ken, 270

MacGregor, William, 270
MacGregor, Roy, 235, 255
Mackenzie Lake, 6
MacDougall, Frank, 159, 161, 163, 168, 173, 214, 229, 251, 252
MacKay, Dr., 99
Mackey, W.M., 6
Madawaska, xxii
Madawaska Lake, 142, 149, 271
Madawaska River & Valley, 1-7, 11, 14, 18, 21, 26, 86, 88, 222, 224, 271, 289, 292
Madawaska (Town), 265, 278, 218
Management of Tolerant Hardwoods in Algonquin Provincial Park, 1973
Martin, Grenville W., 193, 202
Martelle, Ernie, 213
Martin's Siding, 27
Mastine, John, 273-276
Maynooth, 1, 2, 6, 10, 99, 221
McBride, Ben, 36, 228
McClure Township, 31, 107, 143, 284
McCormick, Tom, 230
McIntyre, J., 102
McGregor, Dan, 27, 270
McGregor, Marjorie (McRae), 94, 214, 235-237, 251, 256
McGregor, Sandy, 36, 38, 70, 71, 107, 125, 150
Mckenzie's (Clydegale) Lake, 13
McLean, Mac, 168-172, 215
McLachlin Brothers, xxiii, 27, 80, 86, 270
McLaughlin Township, 86, 106
McNab Township, 25, 27
McNutt, J. Wes, 190
McRae Addition, 188
McRae, Donald M., xvii, xxiii, xxiv, 35, 38, 43, 55, 61, 64, 69, 88, 89, 91, 95-100, 105, 116, 120, 121, 129, 130-132, 137-142, 152-154, 157, 158, 171, 203, 215, 218, 220, 222, 236-241, 243-246, 248-250, 254-257, 263, 265, 270, 273, 278, 288
McRae, Ethel, 26
McRae, Helen, 218, 249, 291

McRae, James C. "Jamie", 264, 268, 274, 279, 292
McRae, John, 73, 132, 154, 155, 157, 158, 201, 209, 214, 256, 257, 263, 273, 282, 285, 287-290, 292
McRae, J.D., 2, 25, 26, 30, 291
McRae, Janet (MacGregor), 29, 84, 92, 218 220, 221, 226-228, 236, 238, 255, 257
McRae, J.S.L., xxiii, 2, 3, 24, 25, 26, 69, 120, 121, 140, 143, 167, 203, 226, 227, 232, 234, 235, 255, 256, 270, 283
McRae, Robert D. "Bob", 116, 119, 132, 135, 136, 138, 139, 141, 148, 149, 154, 155, 157, 158, 196, 203, 223, 239, 257, 265, 266, 268, 273, 274, 276, 279, 283, 289, 290, 292
Meadows, D., 177
Mervart, Dr. J.A., 193
Mickle, Charles (grandfather), 20
Mickle, Charles (father), 20
Mickle, Charles "Charlie" (son), 20
Mickle, Dyment Lumber Company, 21, 22, 31, 32, 36, 80, 216, 278, 294
Migland, Sven, 221
Milldam, 4
Milson Forestry Service, 210
Milton, Bob, 198
Minnesing Lodge, 86, 297
Minneapolis, Minnesota, 11
Morrison, Donald, 243, 249
Munn Lumber Company, xxiii, 15-18, 21, 22, 36, 71, 173, 219
Ministry of Labour, 277
Minskewitz, 242
Munro, Mrs., 218
Murray Brothers Lumber Company, 27, 31, 193, 203, 237, 258
Muskoka District, 16
Muskoka Wood, 80, 105, 170

Napoleonic Wars, 3, 24
National and Provincial Parks Association, 177
Natural Resources Value Inventory System (NRVIS), 199

INDEX

Nepean 26
Nicholas, Jerry, 275
Nightingale Township, 36, 39-41, 44, 46, 53, 55, 63, 69, 71
Nipissing District, xxiii, 1, 17, 226
Noble, Red, 243, 246
Nordin, Vidar, 189, 190
North Branch Madawaska River, 86

O'Brien, J.A., 26
O'Brien, M.J., 26
Odenback, 200
O'Dette, Jack, 178
Office (Basin Depot), 25, 212
Oliver Lumber Company, 286
Omand, D.M., 175, 176, 214
Omanique, 31, 93, 120
Ontario Forest Industries Association, 195
Ontario Hardwoods, 286
Opeongo Lodge, 221
Opeongo Road, xiv, xvii, 1, 5, 6, 113, 220
Operating Plan Cruise (OPC), 194
Osseola, 278
Ottawa River, xvii, xxiii, 1, 3-7, 9, 11, 14, 18, 24, 26, 27, 33, 66, 80, 90, 91, 95, 96, 152, 154, 155, 1654, 173, 212, 219, 224, 228, 235, 243, 250, 251, 255, 258, 260, 264, 271, 289, 294
Ottawa Valley, xxiii, 3-5, 24, 27, 33, 91, 96, 152, 228, 255, 264, 294
Ottawa Arnprior & Parry Sound Railway (OA&PS), xxiii, 1, 3-6, 11, 17, 18, 30, 93, 133, 154, 212, 289
Otterslide Creek, 178

Parks Branch, 173, 176
Parrott, Frank, 197
Parks, John, 217
Paugh Lake, 25
Peace River District, 26
Pembroke, xiv, 1, 4, 27, 30, 36, 80, 86, 89, 95, 123-125, 163, 164, 166, 167, 175, 193, 200, 211, 229, 253, 294
Pembroke Box Company, 27, 80
Pembroke District, 229
Pembroke Lumber Company, 193, 200
Pen Lake, xi, xvii, 13, 38, 63, 67, 68, 239
Peters, Dave, 210
Perley and Pattee, xv, 1, 6, 11, 17, 46, 86
Perry, Albert, 32, 36, 92, 146,
Petawawa River, 16
Phillips, George, 229, 251, 252
Pick, Bob, 213
Pimlot, Douglas, 176, 178, 215, 251
Pinto, Fred, 210
Plonski, Walter, 176
Plunkett, 246
Poaching, 101, 222, 232
Portage du Fort, 204, 287
Post, Geofrey, 218
Post, Dr. Gilbert, 99, 100, 217, 218, 255, 271
Post, O.E., 21, 217, 218, 224
Post Memorial Library, 258, 259, 261
Priddle, George B., 190
Preliminary Master Plan, 175
Protected Places: A History of Ontario's Provincial Park System, 179,180
Proulx Lake, 213
Provincial Board of Health, 80
Provisional Master Plan, 180
Quality Assurance in Hardwood Tree Marking: A Case Study, 203
Quebec City, 3, 66
Quetico Provincial Park, 175, 176, 183
Quyon, 4

Radiant Lake, 200
Rae, Bob, 198
Rainy (Drizzle) Lake, 32,116, 117
Rathburn Company, 31
Raymond, "Doc", 172

Renfrew, 1, 4, 25, 26, 30, 69, 91, 99, 131, 143, 151, 214, 218
Reynolds, Bernard, 190, 197
Rideau Canal, 3
Robarts, John, 182, 215
Roberts, Frank K., 190
Robinson, Bob, 283, 284
Rock Lake, 13, 35, 38, 63-65, 67, 68, 71, 72, 95, 102, 104, 118, 120, 126, 132-147, 149, 151, 153-157, 171, 153-157, 171, 199, 200, 203, 212, 217, 238, 263, 265-267, 272, 286, 289
Rock Lake station, 39, 68, 102
Roche, Phil, 223, 256, 288
Roche, Mrs. Phil, 218
Rutter, Russ, 177

Sabine Township, 2, 10, 31, 105, 107, 110-113, 116, 117, 125, 134, 221, 229, 254, 261, 284
Saint John, 96
Sebastopol, 5
Shalla, Felix, 222, 236
Shalla, Frank, 36, 92, 96, 98, 100, 151, 232
Shalla, Paul, 36, 92, 222, 223, 232
Shaw Brothers, 193
Shurley & Dietrich, 43
Siydock, Andrew, 272
Simpson, John, 195
Sklar, 193
Smith, Ward, 190
Smith Falls, 91
Source Lake, 6, 13, 101
South Algonquin Township, 261
South Tea Lake, 17
Sproule Bay, 13, 16, 18, 107, 108, 113, 167, 219-221, 231
Stalag Luft III, 241
St. Anthony Lumber Company, xv, xxiii, 11-15, 19, 224
St. Elmo, 3, 25
St. Francis Memorial Hospital, 273

Standard Chemical, Iron & Lumber Company, 20
Stelco, 286
Strathearn Group, 286
Stringer, Dan, 217
Stringer, George, 217
Stringer, Jack, 217
Stringer, Jim, 217
Stringer, Joseph, 216, 217
Stringer Lake, 217
Stringer, Omer, 217
Stringer, Wam, 217
Stubbs, Bernie Sr., 43, 93, 226, 289
Sustainable Forest Licences, 196
Swan Lake Forest Research Reserve, 169-173, 205
Swift, Bill, 177, 221

Tayler, Grant, 179
Taylor, Jim, 102
Taylor, Henry, 101, 102
Taylor, Mrs. Henry, 223
Tea Lake, 230
Tea Lake Dam, 17
Temiskaming, 204
Temoix, 287
Thompson-Hyland, 286
Turner, Major, 247
Thomson, Tom, 17, 173
Timber Branch, 175, 176
Timber Management Planning Manual, 198
Timber licences, 169-170
Tomasewski, Felix, xvii, 106, 120, 132, 140, 150, 171
Tough, George, 197
Townsend, E. Ray, 191, 192, 195, 212
Trenton, 16, 17, 43
Tudhope, J.B., 16

United Oil Products, 200

Van, 39, 97

INDEX

Van Meter, Russell, 20, 21, 255
Vance, D.J., 189
Visitor Centre, 169

Wadsworth, James, 5
Wallace, 2, 10, 102, 105, 237, 272
Waters, Dan, 214
Weigh Scale and Scaling, 268, 269
Wellwood Canada, 193
West Smith Lake, 113, 114
Whitefish Lake, xiii, 13, 38, 61, 63, 64, 68, 118, 133, 200, 212, 213, 250, 289
Whitney, 2, 4-6, 11-15, 17-24, 26, 27, 30, 32, 35, 36, 42, 71, 73, 74, 84, 85, 92, 94, 101, 102, 105, 108, 109, 113, 114, 116, 120, 127, 131, 133, 140, 141, 143, 149, 154-156, 166, 167, 171, 200, 202-204, 207, 216-245, 247, 249-261, 263-268, 271-273, 278, 286, 289, 292-294
Whitney, Edwin Canfield, 12, 217
Whitney, James P., 13
Whitton, Jack, 219

Whitton, Charlotte, 219
Wicklow Township, 99
Wilderness Areas Act, 177
Wildlerness Research Station, 86, 176
William Davies Company, 30
Williams, Eric, 244
Wilston, Roy, 252- 254
Wilno, xxiii, 1, 6, 43, 90, 98,
Winters, John, 323
Wolfe, Helena, 26
Wolff, Paul, 94
Workmen's Compensation, 83, 279
Wright, Philemon, 1, 3, 66
Wright, Ruggles, 66
Wrightsville, 3

Yaraskavitch, Joe, 211
Yashinski, Johnny, 275
York River, 25, 30, 174,

About the Author

DONALD L. LLOYD (Don) is a retired geography teacher from Toronto, Ontario. He took his first canoe trip in Algonquin Park in 1945 and since then he has paddled and fished many of the Park's lakes and rivers. To share his experiences and love of Algonquin, Don wrote his best-selling book, *Canoeing Algonquin Park*.

Don has spent many years on the executive of the Algonquin Park Residents' Association. He was one of the original directors of The Friends of Algonquin Park, serving some 23 years. He is presently a member of the Local Citizens Committee for the Algonquin Forest. He is a member of the Ontario Field Ornithologists and a supporter of the following: The Owl Foundation, The Thickson Wood Foundation and The Royal Canadian Geographical Society.

When he first started tripping the Park, evidence of lumbering was commonly seen in the remains of abandoned lumber camps, bridges, dams and sluices. Along portages, purple vetch, daisies and timothy betrayed the fact that lumbermen's horses had passed that way. Curiosity about lumbering and that way of life remains.

After finishing *Canoeing Algonquin Park*, Don asked Bob McRae, a fellow director of The Friends of Algonquin Park, about the possibility of writing a book on the McRae Lumber Company. Working on this project, the author's major reward has been having the opportunity to meet so many wonderful people, all of whom willingly gave of their time, experience and memories and thereby greatly enhanced the book. First and foremost of these are the members of the McRae family, who have patiently educated the author in the ways of lumbering.

Don Lloyd at White Pine Plantation: Hogan Lake Road, June 21, 2005.

The journey in compiling this book has been a long one—but one made delightful by the wonderful people I have met along the way. But be clear on one point "I have had neither apples to polish nor axes to grind."

Source: Keith Fletcher, AFA.